NEUROPROTEOMICS

FRONTIERS IN NEUROSCIENCE

Series Editors
Sidney A. Simon, Ph.D.
Miguel A.L. Nicolelis, M.D., Ph.D.

Published Titles

Apoptosis in Neurobiology
Yusuf A. Hannun, M.D., Professor of Biomedical Research and Chairman, Department
    of Biochemistry and Molecular Biology, Medical University of South Carolina,
    Charleston, South Carolina
Rose-Mary Boustany, M.D., tenured Associate Professor of Pediatrics and Neurobiology,
    Duke University Medical Center, Durham, North Carolina

Neural Prostheses for Restoration of Sensory and Motor Function
John K. Chapin, Ph.D., Professor of Physiology and Pharmacology, State University
    of New York Health Science Center, Brooklyn, New York
Karen A. Moxon, Ph.D., Assistant Professor, School of Biomedical Engineering, Science,
    and Health Systems, Drexel University, Philadelphia, Pennsylvania

Computational Neuroscience: Realistic Modeling for Experimentalists
Eric DeSchutter, M.D., Ph.D., Professor, Department of Medicine, University of Antwerp,
    Antwerp, Belgium

Methods in Pain Research
Lawrence Kruger, Ph.D., Professor of Neurobiology (Emeritus), UCLA School of Medicine
    and Brain Research Institute, Los Angeles, California

Motor Neurobiology of the Spinal Cord
Timothy C. Cope, Ph.D., Professor of Physiology, Wright State University, Dayton, Ohio

Nicotinic Receptors in the Nervous System
Edward D. Levin, Ph.D., Associate Professor, Department of Psychiatry and Pharmacology
    and Molecular Cancer Biology and Department of Psychiatry and Behavioral Sciences,
    Duke University School of Medicine, Durham, North Carolina

Methods in Genomic Neuroscience
Helmin R. Chin, Ph.D., Genetics Research Branch, NIMH, NIH, Bethesda, Maryland
Steven O. Moldin, Ph.D., University of Southern California, Washington, D.C.

Methods in Chemosensory Research
Sidney A. Simon, Ph.D., Professor of Neurobiology, Biomedical Engineering,
    and Anesthesiology, Duke University, Durham, North Carolina
Miguel A.L. Nicolelis, M.D., Ph.D., Professor of Neurobiology and Biomedical Engineering,
    Duke University, Durham, North Carolina

The Somatosensory System: Deciphering the Brain's Own Body Image
Randall J. Nelson, Ph.D., Professor of Anatomy and Neurobiology,
    University of Tennessee Health Sciences Center, Memphis, Tennessee

The Superior Colliculus: New Approaches for Studying Sensorimotor Integration
William C. Hall, Ph.D., Department of Neuroscience, Duke University,
Durham, North Carolina
Adonis Moschovakis, Ph.D., Department of Basic Sciences, University of Crete,
Heraklion, Greece

New Concepts in Cerebral Ischemia
Rick C. S. Lin, Ph.D., Professor of Anatomy, University of Mississippi Medical Center,
Jackson, Mississippi

DNA Arrays: Technologies and Experimental Strategies
Elena Grigorenko, Ph.D., Technology Development Group, Millennium Pharmaceuticals,
Cambridge, Massachusetts

Methods for Alcohol-Related Neuroscience Research
Yuan Liu, Ph.D., National Institute of Neurological Disorders and Stroke,
National Institutes of Health, Bethesda, Maryland
David M. Lovinger, Ph.D., Laboratory of Integrative Neuroscience, NIAAA,
Nashville, Tennessee

Primate Audition: Behavior and Neurobiology
Asif A. Ghazanfar, Ph.D., Princeton University, Princeton, New Jersey

Methods in Drug Abuse Research: Cellular and Circuit Level Analyses
Dr. Barry D. Waterhouse, Ph.D., MCP-Hahnemann University, Philadelphia, Pennsylvania

Functional and Neural Mechanisms of Interval Timing
Warren H. Meck, Ph.D., Professor of Psychology, Duke University, Durham, North Carolina

Biomedical Imaging in Experimental Neuroscience
Nick Van Bruggen, Ph.D., Department of Neuroscience Genentech, Inc.
Timothy P.L. Roberts, Ph.D., Associate Professor, University of Toronto, Canada

The Primate Visual System
John H. Kaas, Department of Psychology, Vanderbilt University
Christine Collins, Department of Psychology, Vanderbilt University, Nashville, Tennessee

Neurosteroid Effects in the Central Nervous System
Sheryl S. Smith, Ph.D., Department of Physiology, SUNY Health Science Center,
Brooklyn, New York

Modern Neurosurgery: Clinical Translation of Neuroscience Advances
Dennis A. Turner, Department of Surgery, Division of Neurosurgery,
Duke University Medical Center, Durham, North Carolina

Sleep: Circuits and Functions
Pierre-Hervé Luoou, Université Claude Bernard Lyon, France

Methods in Insect Sensory Neuroscience
Thomas A. Christensen, Arizona Research Laboratories, Division of Neurobiology,
University of Arizona, Tuscon, Arizona

Motor Cortex in Voluntary Movements
Alexa Riehle, INCM-CNRS, Marseille, France
Eilon Vaadia, The Hebrew University, Jerusalem, Israel

Neural Plasticity in Adult Somatic Sensory-Motor Systems
Ford F. Ebner, Vanderbilt University, Nashville, Tennessee

Advances in Vagal Afferent Neurobiology
Bradley J. Undem, Johns Hopkins Asthma Center, Baltimore, Maryland
Daniel Weinreich, University of Maryland, Baltimore, Maryland

The Dynamic Synapse: Molecular Methods in Ionotropic Receptor Biology
Josef T. Kittler, University College, London, England
Stephen J. Moss, University College, London, England

Animal Models of Cognitive Impairment
Edward D. Levin, Duke University Medical Center, Durham, North Carolina
Jerry J. Buccafusco, Medical College of Georgia, Augusta, Georgia

The Role of the Nucleus of the Solitary Tract in Gustatory Processing
Robert M. Bradley, University of Michigan, Ann Arbor, Michigan

Brain Aging: Models, Methods, and Mechanisms
David R. Riddle, Wake Forest University, Winston-Salem, North Carolina

Neural Plasticity and Memory: From Genes to Brain Imaging
Frederico Bermudez-Rattoni, National University of Mexico, Mexico City, Mexico

Serotonin Receptors in Neurobiology
Amitabha Chattopadhyay, Center for Cellular and Molecular Biology, Hyderabad, India

Methods for Neural Ensemble Recordings, Second Edition
Miguel A.L. Nicolelis, M.D., Ph.D., Professor of Neurobiology and Biomedical Engineering,
    Duke University Medical Center, Durham, North Carolina

Biology of the NMDA Receptor
Antonius M. VanDongen, Duke University Medical Center, Durham, North Carolina

Methods of Behavioral Analysis in Neuroscience
Jerry J. Buccafusco, Ph.D., Alzheimer's Research Center, Professor of Pharmacology
    and Toxicology, Professor of Psychiatry and Health Behavior,
    Medical College of Georgia, Augusta, Georgia

In Vivo Optical Imaging of Brain Function, Second Edition
Ron Frostig, Ph.D., Professor, Department of Neurobiology,
    University of California, Irvine, California

Fat Detection: Taste, Texture, and Post Ingestive Effects
Jean-Pierre Montmayeur, Ph.D., Centre National de la Recherche Scientifique, Dijon, France
Johannes le Coutre, Ph.D., Nestlé Research Center, Lausanne, Switzerland

The Neurobiology of Olfaction
Anna Menini, Ph.D., Neurobiology Sector International School for Advanced
    Studies,(S.I.S.S.A.), Trieste, Italy

Neuroproteomics
Oscar Alzate, Ph.D., Department of Cell and Developmental Biology, University of North
    Carolina, Chapel Hill, North Carolina

NEUROPROTEOMICS

Edited by

Oscar Alzate
University of North Carolina

CRC Press
Taylor & Francis Group
Boca Raton London New York

CRC Press is an imprint of the
Taylor & Francis Group, an **informa** business

CRC Press
Taylor & Francis Group
6000 Broken Sound Parkway NW, Suite 300
Boca Raton, FL 33487-2742

First issued in paperback 2019

© 2010 by Taylor & Francis Group, LLC
CRC Press is an imprint of Taylor & Francis Group, an Informa business

No claim to original U.S. Government works

ISBN-13: 978-1-4200-7625-7 (hbk)
ISBN-13: 978-0-367-38501-9 (pbk)

Library of Congress Cataloging-in-Publication Data

Neuroproteomics / editor, Oscar Alzate.
     p. ; cm. -- (Frontiers in neuroscience)
  Includes bibliographical references and index.
  ISBN 978-1-4200-7625-7 (hardcover : alk. paper)
    1. Proteomics. 2. Nervous system--Diseases--Genetic aspects. I. Alzate, Oscar. II. Series: Frontiers in neuroscience (Boca Raton, Fla.)
    [DNLM: 1. Proteomics--methods. 2. Nervous System Diseases--metabolism. 3. Nervous System Physiological Phenomena--genetics. 4. Proteome--isolation & purification. QU 58.5 N494 2010]
  QP551.N485 2010
  572'.6--dc22
                                          2009031460

Visit the Taylor & Francis Web site at
http://www.taylorandfrancis.com

and the CRC Press Web site at
http://www.crcpress.com

Contents

Series Preface

Our goal in creating the Frontiers in Neuroscience Series is to present the insights of experts on emerging fields and theoretical concepts that are, or will be, in the vanguard of neuroscience. Books in the series cover genetics, ion channels, apoptosis, electrodes, neural ensemble recordings in behaving animals, and even robotics. The series also covers new and exciting multidisciplinary areas of brain research, such as computational neuroscience and neuroengineering, and describes breakthroughs in classical fields like behavioral neuroscience. We hope every neuroscientist will use these books in order to get acquainted with new ideas and frontiers in brain research. These books can be given to graduate students and postdoctoral fellows when they are looking for guidance to start a new line of research.

Each book is edited by an expert and consists of chapters written by the leaders in a particular field. Books are richly illustrated and contain comprehensive bibliographies. Chapters provide substantial background material relevant to the particular subject. We hope that as the volumes become available, the effort put in by us, the publisher, the book editors, and individual authors will contribute to the further development of brain research. The extent to which we achieve this goal will be determined by the utility of these books.

Sidney A. Simon, Ph.D.
Miguel A. L. Nicolelis, M.D., Ph.D.
Series Editors

To Federico and Elizabeth for all their love and support.

Foreword

My first exposure to experimental biology involved a college job, when I worked in the lab of Dr. Norman Anderson in Oak Ridge, Tennessee. He had recently launched his "Molecular Anatomy" (MAN) project, arguably the first step toward contemporary proteomics. I enjoyed my summer, working as a junior lackey in an ultrastructural core facility that supported a component of the project involving centrifugal fractionation of tissue homogenates. (Curiously, our group's main task was to help develop better vaccines for influenza.) I note, 40 years later, that the first citation in Chapter 1 of *Neuroproteomics* is to Norman Anderson. I suppose this anecdote points both to the vision of Dr. Anderson and to the impact that a summer job can have on a college student.

My current research focuses on the arrangement of proteins in excitatory synapses of the rodent forebrain. Over my career I have been repeatedly astonished by the complexity of individual synapses and by the precision with which their numerous components are organized. This embodies one of the joys of neuroscience, but at the same time perhaps its greatest problem: Given the overwhelming complexities that suffuse every level of the nervous system, how can we hope to gain a real understanding of the brain? This goal is one of the few truly difficult problems of natural science. While acknowledging the magnitude of the task, perhaps we should instead marvel that so much has already been discovered.

Neuroscience has advanced thanks to a great deal of hard work, along with clear thinking and a bit of luck; but equally important has been the development of a large and growing technical toolbox. Indeed, it could be argued that the most important advances in neuroscience have rested on the introduction of powerful new tools. Dating back at least to the introduction of the oscilloscope, tools from the physical sciences have had a disproportionate impact on neuroscience. It is in this context that Professor Alzate's book is especially welcome. Neuroproteomics addresses some of the simplest and most straightforward questions about the brain, but these questions remain unanswered. Proteomic tools can offer valuable insights into a wide variety of crucial issues ranging from development to the synaptic plasticity that underlies learning and memory to the basis of neuropsychiatric disease. Recent work in neuroproteomics has also led to entirely novel "systems biology" ways to think about the nervous system (see, for example, http://www.genes2cognition.org).

By assembling a full spectrum of neuroproteomic techniques and applications within a single volume, Dr. Alzate has simplified the practical tasks facing a neurobiologist who needs an introduction to proteomic methods. Moreover, this volume will enable the neuroscience community as a whole to recognize neuroproteomics as an important new tool for the entire field.

Richard Weinberg, PhD
University of North Carolina at Chapel Hill

Preface

Neuroproteomics is gaining momentum. The use of proteomics to elucidate brain phenomena is a natural result of the human's interest in his "thinking organ" and the ever-increasing number of modern techniques for the systematic analysis of proteins and protein interaction networks. Moreover, it is in the brain that the protein interaction networks are at the center stage of the scientific inquiry, as their physico-chemical interactions result in phenomena such as learning, memory, logic interpretation and production of information, visualization, coordination of motion, body orientation, and many others. Neuroproteomics takes advantage of the advent of multiple techniques for isolation, identification, and characterization of large numbers of proteins, and the subsequent analysis of their localization, molecular interactions, and participation in regulatory and metabolic pathways.

Pursuing this broad aim, neuroproteomics uses multiple approaches such as mass spectrometry, electrophoresis, chromatography, surface plasmon resonance, protein arrays, immunoblotting, computational proteomics, and molecular imaging to help elucidate the roles of proteins in the different parts of the nervous system. And this is only the beginning. Many challenges wait ahead for neuroproteomics, including identification of these proteins and their corresponding localization, characterization of protein conformations and post-translational modifications, molecular interactions involving proteins and other molecules, identification of the regulatory and metabolic networks in which these proteins are involved, and ultimately the identification of potential methods to regulate the structures and functions of these protein interaction networks. Finally, neuroproteomics has to explain how this vast amount of protein data changes with time and with external conditions, giving origin to brain functions. This ultimate goal will permit great advances in the understanding of the brain, its functions, its structure, and how deviations from normal states lead to neurological and mental diseases.

This book is the result of several years of work using proteomics to understand biological processes in the nervous system. Proteomics studies large ensembles of proteins and requires many techniques aimed at understanding how proteins interact with each other and with other molecules within the ensemble, and how proteins function in the context of their respective environments and under specific conditions. Investigators using neuroproteomics need to start with very intricate sample preparation, and proceed to mass spectrometry, to HPLC, to 2D-DIGE, to protein arrays, or to protein interaction networks. They need to correlate this information with action potentials, electrical signals, and synaptic activity. Each of these approaches has its own strengths and limitations, and all are needed to achieve a reliable interpretation of neurobiological processes.

This volume provides an overview of what neuroproteomics is, and gives a few examples of what neuroproteomics does. The general idea was to present the principles, the approaches, the difficulties of each technique, and the challenges of the field. Some of the authors have laboratories dedicated to the investigation of neurobiological phenomena, and have integrated proteomics approaches to elucidate relevant biological functions of the nervous system. Other authors are experts in specific approaches that can be used for the study of neuroproteomics.

I would like to acknowledge the help that I received from all the authors. It has been a difficult task to put together all of the information required—in many cases, data were collected specifically for inclusion in this book. I want to extend special thanks to Sid Simon and Miguel Nicolelis for inviting me to undertake this project— it has been an enriching experience. I also want to thank Cristina Osorio for her help in contacting the authors, collecting all of the manuscripts, and for her personal support and assistance with this project, starting from the very beginning. Finally, I want to express my appreciation for the dedication and professionalism of the staff of Taylor & Francis, particularly Barbara Norwitz, Patricia Roberson, and Prudence Board. Thanks to each one of you.

Oscar Alzate, Ph.D.

Editor

Dr. Oscar Alzate currently holds the position of associate professor in the Department of Cell and Developmental Biology at the University of North Carolina (Chapel Hill, North Carolina). He is also the director of the UNC Systems Proteomics Center. His interests include the application of neuroproteomics to the elucidation of molecular pathways and protein interaction networks using animal models for neurodegenerative diseases, particularly Alzheimer's disease. Dr. Alzate, a molecular biophysicist, has developed proteomics laboratories for the Davis Heart and Lung Research Institute at the Ohio State University, the Neuroproteomics Laboratory at Duke University, and the Proteomics Laboratory at the Pontifical Bolivariana University in his native Colombia. His passion is playing with proteins, trying to develop better ways to isolate, identify, and characterize proteins using combinations of all available biophysical techniques.

Dr. Alzate's search for functional protein interaction networks in cell cultures, primary cells, and mouse and human tissue in order to develop models of neurological diseases utilizes differential-display proteomics, mass spectrometry, iTRAQ, protein arrays, MALDI-based tissue imaging, and computational proteomics. In addition to his own research program, he is involved in more than 20 collaborative projects that include the neurobiology of synapses; neuroproteomics of the auditory and visual systems; and Alzheimer's, Parkinson's, and Huntington's diseases, as well as epilepsy and amyotrophic lateral sclerosis.

Contributors

Oscar Alzate, Ph.D.
Department of Cell and Developmental
  Biology
Program in Molecular Biology and
  Biotechnology
University of North Carolina at
  Chapel Hill
Chapel Hill, North Carolina

Malin Andersson, Ph.D.
Department of Pharmaceutical
  Biosciences
Medical Mass Spectrometry
Uppsala University
Uppsala, Sweden

Per Andren, Ph.D.
Department of Pharmaceutical
  Biosciences
Medical Mass Spectrometry
Uppsala University
Uppsala, Sweden

Lutgarde Arckens, Ph.D.
Laboratory of Neuroplasticity and
  Neuroproteomics
Department of Biology
Katholieke Universiteit Leuven
Leuven, Belgium

Carol Haney Ball, Ph.D.
Agilent Technologies
Cary, North Carolina

D. Allan Butterfield, Ph.D.
Department of Chemistry
Center of Membrane Sciences and
  Sanders-Brown Center on Aging
University of Kentucky
Lexington, Kentucky

Richard M. Caprioli, Ph.D.
Department of Biochemistry
Mass Spectrometry Research Center
Vanderbilt University Medical Center
Nashville, Tennessee

Lieselotte Cnops, Ph.D.
Laboratory of Neuroplasticity and
  Neuroproteomics
Department of Biology
Katholieke Universiteit Leuven
Leuven, Belgium

Nedyalka Dicheva, M.S.
UNC Duke Proteomics Center
Program in Molecular Biology and
  Biotechnology
University of North Carolina at
  Chapel Hill
Chapel Hill, North Carolina

Roberto Diez, M.Sc.
GE Healthcare Life Sciences
Piscataway, New Jersey

John F. Ervin
Kathleen Price Bryan Brain Bank
Duke University Medical Center
Division of Neurology
Durham, North Carolina

Michael Herbstreith, B.Sc.
Department of Medicine
Division of Neurology
Duke University Medical Center
Durham, North Carolina

Tjing-Tjing Hu, Ph.D.
Laboratory of Neuroplasticity and
    Neuroproteomics
Department of Biology
Katholieke Universiteit Leuven
Leuven, Belgium

Erich D. Jarvis, Ph.D.
Department of Neurobiology
Duke University Medical Center
Durham, North Carolina

Minoru Kanehisa, Ph.D.
Bioinformatics Center
Institute for Chemical Research
Kyoto University
Kyoto, Japan, and
Human Genome Center
Institute of Medical Science
The University of Tokyo
Tokyo, Japan

Vachiranee Limviphuvadh, Ph.D.
Bioinformatics Center
Institute for Chemical Research
Kyoto University
Kyoto, Japan

Roger D. Madison, Ph.D.
Research Service
Veterans Affairs Medical Center
Department of Surgery
Department of Neurobiology
Duke University Medical Center
Durham, North Carolina

Christine E. Marx, Ph.D.
Research Service
Veterans Affairs Medical Center
Department of Surgery
Department of Neurobiology
Duke University Medical Center
Durham, North Carolina

Mark W. Massing, Ph.D.
Research Service of the Veterans
    Affairs Medical Center
Durham, North Carolina

Mihaela Mocanu, B.S.
UNC Duke Proteomics Center
Program in Molecular Biology and
    Biotechnology
University of North Carolina at
    Chapel Hill
Chapel Hill, North Carolina

Viorel Mocanu, Ph.D.
UNC Duke Proteomics Center
Program in Molecular Biology and
    Biotechnology
University of North Carolina at
    Chapel Hill
Chapel Hill, North Carolina

Cristina Osorio, B.Sc.
Program on Molecular Biology and
    Biotechnology
School of Medicine
University of North Carolina at
    Chapel Hill
Chapel Hill, North Carolina

Carol E. Parker, Ph.D.
UNC Duke Proteomics Center
Program in Molecular Biology and
    Biotechnology
Department of Biochemistry and
    Biophysics
University of North Carolina at
    Chapel Hill
Chapel Hill, North Carolina

Raphael Pinaud, Ph.D.
Department of Brain and Cognitive
    Sciences
University of Rochester
Rochester, New York

Tanea T. Reed, Ph.D.
Department of Chemistry
Eastern Kentucky University
Richmond, Kentucky

Miriam V. Rivas, Ph.D.
Department of Neurobiology
Duke University Medical Center
Durham, North Carolina

Grant A. Robinson, Ph.D.
Department of Surgery
Duke University Medical Center
Durham, North Carolina

Petra Levine Roulhac, Ph.D.
Department of Neurobiology
Duke University Medical Center
Durham, North Carolina

Rukhsana Sultana, Ph.D.
Department of Chemistry
Center of Membrane Sciences and
    Sanders-Brown Center on Aging
University of Kentucky
Lexington, Kentucky

Mao Tanabe, Ph.D.
Human Genome Center
Institute of Medical Science
University of Tokyo
Tokyo, Japan

Liisa A. Tremere, Ph.D.
Department of Brain and Cognitive
    Sciences
University of Rochester
Rochester, New York

Gert Van den Bergh, Ph.D.
Laboratory of Neuroplasticity and
    Neuroproteomics
Department of Biology
Katholieke Universiteit Leuven
Leuven, Belgium

Alexei Vazquez, Ph.D.
The Simons Center for Systems Biology
Institute for Advanced Study
Princeton, New Jersey

Maria R. Warren, Ph.D.
UNC Duke Proteomics Center
Program in Molecular Biology and
    Biotechnology
University of North Carolina at
    Chapel Hill
Chapel Hill, North Carolina

Leonard E. White, Ph.D.
Department of Community and Family
    Medicine
Doctor of Physical Therapy Division
Department of Neurobiology
Center for the Study of Aging and
    Human Development
Duke University Medical Center
Durham, North Carolina

1 Neuroproteomics

Oscar Alzate

CONTENTS

1.1 INTRODUCTION

For the past several years, a large group of collaborators has been working together toward understanding key biological problems related to brain function, brain structure, and the complexity of the nervous system. Problems such as the structure and the function of pre- and post-synaptic densities, the sets of proteins that are regulated by mental processes such as learning and memory formation, the protein networks affected by apoE genotypes in Alzheimer's disease patients, and the structure of synaptic protein complexes in animal models of epilepsy, just to mention a few, have been under scrutiny in these studies. To tackle biological problems using neuroproteomics, we have learned that multiple experimental approaches need to be implemented. Techniques such as 2D-DIGE, mass spectrometry (MS), MS-based tissue imaging, protein arrays, surface plasmon resonance (SPR), protein interaction network analysis, multidimensional liquid chromatography, and many others are now in daily use. This book is intended to provide the reader with an introduction to some of the techniques that are most commonly used in neuroproteomics, and includes some examples of how such techniques are used to understand biological processes. A general overview of these techniques and their scope is discussed in this chapter.

Modern biotechnology is enjoying an explosion of new systematic approaches for the study of biomolecules and their interactions. The advent of genomics, proteomics, metabolomics, peptidomics, lipidomics, glycomics, and all the other "omics" is providing a vast amount of information that is enriching our knowledge of biological systems. In this book, a group of experts in neuroproteomics and its applications present the concept that understanding the dynamics of the proteome of a complex biological system requires the integration of many different experimental approaches.

1.2   NEUROPROTEOMICS

To define neuroproteomics we must start by understanding the term *proteomics*. Although there are many definitions of proteomics, what we mean here by proteomics is the study of a proteome (1), and a proteome is the complete set of proteins of an organ or an organism at a given time and under specific physiological conditions. A proteome is complex and refers to much more than the mere identification of the proteins in the set. In any given proteome, proteins may interact with a certain number of other proteins (or other molecules), determining how the protein functions as part of the whole system. In addition, protein structure and/or function can be altered by changes in the environment, including factors such as temperature, ionic strength, pH, levels of oxidants or anti-oxidants, etc. (2). The study of the proteome should provide information about all of these factors. Proteomics may start by elucidating the "proteome" at a specific time, but it should also determine the "dynamics" of this proteome under all the possible factors that affect the organ or organism. Thus proteomics, by its very nature, is faced with a huge task that requires the collaboration of multiple disciplines including physics, chemistry, biology, and bioinformatics.

Neuroproteomics is the sub-field of proteomics dedicated to answering these same questions about the organs, tissues, and cells that make up the nervous system (3–12). In neuroproteomics, our goals are (*i*) to identify all the proteins of a given tissue, cell type, or organelle under specific conditions at a specific time; (*ii*) to identify the post-translational modifications in all the proteins at that time and under these conditions; (*iii*) to determine how this proteome changes as a function of time (age), environmental changes, genetic factors, and with disease; and (*iv*) to determine how these changes affect the organism as a whole. At the present time, with existing technologies, the current knowledge of protein structure and function, our current knowledge about all possible protein–protein and protein–other molecule interactions, and the financial resources available, the full achievement of all of these goals is not possible. It *is* possible to find a "partial set" of proteins from a given tissue, or to determine how a subset of these proteins changes under the influence of some external factors such as a certain drug, etc. In fact, the complete definition of a proteome as presented above has not yet been determined for any organ or any tissue—not even for a single cell. This is what makes proteomics such an interesting field—despite so much that has been accomplished we realize that there is so much more to do.

In neuroproteomics the different pieces of the nervous system are "fragmented" so that the dynamics of each given sub-proteome can be better understood. Just to mention some examples (and this is not intended to be an exhaustive list), neuroproteomics works at solving the proteome of single neurons or astrocytes grown in cell cultures or from primary brain cells isolated from tissues under several conditions (13–16); at identifying a set of proteins characterizing a brain tumor (17–20); or at determining the set of proteins making up post-synaptic or pre-synaptic densities (21–27). It is also common to try to solve a specific sub-proteome such as the heat-shock response proteome (28–32), or the proteome responding to oxidative stress (33–37). From these examples, it can be seen that specific groups of proteins are targeted for analysis in a way that eventually will lead to solving a single proteome, and possibly being able to determine the dynamics of this proteome. The final goal will be to be able to predict "how" the proteome will evolve when influenced by specific conditions and to use this information to design methods that will modulate the evolution of the proteome. This is the ultimate dream for rational drug design, and molecular manipulation. The accumulation of huge amounts of data all around the world requires the advent of better information systems that will permit this "global" understanding of the dynamics of systems proteomics.

To accomplish the goals described above, proteomics requires the conjunction of many disciplines and techniques. In this book, we describe some of these techniques and give examples of several applications. A short description of the current approaches used in neuroproteomics follows next. For a more detailed description of some these techniques and their applications to proteomics, the reader is referred to specific chapters in this volume and to the literature cited therein.

1.3 EXPERIMENTAL TECHNIQUES CURRENTLY USED IN NEUROPROTEOMICS

1.3.1 MASS SPECTROMETRY

Mass spectrometry (MS) is the workhorse of proteomics. This technique offers tremendous power in terms of sensitivity to detect either digested peptides or intact proteins (38,39). Current developments allow targeting specific protein modifications such as phosphorylation, oxidation, ubiquitination, and others (7,40,41), albeit with varying degrees of difficulty and success. Unlike other approaches, such as x-ray diffraction that require intact proteins either in crystal form or in solution, respectively, MS requires that peptides or proteins be studied as ions in the gas phase.

As described in Chapter 5, matrix-assisted laser desorption/ionization (MALDI) and electro-spray ionization (ESI) are the most common protein and peptide ionization techniques used in MS-based proteomics. The availability of instrumentation and computer programs, together with the availability of protein databases that can be used for automated comparisons with the experimentally derived–MS and MS/MS (gas-phase sequencing) data, makes this the leading technique for protein identification and characterization. In principle, hundreds to thousands of proteins can be identified in a single sample using liquid chromatography (LC) separation coupled on-line with these ionization techniques (LC-MALDI and/or LC/MS/MS). The major obstacles that we face in applying MS to neuroproteomics are limitations in sample availability, which plagues most neuroproteomics experiments.

A recent application of MS to neuroproteomics research is the molecular imaging of brain tissues by MALDI, as described in Chapter 7. This approach, although it has advanced tremendously under the leadership of Dr. Richard Caprioli, is still in its infancy and its applications to neuroproteomics will be mainly for functional analysis of subsets of proteins and for understanding the onset and progression of neurological diseases. A great advantage of MALDI-based tissue imaging is the possibility of creating three-dimensional molecular images of the brain, or brain areas, and then using these images to determine the dynamic evolution of sub-proteomes.

Mass spectrometry is also widely used to characterize protein post-translational modifications (see Chapter 6). Trying to discover protein phosphorylation (42–44), ubiqutination (24), palmitoylation, (45,46), oxidation (see Chapter 10) (33,47–51), and other post-translational modifications (PTMs), and their roles in neurophysiology is a challenging aspect of neuroproteomics. Chapter 6 describes mass spectrometry–based approaches to characterize these and other PTMs relevant to brain function.

1.3.2 TWO-DIMENSIONAL POLYACRYLAMIDE GEL ELECTROPHORESIS

Two-dimensional acrylamide-based gel electrophoresis is a powerful technique that represents a tremendous resource for proteomics studies (see Chapters 3 and 4). Two-dimensional polyacrylamide gel electrophoresis (2D-PAGE) allows the separation of hundreds to thousands of proteins in a single experiment (40,52,53). Separated proteins may be identified by mass spectrometry, or some potential modifications may be analyzed "in" the gel. For example, some phosphorylated proteins can be identified by staining with phospho-specific stains (see Chapter 4) (32) or phospho-Ser/Thr-specific antibodies; oxidized proteins may be identified by Western blotting (see Chapter 10) (33,37,54,55). 2D-PAGE is an extremely useful technique that offers substantial advantages for quick and accurate protein separation and analysis, as well as providing an extremely useful way of "visualizing" a complex sample. Both 1D- and 2D-PAGE can be combined with LC/MS-based separation of gel extracts ("geLC") to provide an additional dimension of separation for complex mixtures.

1.3.3 TWO-DIMENSIONAL DIFFERENCE GEL ELECTROPHORESIS

Differential-display proteomics (comparative proteomics) can be applied using multidimensional liquid chromatography or 2D-gel-based electrophoresis. Differential-display proteomics is very powerful because it offers the advantage of comparing several samples in a single experiment. The samples to be compared may include a control (normal) versus an experimental (disease) sample (see Chapter 4) (6,53,56–59). The experimental sample is selected such that it reflects the effects of a specific condition including the effects of age, a drug, a gene, pH, oxidative stress, heat, etc. The most common approach for differential display is 2D-difference gel electrophoresis (2D-DIGE). This technique permits comparing thousands of fluorescently labeled proteins in a single experiment (see Chapter 4). With this technique, it is possible to determine changes in protein expression, as well as potential post-translational modifications (32,60,61). In addition, different types of fluorophores can be used to probe specific properties of a protein. These may include cysteine, tyrosine, lysine, and histidine modifications and, virtually, any modification for which a specific fluorophore can be found. 2D-DIGE can be used to study post-translational modifications including ubiquitination, phosphorylation, oxidation, and palmitoylation, among other modifications. As with any other technique, 2D-DIGE has limitations and advantages, and it is commonly understood in proteomics laboratories that a single technique will not suffice to answer all possible questions about a particular proteome. Proteins isolated and analyzed by 2D-DIGE can be identified and characterized by MS, or combined with immunoblotting for the analysis of specific subproteomes (3,21,62).

1.3.4   Liquid Chromatography

A workhorse technique that has experienced tremendous advances, liquid chromatography (LC) offers the advantage of separating proteins in a liquid phase (see Chapter 3) (6,63–65). Advances in all the technical aspects associated with LC make this a good complement for any proteomics laboratory. Proteins isolated by LC may be identified and characterized by MS, or they can be run on a 1D- or 2D-PAGE gel for further comparisons (6,63,65,66). In neuroproteomics, the major challenge for LC or LC/MS (see Chapters 3 and 5) is the scarce amount of samples for analysis. Large columns, with large volumes, are therefore not recommended; instead, nanoscale separations, micro-fluidic devices and affinity chromatography with specific antibodies or other molecules to enrich the target molecules may be the methods of choice (67,68). A major disadvantage of these nanoscale separations is the restrictions placed on flow rates as well as the challenges of reducing dead volumes. For on-line LC/MS, there are also restrictions on the selection of buffers and detergents. LC may be combined with differential display proteomics or extended to multidimensional approaches to provide a wider range of proteomics applications (see Chapter 3).

1.3.5   Protein Arrays

Protein arrays are designed to identify protein interactions using a solid surface to capture the proteins of interest, or to characterize a property of a protein of interest (16,41,53,69). This technique can be combined with other approaches such as 2D-DIGE and MS to identify specific proteins that interact with antibodies, peptides, or other suitable molecules properly attached to solid surfaces. In neuroproteomics, we use protein arrays to explore the changes in protein concentration, protein modifications, and protein–protein interactions in nervous systems such as neurons, axons, post-synaptic densities, etc. In this technique, protein lysates are incubated with arrays of antibodies against the protein of interest. (For a detailed discussion of protein arrays, protocols, and application see ref. 69.) The advantage of the arrays is that specific groups of proteins can be targeted for analysis—for instance, mitochondrial proteins, synaptic proteins, or membrane proteins. The molecules immobilized to the solid surface to produce the arrays can be specific for phosphorylated or oxidized proteins. In principle, this approach is comparable to multiplexed Western blots except instead of associating proteins one by one, molecular associations are made against a large number of proteins in a single experiment (69).

1.3.6 IMMUNOBLOT

Western blots (WBs) constitute a unique approach that allows identification of discrete proteins using targeted antibodies against specific proteins (70,71). Widely used in biological research, WB allows characterization of discrete changes in protein expression, and in combination with 2D gels, also allows one to detect changes in protein isoforms, or post-translational modifications (32,37,52,72). The current use of multiplexed WB using pre-labeled fluorescent antibodies permits quantitative characterization of multiple protein changes in a single experiment. Extensive use of WB can be used to determine protein changes in specific systems such as post-synaptic densities (73), mitochondria, etc. This approach is equivalent to a single protein array, or to a low-resolution, non-quantitative molecular image as obtained by MALDI-based tissue imaging (see Chapter 7). Several factors limit the application of immunoblotting for proteomics. These include the low specificity of many antibodies, and the lack of availability of pure antibodies. Moreover, antibodies simply do not exist for some proteins. Currently an antibody proteome project is under way that will provide a valuable tool for neuroproteomics research (http://www.hupo.org/research/hai/) (74,75).

1.3.7 ANALYTICAL ULTRACENTRIFUGATION

Analytical ultracentrifugation (AUC) offers a fast and reliable method for the determination of the molecular weight of a protein, and its hydrodynamic and thermodynamic properties (76,77). AUC is based on the thermodynamic analysis of sedimentation equilibrium, and may be used to determine sample purity, integrity of the structure, and degree of aggregation. The molecular weight determined by AUC is that of the *native* state of the protein, as opposed to the *unfolded* state as determined by gel electrophoresis, or in the gas state as determined by MS. This technique can be used for the study of small molecules (several hundreds of daltons of molecular weight such as small peptides) to multi-million-Da assemblies such as viruses or multicomplex proteins, and organelles (76–80). AUC can be applied to small samples in small volumes as are commonly found in neuroproteomics studies. Sedimentation equilibrium can be used to determine the molecular weights of protein complexes as they exist in solution, including the determination of aggregation states, and to study protein–protein interactions and protein interactions with small molecules (80–82).

1.3.8 SURFACE PLASMON RESONANCE

Surface plasmon resonance (SPR) is a powerful technique to determine protein–protein interactions in solution (83–87). This technique offers the incomparable ability to determine association (*Ka*) and dissociation (*Kd*) constants between two ligands (84,85). A combination of SPR and mass spectrometry allows the identification of protein interaction networks because the proteins bound to a ligand or a group of ligands can be identified by MS (86,87). This technique offers precise quantitative parameters that cannot be determined with protein arrays alone.

1.3.9 CIRCULAR DICHROISM

This technique is largely ignored in the proteomics field, mostly because high-throughput approaches have not been developed. Circular dichroism (CD) spectroscopy offers the advantage of providing a fast and reliable screening for protein secondary and tertiary conformations in solution (88). Many proteins associated with neurological diseases including the amyloid β peptides, α-synuclein, and prion proteins, for instance, form oligomeric conformations, which may be associated with onset and progression of some diseases (see Chapter 9) (89–93). CD is a reliable method for determining the folded conformations of these proteins. Implementing high-throughput CD will provide a method to categorize these proteins and their folded state. For now, neuroproteomics must use CD on discrete samples.

1.4 CHALLENGES

Neuroproteomics faces many challenges (7,8,38,66,94). For many years neuroscientists have been trying to answer questions such as "What is conscience?" "What are dreams made of?" "What are the physical substrates of memory and learning?" "What are the differences between short-term and long-term memories?" and so on. We are confident that neuroproteomics will offer a tool for neuroscientists working on these and many other brain- and mind-associated questions. It is not our expectation that proteomics alone will answer these questions, but instead it will be a powerful tool that will help in the search for the right answers (7,8,66). At the present time, being able to associate protein expression and protein modification with electrophysiological data and with some of the superior functions of the brain will be a good start. This is an area on which many research groups are working, and will be a fantastic beginning of a bright future for neuroproteomics.

Among the many challenges, the one that we face every day and that needs to be solved "up-front" so that data collection can be successful is sample preparation. The common presence of non-proteinaceous components such as nucleic acids, lipids, carbohydrates, and other biomolecules can affect the outcome of a proteomics analysis (see Chapters 3 and 4 for discussions on sample preparation). Many of these "contaminants" may actually be part of a functional proteome. Some of these biomolecules may be part of modulatory mechanisms for enzymatic reactions, including DNA-, RNA-, lipid-, glycolipid-, and carbohydrate–protein interactions. Therefore, elucidating what is a contaminant and what is part of modulatory mechanisms in the cell or organelle is a challenge that needs to be addressed.

Another problem that needs to be addressed is the availability of enough sample for proteomics analysis. For example, when working with post-synaptic densities (PSDs) it is possible to obtain low (tens of micrograms) amounts of proteins. In many cases, these protein lysates have to be cleaned for reliable analysis. During this cleaning process, the amount of protein may decrease between 10% and 40%. Assuming that the PSD contains several hundreds of proteins at any given time, this means that the sample contains on average only a few tens of nanograms of most proteins. Even this "best case" scenario represents the expression levels of the high abundance proteins. Unfortunately, protein identification and characterization by mass spectrometry require at least 20 to 30 ng in most cases. This means that many proteins—and proteins that are usually less abundant—will not be detected, and possibly may not be identified, or may be identified but with insufficient peptide coverage for modification site analysis. In most situations these problems need to be addressed on a case-by-case basis, and individual solutions may be found depending on the availability of animal models or tissues to solve particular situations. We should keep in mind that our goal is not to create catalogs of proteins; we are trying to tackle biological questions relevant to the nervous system. It is also important to keep in mind that in neuroproteomics we are interested in finding functional molecular pathways and not just simply trying to create a catalog of proteins in a tissue or an organelle. The major goal is to determine the *dynamics* of the proteome because "snapshots" alone do not provide enough valuable information about the changes occurring within the system and the other systems associated with it.

Post-translational modifications (PTMs) are another big challenge that neuroscientists need to address in their quest to understand the molecular mechanisms associated with brain functions. Some PTMs have been extensively studied, as is the case with phosphorylation, oxidation, and ubiquitination. The role of these PTMs in some cases has been addressed with optimistic results. Other PTMs such as sumoylation, palmitoylation, and methylation have been less studied but are equally important. However, there are a lot more PTMs than those mentioned above, and the complete picture of their roles in brain functions needs to be addressed (52,95–100). The techniques described in this book (see Chapters 3, 4, 6, and 7) provide some experimental approaches to address some of the questions associated with functional PTMs, but this is just the beginning. A lot more work needs to be done.

The field of neuroproteomics requires a lot more work and a lot of dedication to uncover the protein interaction networks (PINs) (see Chapters 8 and 9) (26,101–108). Currently, PINs are evaluated using genomics data, or by using discrete groups of proteins. However, a global view of the PINs associated with normal- and diseased-brain functions is still not available. Along with the identification of the PINs, it is very important to develop methodologies for their validation, and it is even more important to develop tools that will allow researchers to predict the function and the evolution of these networks under normal and disease conditions, under the effect of aging, or due to environmental factors.

A major push for neuroproteomics has been the as-yet-unfulfilled dream of finding "protein"-based biomarkers for neurodegenerative diseases. Thus far, proteomics has been unable to deliver on this dream. Why? There are many reasons why we still do not have hundreds of true protein biomarkers. We know a lot about individual proteins, but all of the aspects concerning a proteome (as described above) are not yet known even for a *single proteome*, and we know even less about the *dynamics* of proteomes. It will be possible to find true and very useful biomarkers, but this will have to wait a little longer while new technologies and—more importantly—new concepts are developed for determining a whole proteome and how it changes with time and environmental factors.

1.5 ORGANIZATION OF BOOK

The book contains two major parts: the first part describes basic concepts for those principles used in neuroproteomics (Chapters 2 to 8) (Figure 1.1), while the second part is dedicated to illustrating a few examples showing how scientists are using basic principles and techniques to understand molecular mechanisms of neurobiological processes (Chapters 9 to 15) (Figure 1.1). These examples have been selected while keeping in mind the ideas described above, i.e., that neuroproteomics is a set of tools to be used for understanding fundamental neurobiological questions, and not simply to create a catalog of nervous system-associated proteins.

As described earlier, one of the major challenges for neuroproteomics is sample preparation. Chapters 2, 3, and 4 discuss experimental considerations, methods, and concepts associated with sample preparation. Chapter 2 describes the experimental considerations for brain tissue collection and storage. Postmortem tissue is of major interest for research targeting molecular mechanisms of neurodegenerative

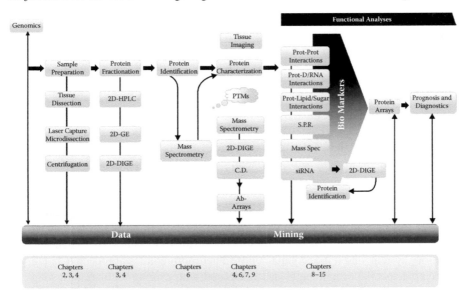

FIGURE 1.1 Neuroproteomics requires many tools of modern biotechnology. As indicated in the upper part of this figure, we are concerned with sample preparation, protein fractionation, protein identification, and protein characterization. Then, we use a combination of techniques such as mass spectrometry, protein arrays, 2D-DIGE, CD, gene silencing, ultracentrifugation, immunoprecipitation, and surface plasmon resonance to elucidate protein function, protein–protein, protein–R/DNA, protein–lipid, and protein–carbohydrate interactions. Using these data in the context of all the data available from public repositories, it is possible to propose functional protein interaction networks, which may be validated by using protein arrays and gene silencing followed by 2D-DIGE and mass spectrometry. The final goal will be the possibility of predicting how a protein network would evolve and how it will react under different experimental conditions. This will result in developing prognosis and diagnosis techniques for neurological diseases, and will allow us to create molecular models for neurological processes.

diseases such as Alzheimer's and Parkinson's diseases. Proper collection, storage, and manipulation of postmortem tissue are very important for reproducible and reliable proteomics analysis. The Bryan Brain Bank at the Department of Medicine and Neurology at Duke University has been banking brain tissues for several years. This bank specializes in brain specimens for neurodegenerative diseases, especially for Alzheimer's disease. Chapter 3 describes modern approaches for multidimensional separations for neuroproteomics. These approaches are mainly based on liquid chromatography and are appropriate for every proteomics method. Chapter 4 is devoted to differential gel electrophoresis (DIGE)-based proteomics. Methods required to prepare proteins for DIGE are described in this chapter, along with experimental considerations for a successful DIGE-based experiment. A discussion of DIGE applications to neuroproteomics is presented in Chapter 4.

The technique most widely used for proteomics is mass spectrometry (MS); three chapters are dedicated to mass spectrometry. Chapter 5 introduces the major concepts of MS for proteomics analysis, including sources, detectors, and analyzers. Chapter 6 presents major applications of mass spectrometry for the analysis of posttranslational modifications of proteins. A new application of MS is MALDI-based tissue imaging. This technique has the potential to become one of the more useful techniques for neuroproteomics. Chapter 7 describes the concepts, techniques, and major applications of MALDI-based tissue imaging. Chapter 8 introduces the concept of protein interaction networks, a new and rapidly evolving field based on the theory of graphs that connects the elements of any group of proteins to other proteins in the system, and how the inter- and intra-connectivity determines the overall behavior of the system (101,104–107).

The second part of the book starts with an example of the identification of a protein interaction network in neurodegeneration (Chapter 9). Chapter 10 analyzes a large amount of data for studies of oxidized proteins in neurodegenerative diseases. This chapter describes the use of 2D gels, mass spectrometry, and 2D Western blots to tackle analysis of protein post-translational modifications. Chapters 11 and 12 present the use of neuroproteomics to study specific aspects of molecular mechanisms in the visual system. Chapter 13 discusses networks of genes associated with the auditory system, and it is accompanied by a neuroproteomics approach (Chapter 14) analyzing specific protein pathways associated with singing regulation in songbirds. The last chapter, Chapter 15, presents several neuroproteomics approaches to understand molecular mechanisms associated with nerve regeneration.

1.6 FINAL REMARKS

This volume is intended as a general introduction to the concepts, techniques, and applications of neuroproteomics. As presented in Figure 1.1, for neuroproteomics studies we have to utilize multiple techniques for sample preparation, protein fractionation, and protein identification. The combination of these approaches results in the creation of extensive catalogs of proteins associated with a tissue or a system; however, these catalogs are useless unless we can identify the function for each protein, the network to which each protein belongs, and the dynamics of the network under experimental conditions. Facing these challenges requires the combination of many approaches, some of which have matured enough to be used in proteomics, and some which are new and still under development. Combining mass spectrometry with surface plasmon resonance, gene silencing, 2D-DIGE, and protein arrays will help to discern these protein interaction networks and their dynamics.

It should be mentioned here that it is not possible to cover all aspects of neuroproteomics in a single volume. The field is growing so fast that it is not even possible to review all the available literature. There are thousands of scientists on all continents contributing to the many different fields of neuroproteomics, including both technical developments and specific applications. In the present volume, we tried to cover us much as possible of the current literature available, and give specific examples that demonstrate exactly what neuroproteomics is, and how and why it is used. I would like to apologize to all of those researchers whose publications were not cited, which is not a disregard for their work, but rather the result of how fast this field is growing, and lack of space in this single volume. Also, we expect that the future will bring applications of techniques such as x-ray crystallography, NMR, and EPR (electron paramagnetic resonance) to neuroproteomics.

Finally I want to thank all the authors who have contributed to this volume. It has been a difficult task that started more than five years ago, when we started designing experiments, preparing samples, collecting data, and trying to validate as much as possible the results obtained by proteomics approaches. Since then, the field has grown considerably and today it is almost impossible to know everything being done in neuroproteomics and its applications. This makes the efforts of all of the authors more valuable as they have tried to give an up-to-date view of particular aspects of this rapidly developing field.

REFERENCES

1. Anderson, N. L. and Anderson, N. G. (1998) Proteome and proteomics: New technologies, new concepts, and new words, *Electrophoresis* 19:1853–61.
2. Creighton, T. E. (1993) *Proteins: Structures and molecular properties* (New York: W. H. Freeman).
3. Tribl, F., Meyer, H. E. and Marcus, K. (2008) Analysis of organelles within the nervous system: Impact on brain and organelle functions, *Expert Rev Proteomics* 5:333–51.
4. Fountoulakis, M. (2004) Application of proteomics technologies in the investigation of the brain, *Mass Spectrom Rev* 23:231–58.
5. Wilson, K. E., Ryan, M. M., Prime, J. E., et al. (2004) Functional genomics and proteomics: Application in neurosciences, *J Neurol Neurosurg Psychiatry* 75:529–38.
6. Tannu, N. S. and Hemby, S. E. (2006) Methods for proteomics in neuroscience, *Prog Brain Res* 158:41–82.
7. Becker, M., Schindler, J. and Nothwang, H. G. (2006) Neuroproteomics—The tasks lying ahead, *Electrophoresis* 27:2819–29.
8. Butcher, J. (2007) Neuroproteomics comes of age, *Lancet Neurol* 6:850–1.
9. Abul-Husn, N. S. and Devi, L. A. (2006) Neuroproteomics of the synapse and drug addiction, *J Pharmacol Exp Ther* 318:461–8.
10. Kim, S. I., Voshol, H., van Oostrum, J. et al. (2004) Neuroproteomics: Expression profiling of the brain's proteomes in health and disease, *Neurochem Res* 29:1317–31.

11. Tribl, F., Marcus, K., Bringmann, G. et al. (2006) Proteomics of the human brain: Sub-proteomes might hold the key to handle brain complexity, *J Neural Transm* 113:1041–54.
12. Marcus, K., Schmidt, O., Schaefer, H. et al. (2004) Proteomics—Application to the brain, *Int Rev Neurobiol* 61:285–311.
13. Boeddrich, A., Lurz, R. and Wanker, E. E. (2003) Huntingtin fragments form aggresome-like inclusion bodies in mammalian cells, *Methods Mol Biol* 232:217–29.
14. Schubert, D., Herrera, F., Cumming, R. et al. (2009) Neural cells secrete a unique repertoire of proteins, *J Neurochem* 109:427–35.
15. Liu, J., Liu, M. C. and Wang, K. K. (2008) Physiological and pathological actions of calpains in glutamatergic neurons, *Sci Signal* 1:tr3.
16. Maurer, M. H. and Kuschinsky, W. (2006) Screening the brain: Molecular fingerprints of neural stem cells, *Curr Stem Cell Res Ther* 1:65–77.
17. Chen, H. B., Pan, K., Tang, M. K. et al. (2008) Comparative proteomic analysis reveals differentially expressed proteins regulated by a potential tumor promoter, BRE, in human esophageal carcinoma cells, *Biochem Cell Biol* 86:302–11.
18. Chumbalkar, V., Sawaya, R. and Bogler, O. (2008) Proteomics: The new frontier also for brain tumor research, *Curr Probl Cancer* 32:143–54.
19. Graner, M. W., Alzate, O., Dechkovskaia, A. M. et al. (2009) Proteomic and immunologic analyses of brain tumor exosomes, *FASEB J* 23:1541–57.
20. Whittle, I. R., Short, D. M., Deighton, R. F. et al. (2007) Proteomic analysis of gliomas, *Br J Neurosurg* 21:576–82.
21. Bai, F. and Witzmann, F. A. (2007) Synaptosome proteomics, *Subcell Biochem* 43:77–98.
22. Cheng, D., Hoogenraad, C. C., Rush, J. et al. (2006) Relative and absolute quantification of postsynaptic density proteome isolated from rat forebrain and cerebellum, *Mol Cell Proteomics* 5:1158–70.
23. Emes, R. D., Pocklington, A. J., Anderson, C. N. et al. (2008) Evolutionary expansion and anatomical specialization of synapse proteome complexity, *Nat Neurosci* 11:799–806.
24. Jordan, B. A., Fernholz, B. D., Boussac, M. et al. (2004) Identification and verification of novel rodent postsynaptic density proteins, *Mol Cell Proteomics* 3:857–71.
25. Li, K. W., Hornshaw, M. P., Van Der Schors, R. C. et al. (2004) Proteomics analysis of rat brain postsynaptic density. Implications of the diverse protein functional groups for the integration of synaptic physiology, *J Biol Chem* 279:987–1002.
26. Li, K. W. and Jimenez, C. R. (2008) Synapse proteomics: Current status and quantitative applications, *Expert Rev Proteomics* 5:353–60.
27. Schrimpf, S. P., Meskenaite, V., Brunner, E. et al. (2005) Proteomic analysis of synaptosomes using isotope-coded affinity tags and mass spectrometry, *Proteomics* 5:2531–41.
28. Arimon, M., Grimminger, V., Sanz, F. and Lashuel, H. A. (2008) Hsp104 targets multiple intermediates on the amyloid pathway and suppresses the seeding capacity of Abeta fibrils and protofibrils, *J Mol Biol* 384:1157–73.
29. Fountoulakis, M. and Kossida, S. (2006) Proteomics-driven progress in neurodegeneration research, *Electrophoresis* 27:1556–73.
30. Herbst, M. and Wanker, E. E. (2006) Therapeutic approaches to polyglutamine diseases: Combating protein misfolding and aggregation, *Curr Pharm Des* 12:2543–55.
31. Herbst, M. and Wanker, E. E. (2007) Small molecule inducers of heat-shock response reduce polyQ-mediated huntingtin aggregation. A possible therapeutic strategy, *Neurodegener Dis* 4:254–60.
32. Osorio, C., Sullivan, P. M., He, D. N. et al. (2007) Mortalin is regulated by APOE in hippocampus of AD patients and by human APOE in TR mice, *Neurobiol Aging* 28:1853–62.

33. Butterfield, D. A. (2004) Proteomics: A new approach to investigate oxidative stress in Alzheimer's disease brain, *Brain Res* 1000:1–7.
34. Ding, Q., Dimayuga, E. and Keller, J. N. (2007) Oxidative damage, protein synthesis, and protein degradation in Alzheimer's disease, *Curr Alzheimer Res* 4:73–9.
35. Forero, D. A., Casadesus, G., Perry, G. and Arboleda, H. (2006) Synaptic dysfunction and oxidative stress in Alzheimer's disease: Emerging mechanisms, *J Cell Mol Med* 10z:796–805.
36. Sultana, R., Perluigi, M. and Butterfield, D. A. (2006) Protein oxidation and lipid peroxidation in brain of subjects with Alzheimer's disease: Insights into mechanism of neurodegeneration from redox proteomics, *Antioxid Redox Signal* 8:2021–37.
37. Sultana, R., Perluigi, M. and Butterfield, D. A. (2006) Redox proteomics identification of oxidatively modified proteins in Alzheimer's disease brain and in vivo and in vitro models of AD centered around Abeta(1-42), *J Chromatogr B Analyt Technol Biomed Life Sci* 833:3–11.
38. Li, K. W. and Smit, A. B. (2008) Subcellular proteomics in neuroscience, *Front Biosci* 13:4416–25.
39. Zetterberg, H., Ruetschi, U., Portelius, E. et al. (2008) Clinical proteomics in neurodegenerative disorders, *Acta Neurol Scand* 118:1–11.
40. Van den Bergh, G. and Arckens, L. (2005) Recent advances in 2D electrophoresis: An array of possibilities, *Expert Rev Proteomics* 2:243–52.
41. Williams, K., Wu, T., Colangelo, C. and Nairn, A. C. (2004) Recent advances in neuroproteomics and potential application to studies of drug addiction, *Neuropharmacology* 47 (Suppl 1):148–66.
42. Grant, S. G. (2006) The synapse proteome and phosphoproteome: A new paradigm for synapse biology, *Biochem Soc Trans* 34:59–63.
43. Jin, M., Bateup, H., Padovan, J. C. et al. (2005) Quantitative analysis of protein phosphorylation in mouse brain by hypothesis-driven multistage mass spectrometry, *Anal Chem* 77:7845–51.
44. Hanger, D. P., Betts, J. C., Loviny, T. L., Blackstock, W. P. and Anderton, B. H. (1998) New phosphorylation sites identified in hyperphosphorylated tau (paired helical filament-tau) from Alzheimer's disease brain using nanoelectrospray mass spectrometry, *J Neurochem* 71:2465–76.
45. Wan, J., Roth, A. F., Bailey, A. O. and Davis, N. G. (2007) Palmitoylated proteins: Purification and identification, *Nat Protoc* 2:1573–84.
46. Yang, J., Gibson, B., Snider, J. et al. (2005) Submicromolar concentrations of palmitoyl-CoA specifically thioesterify cysteine 244 in glyceraldehyde-3-phosphate dehydrogenase inhibiting enzyme activity: A novel mechanism potentially underlying fatty acid induced insulin resistance, *Biochemistry* 44:11903–12.
47. Barnes, S., Shonsey, E. M., Eliuk, S. M. et al. (2008) High-resolution mass spectrometry analysis of protein oxidations and resultant loss of function, *Biochem Soc Trans* 36:1037–44.
48. Butterfield, D. A. and Sultana, R. (2007) Redox proteomics identification of oxidatively modified brain proteins in Alzheimer's disease and mild cognitive impairment: Insights into the progression of this dementing disorder, *J Alzheimers Dis* 12:61–72.
49. Hutson, S. M., Poole, L. B., Coles, S. and Conway, M. E. (2008) Redox regulation and trapping sulfenic acid in the peroxide-sensitive human mitochondrial branched chain aminotransferase, *Methods Mol Biol* 476:139–52.
50. Paige, J. S., Xu, G., Stancevic, B. and Jaffrey, S. R. (2008) Nitrosothiol reactivity profiling identifies S-nitrosylated proteins with unexpected stability, *Chem Biol* 15:1307–16.
51. Pamplona, R., Naudi, A., Gavin, R. et al. (2008) Increased oxidation, glycoxidation, and lipoxidation of brain proteins in prion disease, *Free Radic Biol Med* 45:1159–66.

52. Kim, H., Eliuk, S., Deshane, J. et al. (2007) 2D gel proteomics: An approach to study age-related differences in protein abundance or isoform complexity in biological samples, *Methods Mol Biol* 371:349–91.
53. Lubec, G., Krapfenbauer, K. and Fountoulakis, M. (2003) Proteomics in brain research: Potentials and limitations, *Prog Neurobiol* 69:193–211.
54. Choi, J., Forster, M. J., McDonald, S. R. et al. (2004) Proteomic identification of specific oxidized proteins in ApoE-knockout mice: Relevance to Alzheimer's disease, *Free Radic Biol Med* 36:1155–62.
55. Mori, H., Oikawa, M., Tamagami, T. et al. (2007) Oxidized proteins in astrocytes generated in a hyperbaric atmosphere induce neuronal apoptosis, *J Alzheimers Dis* 11:165–74.
56. Alzate, O., Hussain, S. R., Goettl, V. M. et al. (2004) Proteomic identification of brainstem cytosolic proteins in a neuropathic pain model, *Brain Res Mol Brain Res* 128:193–200.
57. Jacobs, S., Van de Plas, B., Van der Gucht, E. et al. (2008) Identification of new regional marker proteins to map mouse brain by 2-D difference gel electrophoresis screening, *Electrophoresis* 29:1518–24.
58. Pinaud, R., Osorio, C., Alzate, O. and Jarvis, E. D. (2008) Profiling of experience-regulated proteins in the songbird auditory forebrain using quantitative proteomics, *Eur J Neurosci* 27:1409–22.
59. Swatton, J. E., Prabakaran, S., Karp, N. A., Lilley, K. S. and Bahn, S. (2004) Protein profiling of human postmortem brain using 2-dimensional fluorescence difference gel electrophoresis (2-D DIGE), *Mol Psychiatry* 9:128–43.
60. Riederer, B. M. (2008) Non-covalent and covalent protein labeling in two-dimensional gel electrophoresis, *J Proteomics* 71:231–44.
61. Scholz, B., Svensson, M., Alm, H. et al. (2008) Striatal proteomic analysis suggests that first L-dopa dose equates to chronic exposure, *PLoS ONE* 3:e1589.
62. Seshi, B. (2007) Proteomics strategy based on liquid-phase IEF and 2-D DIGE: Application to bone marrow mesenchymal progenitor cells, *Proteomics* 7:1984–99.
63. Haskins, W. E., Kobeissy, F. H., Wolper, R. A. et al. (2005) Rapid discovery of putative protein biomarkers of traumatic brain injury by SDS-PAGE-capillary liquid chromatography-tandem mass spectrometry, *J Neurotrauma* 22:629–44.
64. Kobeissy, F. H., Warren, M. W., Ottens, A. K. et al. (2008) Psychoproteomic analysis of rat cortex following acute methamphetamine exposure, *J Proteome Res* 7:1971–83.
65. Ottens, A. K., Kobeissy, F. H., Wolper, R. A. et al. (2005) A multidimensional differential proteomic platform using dual-phase ion-exchange chromatography-polyacrylamide gel electrophoresis/reversed-phase liquid chromatography tandem mass spectrometry, *Anal Chem* 77:4836–45.
66. Andrade, E. C., Krueger, D. D. and Nairn, A. C. (2007) Recent advances in neuroproteomics, *Curr Opin Mol Ther* 9:270–81.
67. Chen, H. and Fan, Z. H. (2009) Two-dimensional protein separation in microfluidic devices, *Electrophoresis* 30:758–65.
68. Moon, H., Wheeler, A. R., Garrell, R. L., Loo, J. A. and Kim, C. J. (2006) An integrated digital microfluidic chip for multiplexed proteomic sample preparation and analysis by MALDI-MS, *Lab Chip* 6:1213–9.
69. Fung, E. (2004) *Protein arrays: Methods and protocols* (Totowa, NJ: Humana Press).
70. Dickson, C. (2008) Protein techniques: Immunoprecipitation, in vitro kinase assays, and Western blotting, *Methods Mol Biol* 461:735–44.
71. Gallagher, S. and Chakavarti, D. (2008) Immunoblot analysis, *J Vis Exp* Jun 20.
72. Barnouin, K. (2004) Two-dimensional gel electrophoresis for analysis of protein complexes, *Methods Mol Biol* 261:479–98.

73. Ehlers, M. D. (2003) Activity level controls postsynaptic composition and signaling via the ubiquitin-proteasome system, *Nat Neurosci* 6:231–42.

74. Persson, A., Hober, S. and Uhlen, M. (2006) A human protein atlas based on antibody proteomics, *Curr Opin Mol Ther* 8:185–90.

75. Saerens, D., Ghassabeh, G. H. and Muyldermans, S. (2008) Antibody technology in proteomics, *Brief Funct Genomic Proteomic* 7:275–82.

76. Schachman, H. K. (1959) *Ultracentrifugation in biochemistry* (New York: Academic Press).

77. Van Holde, K. E. (1975) *Sedimentation analysis of proteins* (New York: Academic Press).

78. Creeth, J. M. and Knight, C. G. (1967) The macromolecular properties of blood-group substances. Sedimentation-velocity and viscosity measurements, *Biochem J* 105:1135–45.

79. Creeth, J. M. and Pain, R. H. (1967) The determination of molecular weights of biological macromolecules by ultracentrifuge methods, *Prog Biophys Mol Biol* 17:217–87.

80. Teller, D. C. (1973) Characterization of proteins by sedimentation equilibrium in the analytical ultracentrifuge, *Methods Enzymol* 27:346–441.

81. Geerlof, A., Brown, J., Coutard, B. et al. (2006) The impact of protein characterization in structural proteomics, *Acta Crystallogr D Biol Crystallogr* 62:1125–36.

82. Stafford, W. F., 3rd (2009) Protein-protein and ligand-protein interactions studied by analytical ultracentrifugation, *Methods Mol Biol* 490:83–113.

83. Ahmed, F. E. (2008) Mining the oncoproteome and studying molecular interactions for biomarker development by 2DE, ChIP and SPR technologies, *Expert Rev Proteomics* 5:469–96.

84. Jonsson, U., Fagerstam, L., Ivarsson, B. et al. (1991) Real-time biospecific interaction analysis using surface plasmon resonance and a sensor chip technology, *Biotechniques* 11:620–7.

85. Myszka, D. G. (1997) Kinetic analysis of macromolecular interactions using surface plasmon resonance biosensors, *Curr Opin Biotechnol* 8:50–7.

86. Visser, N. F. and Heck, A. J. (2008) Surface plasmon resonance mass spectrometry in proteomics, *Expert Rev Proteomics* 5:425–33.

87. Yuk, J. S. and Ha, K. S. (2005) Proteomic applications of surface plasmon resonance biosensors: Analysis of protein arrays, *Exp Mol Med* 37:1–10.

88. Greenfield, N. J. (2006) Using circular dichroism spectra to estimate protein secondary structure, *Nat Protoc* 1:2876–90.

89. Kahle, P. J. and Haass, C. (2004) How does parkin ligate ubiquitin to Parkinson's disease? *EMBO Rep* 5:681–5.

90. Matus, S., Lisbona, F., Torres, M. et al. (2008) The stress rheostat: An interplay between the unfolded protein response (UPR) and autophagy in neurodegeneration, *Curr Mol Med* 8:157–72.

91. Sahara, N., Maeda, S. and Takashima, A. (2008) Tau oligomerization: A role for tau aggregation intermediates linked to neurodegeneration, *Curr Alzheimer Res* 5:591–8.

92. Uehara, T. (2007) Accumulation of misfolded protein through nitrosative stress linked to neurodegenerative disorders, *Antioxid Redox Signal* 9:597–601.

93. Uversky, V. N. (2008) Alpha-synuclein misfolding and neurodegenerative diseases, *Curr Protein Pept Sci* 9:507–40.

94. Johnson, M. D., Yu, L. R., Conrads, T. P. et al. (2005) The proteomics of neurodegeneration, *Am J Pharmacogenomics* 5:259–70.

95. Chun, W. and Johnson, G. V. (2007) The role of tau phosphorylation and cleavage in neuronal cell death, *Front Biosci* 12:733–56.

96. Ferrer, I., Martinez, A., Boluda, S., Parchi, P. and Barrachina, M. (2008) Brain banks: Benefits, limitations and cautions concerning the use of post-mortem brain tissue for molecular studies, *Cell Tissue Bank* 9:181–94.

97. Gutstein, H. B., Morris, J. S., Annangudi, S. P. and Sweedler, J. V. (2008) Microproteomics: Analysis of protein diversity in small samples, *Mass Spectrom Rev* 27:316–30.

98. Ogasawara, H., Doi, T. and Kawato, M. (2008) Systems biology perspectives on cerebellar long-term depression, *Neurosignals* 16:300–17.

99. Routtenberg, A. (2008) The substrate for long-lasting memory: If not protein synthesis, then what? *Neurobiol Learn Mem* 89:225–33.

100. Sunyer, B., Diao, W. and Lubec, G. (2008) The role of post-translational modifications for learning and memory formation, *Electrophoresis* 29:2593–2602.

101. Bullmore, E. and Sporns, O. (2009) Complex brain networks: Graph theoretical analysis of structural and functional systems, *Nat Rev Neurosci* 10:186–98.

102. Grant, S. G. and Husi, H. (2001) Proteomics of multiprotein complexes: Answering fundamental questions in neuroscience, *Trends Biotechnol* 19:S49–54.

103. Stelzl, U. and Wanker, E. E. (2006) The value of high quality protein-protein interaction networks for systems biology, *Curr Opin Chem Biol* 10:551–8.

104. Volonte, C., D'Ambrosi, N. and Amadio, S. (2008) Protein cooperation: From neurons to networks, *Prog Neurobiol* 86:61–71.

105. Venkatesan, K., Rual, J. F., Vazquez, A. et al. (2009) An empirical framework for binary interactome mapping, *Nat Methods* 6:83–90.

106. Yu, H., Braun, P., Yildirim, M. A. et al. (2008) High-quality binary protein interaction map of the yeast interactome network, *Science* 322:104–10.

107. Yook, S. H., Oltvai, Z. N. and Barabasi, A. L. (2004) Functional and topological characterization of protein interaction networks, *Proteomics* 4:928–42.

108. Wuchty, S., Oltvai, Z. N. and Barabasi, A. L. (2003) Evolutionary conservation of motif constituents in the yeast protein interaction network, *Nat Genet* 35:176–9.

2 Banking Tissue for Neurodegenerative Research

John F. Ervin

CONTENTS

2.1 INTRODUCTION

Human brain banking has become an essential part of the research landscape in neurodegenerative disorders and neurobiology. The demand for high quality banked tissue has been on a steady rise for quite some time. Advanced research studies, including proteomics, metabolomics, m-RNA micro arrays, and genomics, are fast becoming the standard in neuroscience investigations. Since many investigators study human diseases or biological processes, it is therefore not a surprise that human tissue is in high demand to verify findings from animal models of disease. Many leading neuroscientists are focusing a large component of their research on techniques that require the collection of the highest quality of human brain tissue. The Kathleen Price Bryan Brain Bank (KPBBB; http://adrc. mc.duke.edu/BB.htm) at Duke University Medical Center (DUMC) in Durham, North Carolina, has over 20 years of experience with this process (1). Successful human brain banking requires not only attention to the users' needs for the highest quality of tissue, but it is also imperative for brain bankers to ensure that the donor's wishes are honored. There have been great strides on a national level to facilitate the availability and distribution of these resources to the ever-growing demand in the neuroscience community. The National Institute on Aging (NIA) through the National Alzheimer's Coordinating Center (NACC) has created an infrastructure and informatics network to support collaboration among the individual NIA-funded Alzheimer's Disease Centers (ADCs) and to serve as a resource for the neuroscience research community. The banked tissue thus obtained is an invaluable resource available to qualified researchers. This chapter describes general concepts concerning proper acquisition, storage, and distribution of brain tissue for neurodegenerative research.

2.2   TISSUE PROCUREMENT AND STORAGE

Although each ADC is required to report specific neuropathological data concerning the tissue collected, each brain bank often varies in its procedures for procuring its tissue for research. A protocol has not been established or accepted that instructs how each bank should go about its task of tissue collection. Historically, each bank may collect tissue according to different protocols in order to accommodate the specific needs of the investigators who most often use its tissue (2,3). The differences in collection and storage protocols among ADC brain banks today are often driven by the availability of resources and funding (4). Neuroscience investigators involved with the NACC have become more organized in their efforts to amplify the usefulness of the individual brain banks that have collected tissue over the last several decades. However, there has been some discussion about how modern brain banks should be collecting and storing tissues in a way that provides high quality tissue for a wider range of applications (5).

In addition to brain banks that are part of the NACC and focus on Alzheimer's disease (AD), there are other neurodegenerative investigative groups that have established human brain banking resources. Parkinson's, Huntington's, amyotrophic lateral sclerosis (ALS), and Creutzfeldt–Jakob disease (CJD) are also areas of research that need human tissue to study (6,7). Although a majority of brain banks focus on neurodegenerative research, there are several banks collecting tissue from individuals with psychiatric disorders, brain tumors, developmental anomalies and malformations, and brain tissue from acquired immunodeficiency syndrome (AIDS) patients who are infected with human immunodeficiency virus (HIV) (8,9). Each of these types of brain banks implements different protocols with a focus on collecting the regions most affected in each disease process while emphasizing any special safety and banking precautions that are needed for the inherent risks working with infectious material (10,11). Regardless of the type of tissue banked, anyone who works with fresh human tissue should take the necessary safety steps and follow universal precautions for protection against the spread of infectious diseases.

The KPBBB has been collecting tissue since 1985 with an emphasis in Alzheimer's disease and other dementing illnesses. It takes a large amount of effort and coordination between a variety of committed personnel with highly specialized skills in order to successfully retrieve, bank, and distribute quality tissue for today's neuroscience investigators. Having the proper facilities and trained personnel in place is not a trivial matter when running a successful brain bank. The personnel and facilities that are required often cross over different departments and disciplines within a hospital. Coordinating and managing a working relationship among multiple entities within a hospital setting can be a very laborious task. There are always funding and cost issues that must be worked out by the collaborating parties. There are safety concerns, a growing number of legal issues that must be fully addressed, and strict adherence to the federal law is paramount when banking and distributing human tissue. Without a knowledgeable liaison involving the different components, it would be very difficult to achieve a successful working relationship that is agreeable to all those involved. The quality of tissue collected has a direct relationship to the effort put forth by each and every person involved with coordinating the chain of events that culminates with the banking of tissue that can be used for a variety of scientific experiments.

The most useful specimens that are banked are derived from patients' cases that have been followed clinically and have a documented medical history that can be used in conjunction with the neuropathological diagnoses. The process for procuring the highest quality tissue often starts many years before the tissue is actually banked. The neuropathology core is one component of a multidisciplinary team. The Joseph and Kathleen Bryan Alzheimer's Disease Research Center at Duke University consists of a clinical core where patients are seen and treated at the clinic for memory disorders. The clinical core facilitates the enrollment of their patients in a variety of clinical studies as well as providing education for patients and family members about the autopsy donation program.

Once a person decides to participate and become a tissue donor, the autopsy nurse coordinator establishes a relationship with the donor and the donor's family. Quite often, many years may transpire between enrollment and tissue procurement. At the time of enrollment, there are many legal documents that must be explained and signed. There is a special relationship formed by the donor, the donor's family, and the coordinating nurse. These relationships are built on many discussions about the donor and the family that can be quite personal in nature. The nurse coordinator is a trusted source of information about what will happen to a loved one who has decided to enroll in the autopsy program. These relationships are a fundamental part of a successful donation program and they have translated into a high rate of participants who are comfortable and willing to get follow-up exams to monitor their progress as they age. It is important to remain in contact with the donor and his or her family, so the nurse coordinator frequently makes house visits and may administer a neuropsychological field test if the donor is not able to come to the clinic. However, in the late stages of AD, the donor's dementia may have progressed to the point that a neuropsychological evaluation cannot be performed. As donors age or they enter an advanced disease state, it becomes even more critical for the donor's family to contact the nurse coordinator so that when the last phase of the process begins everyone is prepared.

A call from the next of kin or nursing facility usually notifies the nurse coordinator that the donor has reached a point where he or she may expire in the next few days. If possible, any perimortem observations are recorded and the autopsy nurse coordinator will enter them into the medical record. Once the nurse coordinator has been notified of a death several things must then be done in preparation for the autopsy. Arrangements for the body to arrive at the hospital by a reliable transportation service will be required. Any available medical records and a signed consent for autopsy must be given to the pathologist assigned to the case. The brain bank and autopsy technicians are notified of the death as soon as possible in order to prepare items for tissue collection and transport to the autopsy suite. If a donor lives too far away to transport the body to DUMC for an autopsy to begin in a timely manner, other arrangements can be made. Prior arrangements can be made for the tissue to be retrieved and shipped to the brain bank from almost anywhere in the United States. The nurse coordinator can help a family get in contact with a facility and a neuro-pathologist who are willing to help retrieve and ship the tissue to the brain bank. The autopsy program nurse coordinator is a critical part of the team that facilitates a smooth transition from donor enrollment to successful performance of the autopsy.

When the autopsy coordinator arranges for the transportation of the body to DUMC, it is optimal for the body to be cooled down prior to arriving at the autopsy suite. Depending on the expected delay to the hospital, the body may have been refrigerated before delivery to the autopsy suite. The head would have been chilled using a bag of wet ice in order to start cooling the brain. The autopsy technicians will begin removing the brain as soon as the pathologist is finished with the external exam. Ventricular cerebral spinal fluid (CSF) is taken using a large needle inserted through the lateral side of the brain. The CSF is placed on ice for transport back to the lab where it will be aliquoted into 1 mL vials and stored in a −80°C freezer. Upon removal from the cranium, the brain is weighed and grossly inspected. The specimen is placed on wet ice in order to further cool the tissue. It is standard procedure for the brain to be bisected through the corpus callosum, cerebellum, and brain stem. This divides the brain into left and right hemispheres. One hemisphere of the brain will be submerged in a tissue fixative such as 10% phosphate-buffered formalin for a mini-mum of one week. The specimen can be dissected and paraffin-embedded after one week of fixation. The embedded tissues are used for preparation of routine histologic sections and a battery of immunohistological stains to detect plaques, tangles, Lewy bodies or other inclusions. Before any tissue is used for research it must be evaluated by a neuropathologist and the data catalogued in the research database. The fixed tissue can remain submerged in formalin and kept at room temperature for long-term storage. The other hemisphere may be fresh-frozen and kept free of any chemical processes. The cerebellum and brain stem may be removed from the neocortex at the level of the third nerve. This cut will expose the substantia nigra (Figure 2.1) so that it can be removed from the brain stem and stored separately.

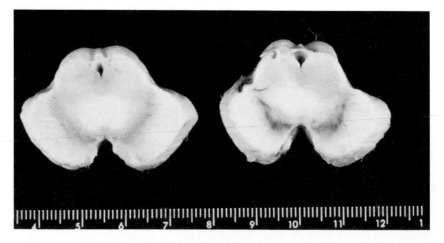

FIGURE 2.1 Substantia nigra: abnormal left; normal right.

The fresh-frozen hemisphere will be coronally sliced 1- to 2-centimeters thick from frontal pole to occipital pole. Cutting the hemisphere into coronal slices allows access to internal structures that may be dissected out and stored. Commonly dissected structures include the hippocampus (Figure 2.2A), amygdala, and basal ganglia (Figure 2.2B). Dissected pieces are placed in labeled aluminum foil and placed into a container of liquid nitrogen to quickly freeze the tissue. The coronal slices are placed in numbered freezer bags beginning with the frontal lobe and ending with the

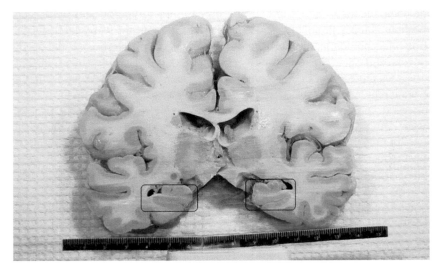

FIGURE 2.2A Hippocampus shown inside the small rectangles.

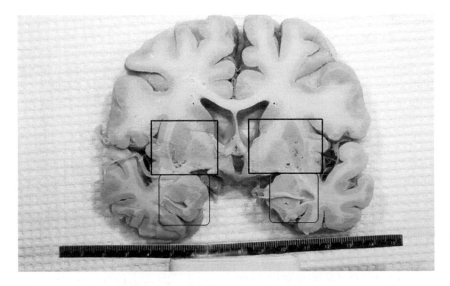

FIGURE 2.2B Amygdala within small rectangles; basal ganglia within the large rectangles.

occipital lobe. These bags are placed on wet ice for transportation back to the labora-tory. The cerebellum and brain stem are placed in separate bags to be transported and frozen with the coronal slices. The bags are frozen between chilled steel plates in an ultra low −80°C freezer. Long-term storage of frozen specimens at most banks is in ultra low −80°C freezers.

In advance of an autopsy, there are three things that must be considered. How will the tissue be dissected, preserved, and stored? As discussed above, not all brain banks support the same research interests. These differences will influence each of the three components of the brain banking process. Each type of bank will need to establish an autopsy protocol based on what is important for its research focus. The KPBBB has predominantly focused on banking Alzheimer's disease and aged-matched control tissues. However, we also have several other types of neurodegen-erative diseases represented in our collection. Each neurodegenerative disease has a specific brain region that is most affected. These regions are what investigators most often request from the bank for their research. For example, in Parkinson's disease, the substantia nigra is most often sought after. In Huntington's disease, the basal ganglia are of interest, specifically including the caudate nucleus. The hippocampus is the most requested region associated with Alzheimer's disease research. In addi-tion to dissecting each of these areas, we also collect other structures including the amygdala. We enroll normal controls, which would have the same regions retrieved at the time of autopsy. Due to the availability of these tissues in our collection, we have distributed tissue for a wide variety of studies over the years. No matter which areas are chosen for special dissection, it is best to retrieve frozen tissue as consis-tently as possible to maximize its usefulness in comparative studies.

2.3   OTHER CONSIDERATIONS

There are ongoing discussions in the neuroscience community as to what methods a brain bank should employ in order to maximize the usefulness of its frozen tissue (5,12). Some banks dissect different brain regions and also use a different method of freezing. There are methods of freezing other than using liquid nitrogen or chilled metal plates in a −80°C freezer. Some banks use an isopentane/dry ice slushy mix-ture, while others use only the vapor from the liquid nitrogen. Some banks cryopro-tect the samples before freezing and some do not. Selecting tissue for comparative studies has many variables that come into play when an investigator requests tissue from a brain bank. The method of freezing can be an issue depending on the project. We are asked to select disease cases with matching normal controls using gender, age at death, apolipoprotein E (*APOE*) genotype, and specific neuropathy diagnosis. In addition to these variables, most investigators insist on using the shortest post-mortem interval (PMI) possible. All of these details are databased and can be que-ried in order to find appropriate matches. However, the number of specimens that can be selected for a project decreases with each variable. Tissue that is prepared using similar methods by different neurodegenerative brain banks allows investigators to request tissue from several brain collections.

The advancement of scientific techniques used in gene expression studies has been driving the debate concerning the methods of banking diseased neurological tissue. There has been published research to show that biomolecule stability is highly variable. These studies show that the stability of proteins and RNA transcripts is not guaranteed using screening criteria such as PMI (13–15). PMI is the time it takes to get the tissue in the freezer after death. Cooling the tissue as soon as possible seems to be very beneficial in preserving intact RNA and proteins. However, it appears that individual proteins and RNA transcripts may react differently under the same conditions. These studies point to a lack of predictability of biomolecule integrity using measurements such as PMI. There are some biomolecules that can withstand a wider range of PMIs and some that cannot. It is expected that some biomolecules may not be recoverable due to factors that promote degradation no matter how optimal the conditions are during tissue retrieval. At this time it is prudent to evaluate the post-mortem integrity of the molecules of interest before conclusions can be drawn about the role of these molecules in the pathogenesis of any disorder.

REFERENCES

1. Hulette, C. M., Welsh-Bohmer, K. A., Crain, B. et al. (1997) Rapid brain autopsy. The Joseph and Kathleen Bryan Alzheimer's Disease Research Center experience, *Arch Pathol Lab Med* 121:615–8.
2. Cochran, E. J., Gostanian, O. M. and Mirra, S. S. (1995) Autopsy practices at CERAD and Alzheimer disease center sites: A survey of neuropathologists, *Alzheimer Dis Assoc Disord* 9:203–7.
3. Tourtellotte, W. W., Rosario, I. P., Conrad, A. and Syndulko, K. (1993) Human neuro-specimen banking 1961–1992. The National Neurological Research Specimen Bank (a donor program of pre- and post-mortem tissues and cerebrospinal fluid/blood; and a collection of cryopreserved human neurological specimens for neuroscientists), *J Neural Transm Suppl* 39:5–15.
4. Hulette, C. M. (2003) Brain banking in the United States, *J Neuropathol Exp Neurol* 62:715–22.
5. Vonsattel, J. P., Del Amaya, M. P. and Keller, C. E. (2008) Twenty-first century brain banking. Processing brains for research: The Columbia University methods, *Acta Neuropathol* 115:509–32.
6. Cruz-Sanchez, F. F., Moral, A., de Belleroche, J. and Rossi, M. L. (1993) Amyotrophic lateral sclerosis brain banking: A proposal to standardize protocols and neuropathological diagnostic criteria, *J Neural Transm Suppl* 39:215–22.
7. Reynolds, G. P. and Pearson, S. J. (1993) Neurochemical-clinical correlates in Huntington's disease—Applications of brain banking techniques, *J Neural Transm Suppl* 39:207–14.
8. Haroutunian, V. and Pickett, J. (2007) Autism brain tissue banking, *Brain Pathol* 17:412–21.
9. Schmitt, A., Bauer, M., Heinsen, H. et al. (2007) How a neuropsychiatric brain bank should be run: A consensus paper of Brainnet Europe II, *J Neural Transm* 114:527–37.
10. Bell, J. E. and Ironside, J. W. (1997) Principles and practice of 'high risk' brain banking, *Neuropathol Appl Neurobiol* 23:281–8.
11. Morgello, S., Gelman, B. B., Kozlowski, P. B. et al. (2001) The National NeuroAIDS Tissue Consortium: A new paradigm in brain banking with an emphasis on infectious disease, *Neuropathol Appl Neurobiol* 27:326–35.
12. Webster, M. J. (2006) Tissue preparation and banking, *Prog Brain Res* 158:3–14.

13. Crecelius, A., Gotz, A., Arzberger, T. et al. (2008) Assessing quantitative post-mortem changes in the gray matter of the human frontal cortex proteome by 2-D DIGE, *Proteomics* 8:1276–91.

14. Ervin, J. F., Heinzen, E. L., Cronin, K. D. et al. (2007) Postmortem delay has minimal effect on brain RNA integrity, *J Neuropathol Exp Neurol* 66:1093–9.

15. Lipska, B. K., Deep-Soboslay, A., Weickert, C. S. et al. (2006) Critical factors in gene expression in postmortem human brain: Focus on studies in schizophrenia, *Biol Psychiatry* 60:650–8.

3 Multidimensional Techniques in Protein Separations for Neuroproteomics

Carol Haney Ball and Petra Levine Roulhac

CONTENTS

3.1 SUMMARY

Proteomic analysis of brain tissue is becoming an integral component of neuroscience research. A proteomic analysis typically involves protein separation, protein identification, and protein characterization. Global characterization of proteins is providing new insights into biological structures such as synapses, axons, and dendrites. Additionally, proteomics has been applied in the investigation of various neurological disorders and has resulted in the detection and identification of a large number of disease-related proteins. The field of neuroproteomics faces special challenges given the complex cellular and subcellular architecture of the brain including sample preparation and limited sample amounts. This chapter presents an overview of current separation-based proteomic technologies, and the advantages and disadvantages of these technologies, and reviews the recent use of these approaches in neuroscience research.

3.2 INTRODUCTION

The increasing use of proteomics has created a basis for the development of new strategies for the rapid identification of protein profiles in living organisms. It has also become evident that proteomics has potential applications other than protein and peptide identification. Neuroproteomic studies have provided information about protein function, subcellular localization, activity, interaction partners, biochemical pathways, and molecular networks. Despite the advance of proteomic technologies, the brain poses particular challenges to studying protein function. There is a huge level of heterogeneity, with complex neuronal morphologies and the existence of unique subcellular structures, such as dendrites, dendritic spines, and axons. There are billions of neurons in the brain. Thus, intercellular and intracellular signal transduction plays a major role in brain function. It has been estimated that approximately one fifth of human genes encode proteins involved in signal transduction (1). It is likely that a thorough understanding of neuronal function will require identification and characterization of the complex neuronal proteome, and that future proteomic approaches will complement pharmacogenomics and may result in the development of therapeutic agents for neurological disorders.

Recent advances in proteomic technologies offer significant potential not only for gaining a better understanding of brain function, but also for achieving more effective treatments for neurological disorders such as Alzheimer's disease. A study by Kislinger et al. used multidimensional protein identification technology (MudPIT) and comparative proteomics to analyze organ- and organelle-specific protein expression in the mouse brain. The authors identified 4388 proteins in the brain. Of the 4388 proteins identified, 1336 localized to the cytosol, 1075 to mitochondria, 907 to nuclei, and 1040 to membrane fractions (2). Protein abundance patterns in the brain can also be localized by using imaging mass spectrometry or other high-throughput proteomic strategies (3–6) (see Chapters 4–7). The proteomic profiling of different regions or subcellular compartments of healthy and diseased brains promises new insights into the molecular basis of brain function and the pathogenesis of brain-related diseases (3,6).

Protein–protein interactions are the basis on which the cellular structure and function are built, and interaction partners are an immediate lead into the biological function that can be exploited for therapeutic purposes. One way to increase the understanding of the mechanisms of brain-related proteins is to identify interacting partners and to establish the location of individual proteins in a cellular pathway. Methods to explore interaction partners of a protein are affinity chromatography in conjunction with mass spectrometry and Western blotting (7,8), and the use of surface plasmon resonance (SPR) combined with protein identification by mass spectrometry (SPR-MS) (9). An early proteomic analysis of brain multiprotein complexes was the purification and identification of the molecular constituents of the N-methyl-D-aspartate (NMDA) receptor-adhesion protein signaling complexes (10). The NMDAR multiprotein complex (NRC) was shown to comprise 77 proteins organized into receptor, adaptor, signaling, cytoskeletal, and novel proteins, of which 30 are implicated from binding studies and another 19 participate in N-methyl-D-aspartate receptor (NMDAR) signaling (10). Future interaction neuroproteomic studies may reveal how protein–protein interactions contribute to the proliferation of neurodegenerative diseases and how they affect signal transduction pathways.

Proteomics as a global analysis of proteins has opened a wide range of new opportunities to study distinct subcellular structures in the brain. One such structure is the synapse. Chemical synapses are specialized asymmetric cell–cell contacts between neurons. Synaptic transmission involves an intricate network of synaptic proteins that form the molecular machinery responsible for transmitter release, activation of transmitter receptors, and signal transduction cascades. Recently, a number of proteomic studies have been performed on synaptic subdomains, including postsynaptic density (PSD) and synapse protein complexes (11–14). A proteomic analysis of the PSD has been instrumental in the formulation of hypotheses that may explain the molecular basis of neuropathogenic events. A recent study examined the alteration of the hippocampal PSD in response to morphine using the isotope coded affinity tag (ICAT)-based relative quantitation approach (13). A total of 102 proteins were identified. Future proteomic studies should yield insights into the structure of the synapse.

The analysis of proteins in the brain is extremely challenging due to the complex nature, heterogeneity, and wide dynamic concentration range (15,16). To perform a reliable analysis there are a variety of multidimensional approaches available (17,18). There is no universal procedure, but rather components selected to answer the questions important to the researcher. Mass spectrometric detection is an integral part of the multidimensional approach (see Chapters 5 and 6) (19). Prior to embarking on the experiment the researcher must decide on the details such as the origin of the samples, the time point or points needed, and the dynamic range required. Identifying as many proteins as possible in a sample is a common goal (20,21). However, in neuroscience many studies are designed from a hypothesis-driven approach focusing on identification of functional protein complexes (22). It is important to remember that any given sample will reflect only a snapshot of the ever-changing biological system.

The good news is these challenges can be addressed by the variety of techniques available. For that reason, multiple stages of separations are usually employed to answer important questions surrounding protein identification, system function, and pathway construction. These stages are orthogonal and can provide clean-up, fractionation, and specificity. In its simplest form, the multidimensional workflow can be defined as target collection, sample preparation (one step or several), separations (one step or several), and detection. The actual workflow will be much more complex. The number of separation stages is determined by the complexity of the sample and the goals of the research. A recent publication by Hoffman et al. compared various multidimensional separation strategies (23). They report that a 1D experiment will yield about 100 protein identifications. A 2D experiment (1-DE gel, 20–40 fractions analyzed by high performance liquid chromatography [HPLC]) and a 3D (isoelectric focusing [IEF], 1-DE gel, 100–150 fractions on HPLC) experiment will yield 800–1000 and more than 2000 proteins, respectively. Lastly, a 4D experiment (depletion, IEF, 1-DE gel, 100–150 fractions on HPLC) yields more than 2800 protein identifications. This chapter focuses on the separation choices available to the neuroscience researcher.

3.3 SAMPLE PREPARATIONS

Sample preparation is very important to a successful separation. Sample preparation methods can range from simple solubilizations to complex extractions. Isolating the protein component of the target tissue, complex, or organelle involves releasing the protein from the cells, breaking protein–lipid interactions, and solubilizing the proteins in a suitable buffer for protein fractionation and analysis (24). Often nucleic acids, lipids, polysaccharides, and a variety of cellular debris are removed prior to proteomic analyses. Cañas et al. reviewed the techniques, strategies, and pitfalls of various sample preparation techniques (25). Unique sample preparations are necessary for all samples from composite sections of brain tissue, specific brain areas, areas of cellular similarity, and subcellular components (organelles, cytoplasm, membranes, target cellular compartments, and protein complexes such as synaptosomes) (11,26,27). Some sample preparation techniques are listed in Table 3.1 and include laser capture microdissection (28–32), centrifugation (33–35), and protein precipitation (36–38).

Laser capture microdissection (LCM) is an indispensable tool in neuroscience. LCM is a method for isolating highly pure cell populations from a heterogeneous tissue sample under direct microscopic visualization (28,31). LCM technology can selectively dissect the cells of interest or can isolate specific cells by cutting away unwanted cells for histologically pure cell populations. LCM has been proven to be capable of isolating specific cells directly from tissue slides with a resolution as small as 3–5 mm in diameter (28,29,31). Espina et al. reviewed current LCM technology, with an emphasis on troubleshooting advice derived from LCM users (29).

Centrifugation and protein precipitation are the most commonly used first steps in sample preparation. Protein precipitation involves the removal of high abundance proteins or sample fractionation (38). One of the limitations of protein precipitation for removal of the high abundance proteins is the simultaneous removal of target proteins, which are bound with the high abundant proteins being removed (such as serum albumin). Centrifugation is also one of the most widely used methods for organelle isolation. Samples are spun at high speeds and the resulting force causes the separation of various cellular components based on their specific density (39). Several other techniques that exploit various physical properties (e.g., electrical charge for free flow electrophoresis) have been applied to study complex organelles.

TABLE 3.1
Techniques for Sample Preparation

Technique	Applications	Advantages
Laser capture microdissection	Tissue fractionation	Individual populations of various cell types can be isolated
Centrifugation	Subcellular fractionation	Easy to use and amenable to various proteomic platforms
Protein precipitation	High abundance protein depletion	Able to remove high abundance proteins, contaminants, and debris

However, the advantage of centrifugation is that it is easily set up and ideally combined with analytical proteomic techniques.

3.4 INTACT PROTEIN OFF-LINE MULTIDIMENSIONAL APPROACHES

Off-line separations are frequently used as the first stage of a multidimensional separation scheme following the physical sample preparation steps. If the sample is extremely complex or if a higher level of specificity is needed, a second off-line step can be added to the experimental design. The researcher is reminded that each added dimension gives more specificity, but there is a potential for protein loss and time will be added to the experiment. The goals of the experiment will define the appropriate number of dimensions needed.

The purpose of the first stage of a multidimensional experiment is to simplify a complex mixture. Brain samples are some of the most complex and are rarely used for direct analysis without prior fractionation. Several fractionation choices of intact proteins are shown in Table 3.2. The off-line approaches include two-dimensional gel electrophoresis (2-DE), one-dimensional gel electrophoresis (1-DE), difference gel electrophoresis (DIGE), off-gel electrophoresis (OGE), and a number of affinity separations that have varying degrees of selectivity.

2-DE remains the most common method used to simplify a complex protein mixture (40,41) (see Chapter 4 for a detailed discussion of 2D-DIGE-based proteomics).

TABLE 3.2
Off-Line Protein Fractionation Techniques

Technique	Process	Application
2-DE (2-dimensional gel electrophoresis)	Soluble protein separation in gel by isoelectric focusing (IEF) followed by separation based on molecular weight (SDS-PAGE)	Fractionation of proteins, through highly simplified protein spots
1-DE (1-dimensional gel electrophoresis)	Soluble protein separation in gel by isoelectric focusing (IEF) or molecular weight (SDS-PAGE)	Low-resolution separation of proteins. Generally used prior to subsequent separation steps
DIGE (differential gel electrophoresis)	Fluorescent dye labeled soluble proteins separated by isoelectric point followed by separation based on molecular weight	Quantitative 2-D fractionation used for accurate relative expression differences of soluble proteins
OGE (off-gel electrophoresis)	Soluble protein separation in a gel/buffer combination. Isoelectric focusing without ampholytes	Separation of intact proteins. Used to achieve moderately simplified protein mixtures
Target affinity enrichment	Protein binds to an immobilized substance selective for the protein of interest. Protein binding is reversible	Known protein complexes MARS, oxidations, glycosylations, phosphorlyations, RNA/DNA. For highly specific protein fractionation

A report on the study of the mouse brain proteome using 2-DE was recently published (42). 2D-DIGE has the advantage of simultaneously providing separations and visualization that can be used in detecting protein expression differences. These gels separate the proteins using two orthogonal physical properties of the proteins. The first dimension separates based on the pI of the protein and different pH ranges can be used. The most general is a broad pH range (pH 3–11). These gels have the advantage of separating a wide range of different proteins. Smaller pH ranges (pH 4–7, pH 6–9, and pH 6–11, among many other possibilities) can be used to provide better resolution of a subset of the proteins in the sample. Larger format gels also provide increased resolution (43). The higher the resolution, the easier it is to visualize differences between the gels in terms of proteins present and establish quantitative expression levels. The trade-off on the narrow pH range strips is that fewer proteins are isolated compared to the wide pH range strip. This could mean that multiple strips must be run, which greatly increases the sample needed and the time required to complete the experiments.

In the second dimension, the proteins are separated based on molecular weight. A polyacrylamide gel with sodium dodecyl sulfate controls the separation (SDS-PAGE). The combination of 1-DE and 2-DE gels with mass spectrometry has improved the accuracy of protein identification and molecular weight determination. Limitations of the 2-DE technique include the concentration dynamic range, suitable molecular weight range, suitable pI range, and protein hydrophobicity (44,45). 2-DE is mostly applicable to the separation of soluble proteins. Not all membrane and other hydrophobic proteins will migrate properly into or separate on a 2-DE gel. There are multiple staining techniques that can be used for visualization, each having different lower detection limits. Proteins with pI's below 4 and above 10 or with molecular weights less than ~10 kDa or above ~100 kDa are not effectively isolated in 2-DE. Running a 2-DE gel is time consuming and there are a number of steps that follow the separation. After a 2-DE separation, the spots of interest are excised, de-stained, and digested. The time is well spent because the resulting samples are sufficiently purified for direct application to a matrix-assisted laser desorption ionization (MALDI)/MS or liquid chromatography (HPLC)/MS analysis. The primary use of 2-DE gel analysis is the fractionation of many proteins from a heterogeneous mixture. Lubec et al. identified 110 proteins from the microsomal and cytosolic fractions from the human frontal cortex. They also categorized the identified proteins by function (46). It is rare for an experiment to call for all of the spots to be analyzed. Intact proteins can be removed from the gel for further analysis by several methods, including electroblotting onto nitrocellulose or polyvinylidene fluoride (PVDF) membranes (47).

A simpler approach to separation is a 1-DE gel experiment that uses either isoelectric focusing or the SDS-PAGE separation. The gel bands can be fractionated again (48) or excised, de-stained, and in-gel digested as in 2-DE. Because the fractions are complex protein mixtures in 1-DE, additional dimensions are generally performed such as an ion-exchange/HPLC chromatographic method. Stevens et al. identified 112 proteins with high confidence from the rat forebrain by a multidimensional approach which included three stages of sample preparation (1-DE gel, strong cation exchange [SCX], and HPLC) (34). Similar to an intact protein separation, Wu et al. used strong anion exchange HPLC as the first dimension. They subsequently collected the strong anion exchange HPLC fractions and separated them on a 1-DE SDS-PAGE gel. The differentially expressed protein bands were excised and digested. Lastly, the resulting peptides were analyzed by both MALDI and electrospray ionization (ESI) mass spectrometry. This method was applied to the study of the Alzheimer's disease relevant protease, beta amyloid cleaving enzyme. Wu et al. identified four proteins that seem to be part of the protease functional mechanism (49).

DIGE offers a solution to the reproducibility and differential expression issues of 2-DE (50,51). In this method two protein samples are labeled with cyanine dyes (Cy3 or Cy5). The samples are then mixed, separated, and visualized on a single gel (see Chapter 4). The proteins from each sample migrate to the same position for precise identification and the fluorescent signals give a more accurate relative quantification and wider concentration dynamic range for differential comparisons. The DIGE method is still limited to soluble proteins of appropriate pI and molecular weight. Kakisaka et al. used a 4-D experiment which included protein depletion followed by anion exchange fractionation prior to running the DIGE (a quantitative 2-DE gel) and HPLC/MS experiment to identify cancer biomarkers (52). In unfractionated, undepleted plasma, 290 spots were observed in the DIGE experiment. In depleted plasma with five fractions separated by anion exchange, a total of 1200 spots were observed.

Recently, off-gel fractionation (OGE) procedures for the separation of intact proteins by pI has been introduced (53–55). The separation is analogous to an isoelectric focusing 1-DE gel, with the difference that the separated proteins are present in the solution phase at the end of the experiment for direct analysis by HPLC/MS. This is important to the researcher because it provides greater flexibility in the following separation stage choices. Similar to a 1-DE gel experiment, the OGE separation of proteins is not high resolution. As a result, the fractions are amenable to various multidimensional separation strategies including (i) off-line reversed phase chromatography with fraction collection (Figure 3.1) for MALDI detection; (ii) on-line reversed phase chromatography (Figure 3.2) with electrospray MS; (iii) digestion followed by on-line SCX-reversed phase HPLC (RPLC)-MS; (iv) digestion followed by peptide off-gel fractionation; or (v) HPLC-MS. Another positive feature is that the process does not involve the use of ampholytes, which interfere with direct analysis by mass spectrometry. One major limitation of the technique is that visualization is not available. Thus, OGE is not useful for looking at differential protein expression. Also, like 1-DE, the OGE procedure is limited to soluble proteins with pI's <9 and molecular weights between 10 kDa and 150 kDa.

There are a wide variety of affinity-based separations. In all cases, the key is the reversibility of the binding reaction. The most specific utilize an antibody or other compound which binds a specific target protein or complex (6). For affinity-based separations, the binding substance is immobilized on a surface, the sample is run through the column, and the target protein or complex is retained, washed, and then eluted (56). These affinity methods require knowledge of the target protein and the appropriate affinity material (antibody, RNA or DNA fragment, aptamers, etc.). Burre et al. successfully isolated synaptic vesicles through bead-bound SV2 antibody and identified the associated soluble and membrane proteins (57).

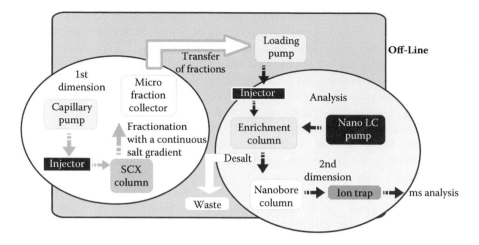

FIGURE 3.1 Two-dimensional liquid chromatography off-line workflow. The first dimension (*left*) typically includes protein fractionation by ionic strength using a strong cation exchange (SCX) column. This stage fractionates proteins using a continuous salt gradient, which are collected and manually transferred to a loading pump for the second dimension. The second dimension (*right*) uses a "perpendicular" property that creates another layer of fractionation, from which the proteins will be taken to further analysis (such as mass spectrometry).

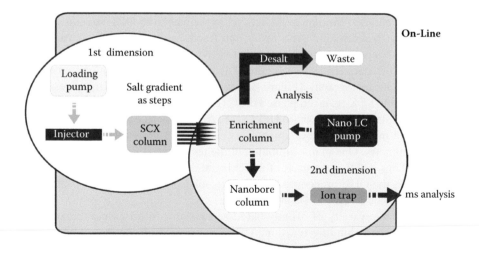

FIGURE 3.2 Two-dimensional liquid chromatography on-line workflow. A fundamental difference in concept is applied to the on-line method. In this case instead of manually "loading" the first fractionation into the loading pump, the analytes are sent directly to a C18 enrichment column; after sample enrichment a valve is used to flush the analyte back into a nanobore column, where organic solvents are used to elute the sample directly into the mass spectrometer.

Another, although less specific, fractionation method is the multiple affinity removal system (Figure 3.3). In the case of multiple affinity removal systems, high abundance proteins are retained and the less abundant proteins are collected in the flow-through. There are several sources of columns which remove anywhere from 6 to the top 100 proteins (58,59). The depletion methods remove the high abundance proteins from plasma or cerebrospinal fluid (CSF) but are not applicable to the analysis of tissues. Running a depletion step effectively increases the number of proteins that can be visualized (Figure 3.4). Figure 3.4A shows a 2-DE gel without depletion. Notice that few protein spots can be seen other than the high abundance proteins. Compare this to Figure 3.4B which shows the enhanced number of detected proteins. A common application for this approach is protein profiling for maximum protein identification or for biomarker discovery. In the depletion method only two fractions

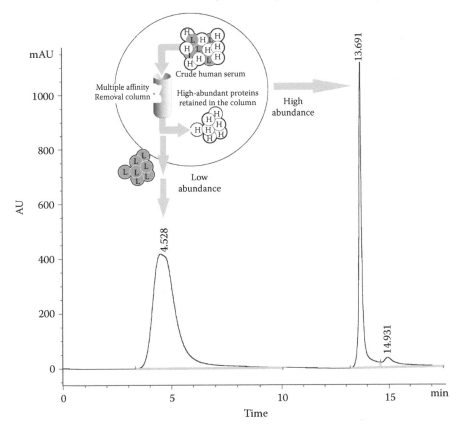

FIGURE 3.3 Separation of high abundance and low abundance proteins using a depletion column. Very complex fluids such as serum and cerebrospinal fluid are extremely rich in small numbers of proteins that account for most of the protein contents in the sample, leaving the low abundance proteins difficult to detect. To improve fractionation of these types of samples immuno-depletion columns are regularly used. These columns retain the high abundance proteins while allowing the low abundance proteins to pass through. The high abundance proteins are then removed, and collected for further analysis if necessary.

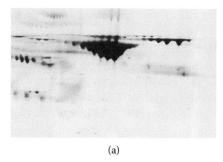

(a)

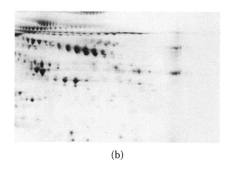

(b)

FIGURE 3.4   (a) Serum samples previous to high abundant depletion. (b) Same sample after the high abundant proteins have been depleted.

are collected (Figure 3.3). The flow-through fraction contains the low abundance proteins while the bound fraction contains the high abundance proteins. The bound fraction is eluted and the column can be used multiple times. The fraction can also be used for downstream analysis. The flow-through fraction, which contains a high number of proteins, is ready for subsequent analysis either as intact proteins or digested peptides. For highly complex samples, a 2-DE gel is frequently performed on the flow-through. A recent study by Burgess et al. used a multidimensional approach to identify proteins released into the CSF following brain injury (60). In this four-stage approach, the CSF was first depleted of high abundance proteins by a multiple affinity column separation. The flow-through from the affinity column separation was subjected to off-gel electrophoresis, followed by 1-DE SDS-PAGE. The protein bands were excised from the gel, digested, separated by HPLC, and identified by MS/MS. Two hundred ninety-nine proteins were identified of which 172 proteins were not previously known to exist in CSF. Most of the proteins identified were intracellular, suggesting that they were associated with damaged cells.

There are other affinity separations that have more specificity than the high abundance removal system but less specificity than target affinity separations. These columns fractionate a particular class of proteins rather than one specific protein or protein complex (61). These classes include membrane proteins, glycoproteins, phosphoproteins, and others (8). Considerable efforts have gone into developing methods to fractionate membrane and other hydrophobic proteins. Heparin affinity columns show affinity for several classes of proteins including membrane proteins

(62). Benzyldimethyl-n-hexadecylammonium chloride-(16 BAC)/SDS-PAGE gel is a cationic detergent used in the first dimension of a 2-DE separation to help solubilize membrane and other hydrophobic proteins. A study by Bierczynska-Krzysik et al. used a modified 16-BAC method and subsequent SDS-PAGE separation to identify 106 proteins from 187 spots (63). This study focused on the identification of insoluble and transmembrane brain proteins. Another technique useful for separating membrane proteins is blue native–PAGE (BN-PAGE). BN-PAGE is a gel-based separation that uses Coomassie blue dye anionic binding to help solubilize the membrane proteins. An advantage of this procedure is that in many cases intact protein complexes may be isolated (64,65).

Glycoproteins are of increasing interest because of the growing use of protein drugs, many of which are glycosylated. Heterogeneity of the glycoproteins complicates the analysis. When working with mixtures of proteins, some of which are glycosylated and others that are not, there is a need to selectively extract these modified proteins. Lectin columns are routinely used for this purpose, and more recently a highly specific affinity method has been described (66,67). The isolated glycoproteins are digested and the glycopeptides are analyzed to identify the site and structure of the glycan (68). Subsequently, the glycopeptides can be digested with exoglycosidase to release the oligosaccharides which are then profiled using a graphitized carbon column and mass spectrometry (69).

Phosphoproteins are of major importance in biological systems as they signal gene expression, cell adhesion, cell cycle, proliferation, and differentiation, to name a few. Many proteins undergo phosphorylation but only at very low levels compared to the non-phosphorylated proteome. In complex mixtures the phosphoproteins are represented at trace levels. Methods for isolating phosphoproteins include chromatofocusing, ion exchange chromatography, immobilized metal affinity chromatography (IMAC), metal oxide affinity chromatography (MOAC), and highly specific antibody immunoaffinity techniques (70,71).

3.5   INTACT PROTEIN ON-LINE MULTIDIMENSIONAL APPROACHES

There are relatively few on-line techniques for the identification of proteins. As a general rule off-line fractionation followed by MS analysis or digestion with further dimensions of chromatography are used. Complications that affect the detection of intact proteins and corresponding accuracy of the identifications include the inherent heterogeneity of biological molecules and the purity of the protein entering the detector. Common detection techniques are UV, laser-induced fluorescence, and MS.

A number of techniques are used for on-line separation of proteins including ion exchange chromatography/reversed phase HPLC (IEC/HPLC), size exclusion/reversed phase HPLC (SEC/HPLC), size exclusion/capillary electrophoresis (SEC/CE) (Figures 3.1 and 3.2), reversed phase HPLC with CE or capillary zone electrophoresis (CZE), and others (72). To be an effective on-line technique, the solvents and injection volumes must be compatible and the first dimension run must be slower and have lower peak resolution than the second dimension separation. When this cannot be accomplished it is generally possible to set up an off-line process instead. An additional consideration is that in all cases the on-line technique involves specialized (and often complex) chromatographic setups for both the instrumentation and the separation procedures.

Ion exchange chromatography (IEC)/HPLC is probably the best known application (73). It has been applied to both protein and peptide analyses. Ion exchange and reversed phase chromatography are highly compatible and provide separations based on orthogonal physical parameters. The solvents, concentrations, and injection volumes are easily matched, and commercially available systems make this a viable choice. The IEC run is measured in minutes, sometimes longer if the sample is complex, while the HPLC is fast, measured in seconds. Strong cation exchange (SCX) and strong anion exchange columns have been used in the first dimension in the analysis of intact proteins. Detection is commonly reported using UV and MS.

Capillary electrophoresis is a well-known separation technique for proteins as well as peptides (74,75). The multidimensional technique is somewhat more difficult to perform because the peak volume in HPLC is not compatible with the injection volume in CE. A method of introducing the sample from the HPLC with a stacking injection on the CE can be used to link the two dimensions. Size exclusion as a first dimension with capillary electrophoresis as the second dimension has also been demonstrated. This coupling has the same logistical issues with injection volume and peak volumes as discussed above. In addition, SEC solvent salt concentration must be low to be compatible with the CE. Specialized gating systems work for both 2-D schemes, but they require considerable expertise to construct and operate. Size-exclusion chromatography followed by RP-HPLC (SEC-HPLC) can be used in an on-line experiment and the two dimensions have compatible flow rates and matrices (76). The technique does not handle complex mixtures of proteins very well, is relatively labor intensive, requires custom instrument design which is not commercially available, and involves a multistep process including heart cutting and stop-flow sequences. (Heart cutting refers to isolating unresolved solutes from one separation and taking the mixture to another column where they will be resolved). Additionally, the SEC eluent is not compatible with either UV or MS detection without desalting. Happily, the combination with HPLC provides this necessary desalting step on-line. Advantages of the technique include reproducibility, stability, and improved resolution gained by coupling with the HPLC column (77).

3.6   PEPTIDE MULTIDIMENSIONAL APPROACHES AND QUANTITATIVE ANALYSES

Multidimensional peptide analysis is by far the most common approach used in the identification of proteins. Differential expression is one of the driving forces for many hypothesis-driven proteomic studies. Recent advances in isotopic labeling of peptides provide accurate relative and absolute quantitative results. By applying a multidimensional separation with chemically modified proteins a new era in differential expression studies has arrived (78).

Both on-line and off-line multidimensional schemes have been used successfully. Most of the off-line fractionation techniques for intact proteins take the collected proteins and digest them into peptides for subsequent separations. In 1-DE and 2-DE gel experiments, digestion effectively releases the protein from the gel in the form of peptides. Table 3.3 lists some of the techniques that are used after a digestion step. The advantages of on-line separations are the ability to automate the analysis and to reduce sample loss through liquid handling. The advantages of off-line separation techniques include the removal of column/column incompatibility and the flexibility of the mass spectrometry ionization choice (MALDI or electrospray). Widely used methods include 2-D strong cation exchange-reversed phase HPLC separation (SCX/HPLC or MudPIT), HPLC coupled to capillary electrophoresis (HPLC/CE), and a number of affinity-based separations such as selective phosphopeptide isolation using immobilized metal affinity chromatography (IMAC), and immunoaffinity removal of glycopeptides. Quantitative analyses include isotope coded affinity tag (ICAT), absolute quantitative analysis (AQUA), and isotope tagging reagents for absolute quantification (iTRAQ).

The SCX/HPLC technique has been widely used with highly complex peptide mixtures. The advantages of this technique include minimal sample handling, ease of use, and automation. The disadvantage is that identifications may be missed due to the co-elution of multiple peptides. The first dimension (SCX) can be run off-line with the collected fractions run on HPLC/MS or MALDI/MS. A comparison of the

TABLE 3.3
Multidimensional Peptide Analyses and Quantitative Methods

Technique	Process	Application
SCX/HPLC (strong cation exchange/HPLC)	On-line elution of tryptic digests based on ionic strength followed by reversed phase HPLC	Highly complex protein digest mixtures
HPLC/CE	On-line elution based on hydrophobicity and charge/mass ratio	Low level, small volume protein identification of digests
HPLC/CZE	On-line elution based on hydrophobicity and size	Low level, small volume protein
Class enrichment	Affinity interaction of peptide	Preferential removal of modified peptides
ICAT (isotope coded affinity tag)	Biotin derivatized cysteine residues selectively isolated by avidin column chromatography	Relative quantification and identification of proteins based on isotopic labels
AQUA (absolute quantitative analysis)	Isotopically labeled peptides quantified by the presence of a stable isotopically labeled internal standard	Absolute quantification of peptide concentration
iTRAQ (isotope tagging reagents for absolute quantification)	Isobaric labeling of primary amines	Comparison of up to four samples quantitatively in MS/MS mode

on-line/off-line performance is given by Nagele et al. (79). Their results show more proteins identified in the off-line approach (144 off-line, 101 on-line). The cost of the added information is time and labor. The on-line technique is referred to as MudPIT (multidimensional protein identification technology) because rather than selected spots from a gel, the entire complex mixture of peptides is analyzed (80–83). This "shotgun" approach is useful when the goal of the researcher is to identify as many proteins in the sample as possible with as little sample handling and prefractionation as possible (84,85).

The first dimension of the SCX/HPLC method relies on binding the positively charged amines of the peptide with an anionic group bound to the column packing. Peptides carry a positive charge in acidic solutions and this assists in retention. However, the solution should not have a pH less than 3. To elute the peptides, increasing concentrations of ionic strength buffers are injected in a stepwise manner onto the column to disrupt the peptide/sorbent bond. The buffer cations compete for binding sites on the column and as a result loosely bound peptides (singly charged) are eluted first. By increasing the ionic strength of each injection, peptides of increasingly strong positive charge (multiply charged) are eluted.

In the original MudPIT design the SCX phase and the reverse phase can be packed into the same column. A modification to the design has the SCX column and the Reversed phrase column separated by a switching valve (86). In the two-column approach, a small Reversed phrase column is placed between the SCX and HPLC columns via a switching valve to trap peptides as they elute from the SCX column (87). The run times of both designs are similar and both designs can be easily automated. The advantage of the two-column approach is that the SCX fractions are desalted prior to the Reversed phrase analysis. The presence of salts can interfere with the performance of the electro-spray ionization source. After trapping and washing, the small column is placed in line with the reverse phase analytical column. Peptides are eluted from these columns with a gradient of increasing organic mobile phase. The entire experiment involves sequential buffer injections with each injection triggering a full HPLC run. The number of salt injections can be large (from 10–25) and the HPLC run times can be in excess of an hour. In all cases, the on-line analyses will be performed at low flow rates on nanobore columns with electrospray MS/MS.

The advantages of the SCX/HPLC experiment are that the sample preparation is minimized and many proteins can be identified in one experiment. There are no time-consuming gel electrophoresis fractionation steps prior to digestion. Most importantly, the entire SCX/HPLC experiment can be automated. There are several disadvantages to this technique. First, the elution of the peptides from the SCX column is not precise. That is, the same peptide may elute across several of the SCX injections and appear in several of the HPLC runs. This reduces the amount of that peptide in any one HPLC run and thus decreases the sensitivity attainable. Also, most peptides are singly charged, even in acidic solutions, so the first few buffer injections contain the majority of peptides. This results in samples that are too complex. Breakthrough on the trapping column can occur if the sample is too concentrated for the capacity of the column. Di- and tri-peptides are usually not trapped.

When complex mixtures of proteins are digested, the complexity grows by more than 10-fold because each protein yields many peptides. Given this complexity it is not unusual to use an approach with more than two dimensions. A comparison of SCX/HPLC and off-gel/HPLC of complex protein digests provides a good example of the choices on which procedure and how many stages of separation to apply. In a 2-D experiment on *Escherichia coli* lysate, 900 proteins were identified with high confidence in 24 OGE fractions, and 500 proteins were identified with high confidence in 36 SCX fractions. A comparison between two 3-D experiments on plasma yields similar results. The workflow for the two 3-D experiments on plasma is as follows: depletion (off-line), digestion, SCX (off-line), and HPLC/MS or depletion (off-line), digestion, OGE (off-line), and HPLC/MS. Seventy-nine proteins were identified in 36 SCX salt fractions and 183 proteins were identified in 24 OGE fractions (88).

Reverse phase HPLC can be used in combination with capillary electrophoresis (CE) or capillary zone electrophoresis (CZE) for separation of intact proteins or protein digests. The separation is based on charge or size, respectively. This process works well as either an on-line or off-line technique (89–92). The on-line approach has significant challenges because the optimum HPLC operation and CE operation do not couple in a straightforward manner. The researcher must consider the buffer effects, sample concentrations, and the coordination between the chromatographic run time and the electrophoretic migration time. Fast CZE is ideally suited for the second dimension following an HPLC separation since the entire CZE analysis can be completed in seconds. CE injection volumes are in the nanoliter range; therefore, sampling from the high flow-rate HPLC requires specialized equipment. For most of the analyses the peptides are labeled and detected by fluorescence. This is a high sensitivity detection method and ideally suited for CE and CZE. Although there are significant challenges to setting up the on-line approach, it has the advantages of automation, high peak capacity, and speed. In the off-line approach, fractions are collected at the HPLC and then run in a 1-D fashion on the CE. While this creates a manual bottleneck, it is a simpler design to implement.

Affinity chromatography provides far more specificity in one dimension of the separation. In some cases the affinity is for a particular class of compounds (i.e., phosphopeptides), and in others it is specific to one individual protein. In hypothesis-driven design, there is often the need to isolate a target protein, often at trace levels, from a much more complex mixture. There are a wide variety of methods to preferentially isolate low abundance modified peptides from complex digest mixtures. Phosphoproteins are extremely important in brain research due to their involvement in signaling pathways and other biological activities. To complicate matters, the phosphate group is labile under most MS detection methods. Phosphopeptides can be isolated by immobilized metal affinity chromatography (IMAC) or immunoaffinity columns (22). The immobilized metal affinity columns have Fe(III), Ga(III), or TiO_2 metals bound to the sorbent. The columns can be used to isolate phosphopeptides or phosphoproteins. This on-line technique is compatible with HPLC/MS. Detection limits for proteins are in the low picomolar range while peptides can be detected in the femtomolar range (93). One limitation of the technique is its lack of specificity because histidine-containing peptides also bind to the IMAC columns. Immunoaffinity columns for phosphopeptide isolation provide greater selectivity but are less robust and require highly specialized conditions for optimal performance. Several antibodies are used including anti-phosphoserine, -tyrosine, and -threonine.

Glycosylated proteins are implicated in many biological processes. The importance of the heterogeneity of the glycans has been found to be significant. Lectin affinity columns have been used to capture glycoproteins and are also very effective in enriching glycopeptides (94). The fractions can be collected for off-line analysis by MALDI or coupled on-line to HPLC with UV or MS detection.

An important area of proteomics involves the need for quantification (95). Most proteomics technologies to date can measure relative quantification. The search for biomarkers or the extent of post-translational modification necessitates the ability to accurately express the changes in the amount of protein in the context of a second cellular state or control sample. Table 3.3 lists a few of the quantitative approaches that may be used in proteomic studies. While not technically an added dimension of separation, quantification represents an important added dimension of information that can be included in to a multidimensional experiment; it also may obviate the need to use 2-D gels to look for differentially expressed proteins.

ICAT is one of the first of these quantitative approaches (96). This technique compares the relative quantity of proteins in two different samples for expression analyses. In the method, each protein sample is labeled with isotopically different affinity tags (C^{13}- and D-linked biotin), which bind specifically to cysteine residues. Once the labeling is complete, the two samples are combined and digested with trypsin. The mixture is run through an avidin column and the cysteine-containing peptides are retained. The retained peptides are eluted and analyzed by reversed phase HPLC/MS. The major advantage to this technique is the ability to get accurate information on the relative quantity of the same peptides from each sample. A second advantage is that the peptide mixture is greatly simplified. The simplified peptide mixture improves the probability that spectra from low abundance peptides can be detected because there are fewer co-eluting species. In essence, the sample is enriched. However, there are a couple of limitations to the technique. Most importantly, proteins without cysteine are not captured. When a protein has few cysteine residues, identification is heavily dependent on locating those cysteines. It is likely that some of these proteins may be missed or erroneously identified. Lastly, the procedure is complicated and labor intensive. The technique requires three manual off-line steps (i.e., digestion, biotin labeling, avidin chromatography) followed by ion exchange and HPLC/MS.

A technique termed AQUA, "absolute quantification," uses a synthetic copy of a target peptide that has been isotopically labeled for use as an internal standard. The internal standard is mixed into the protein mixture during digestion and a selected monitoring MS experiment is performed to measure the absolute quantity of that peptide, and hence the protein. The procedure is used for differential expression analysis of known target proteins (97–99). The benefits of this method are the absolute quantitative results and the ability to examine the extent of post-translational modifications. Cheng et al. used a multidimensional approach to identify and quantify differentially expressed proteins associated with the postsynaptic density of rat forebrain and cerebellum (100). In the first dimension ICAT labeling was used to identify differentially expressed proteins. This experiment resulted in the identification of 296 proteins, of which 43 showed a statistically significant expression difference between the two brain regions. Appropriate peptides from 32 of those proteins were synthesized and used as internal standards for absolute quantification using the AQUA method.

A third labeling technique, iTRAQ, uses isobaric tags (methylpiperazine acetic acid *N*-hydroxysuccinimide ester), which react with primary amino groups (N-terminus and lysine residues). As a result all proteins are labeled. Furthermore, up to four samples can be labeled providing multiple sample comparisons in a single run. The analysis is performed by HPLC/MS/MS and the relative abundances can be calculated (101). One advantage of this technique is that no extra chromatographic step for isolation of the derivatized peptides is necessary (as in ICAT). Also, with all proteins having multiple labeled sites, there is greater confidence in the protein identifications because multiple peptides may be found for each protein. Lastly, with more peptides for each protein the relative area results can be evaluated for consistency. Ogata et al. used iTRAQ to quantitatively examine protein differences between male and female CSF (102).

3.7 CONCLUSIONS AND FUTURE DIRECTIONS

Currently there are two main research classes of proteomics applied in neuroscience: profiling and functional proteomics. Profiling encompasses maximum protein identifications and differential expression either with total protein identification or biomarker identification. Kobeissy et al. used a multidimensional approach to identify biomarkers of traumatic brain injury for potential diagnosis and treatment (103). Functional proteomics focuses on understanding the structure of protein clusters of known function. The identified proteins can then be classified as participating in that function by association (104–107) (see also Chapter 9). Some of the identified proteins in the complex may already be classified in that function. This confluence of information helps confirm existing hypotheses and builds a more detailed understanding of biological interactions.

Both of these research goals, as well as others such as genomic/proteomic correlations, occupy important areas of active research (108,109). Thus, all of the multidimensional proteomic technologies are useful and valid provided they are used in a way that maximizes the advantages and minimizes the limitations. To be successful the experiment requires intelligent design, which includes a complete workflow definition including sample choice, sample preparation, separations, detection, protein identification, quantification, and biological significance.

Reproducibility of results remains a concern, in part due to factors related to the sample preparation and analysis as well as human factors such as disease progression, differences in individuals, age, sex, and many other environmental and physical factors. The large amount of data generated also present a challenge, but numerous search engines are available (110–112). A number of new software programs are available that aid in the statistical evaluation of data (113,114) and pathway analysis (115,116).

Regardless of the current limitations, protein identification, protein expression analysis, and biomarker discovery are now firmly entrenched in neuroscience. Sample preparation remains a major area where improvements must be made to deal with the complexities of gathering reproducible samples from brain tissues. The need for consistency and precision is the cornerstone of making accurate conclusions. It is well recognized that reducing the complexity of a sample through fractionation is a key component of the experimental design. Fractionation yields more protein identifications, better expression comparisons, and greater access to lower abundance proteins. On-line multidimensional separations continue to improve in terms of speed, sensitivity, reproducibility, and automation.

There are many potential applications for proteomics in neuroscience. Such applications include the determination of the brain proteome, the identification of disease-related proteins implicated in neurological disorders, comparative protein expression profiling, post-translational modification (PTM) profiling, and mapping protein–protein interactions. All of these applications have strengths and limitations. One of the major challenges ahead for the neuroscience researchers is to determine the most appropriate applications of the various proteomic platforms for their research goals.

High-throughput proteomic technologies specifically for the identification and characterization of protein modifications have to be created. Post-translational modifications are features of proteins that affect activity and localization. Modified proteins activate, target, and have important roles in determining turnover and enzyme activity. Current high-throughput technologies such as automatic peptide mass fingerprinting data ignore unknown or modified peptides. This may result in the loss of many interesting proteins. High-throughput proteomic technologies with greater sensitivity also need to be developed to identify low abundance proteins as protein amplification methods are not available. This is very important for the applicability of proteomics in clinical settings. Proteomics technologies are advancing rapidly and addressing fundamental questions in neuroscience involving brain function and behavior.

ABBREVIATIONS

2-dimensional gel electrophoresis (2-DE)
Absolute quantitative analysis (AQUA)
Benzyldimethyl-n-hexadecylammonium chloride (16 BAC)
Blue native-polyacrylamide gel electrophoresis (BN-PAGE)
Capillary electrophoresis (CE)
Capillary zone electrophoresis (CZE)
Cerebrospinal fluid (CSF)
Difference in gel electrophoresis (DIGE)
Electro-spray ionization (ESI)
High performance liquid chromatography (HPLC)
Immobilized metal affinity columns (IMAC)
Ion exchange chromatography (IEC)
Isotope coded affinity tag (ICAT)
Isotope tagging reagents for absolute quantification (iTRAQ)
Laser capture microdissection (LCM)
Mass spectrometry (MS)
Matrix assisted laser desorption ionization (MALDI)
Metal oxide affinity chromatography (MOAC)

Multidimensional protein identification technology (MudPIT)
N-methyl-D-aspartate (NMDA)
N-methyl-D-aspartate receptor (NMDAR)
Off-gel electrophoresis (OGE)
Postsynaptic density (PSD)
Post-translational modification (PTM)
Reversed phase (RP)
Reversed phase liquid chromatography (RPLC)
Size exclusion chromatography (SEC)
Sodium dodecyl sulfate polyacrylamide gel electrophoresis (SDS-PAGE)
Strong cation exchange (SCX)
Tandem mass spectrometry (MS/MS)

REFERENCES

1. Blume-Jensen, P. and Hunter, T. (2001) Oncogenic kinase signalling, *Nature* 411:355–65.
2. Kislinger, T., Cox, B., Kannan, A. et al. (2006) Global survey of organ and organelle protein expression in mouse: Combined proteomic and transcriptomic profiling, *Cell* 125:173–86.
3. Johnson, M. D., Floyd, J. L. and Caprioli, R. M. (2006) Proteomics in diagnostic neuropathology, *Journal of Neuropathology and Experimental Neurology* 65:837–45.
4. Laurent, C., Levinson, D. F., Schwartz, S. A. et al. (2005) Direct profiling of the cerebellum by matrix-assisted laser desorption/ionization time-of-flight mass spectrometry: A methodological study in postnatal and adult mouse, *Journal of Neuroscience Research* 81:613–21.
5. Petyuk, V. A., Qian, W. J., Chin, M. H. et al. (2007) Spatial mapping of protein abundances in the mouse brain by voxelation integrated with high-throughput liquid chromatography-mass spectrometry, *Genome Research* 17:328–36.
6. Schwartz, S. A., Weil, R. J., Thompson, R. C. et al. (2005) Proteomic-based prognosis of brain tumor patients using direct-tissue matrix-assisted laser desorption ionization mass spectrometry, *Cancer Research* 65:7674–81.
7. Joseph, T., Lee, T. L., Ning, C. et al. (2006) Identification of mature nocistatin and nociceptin in human brain and cerebrospinal fluid by mass spectrometry combined with affinity chromatography and HPLC, *Peptides* 27:122–30.
8. Lee, W. C. and Lee, K. H. (2004) Applications of affinity chromatography in proteomics, *Analytical Biochemistry* 324:1–10.
9. Buijs, J. and Franklin, G. C. (2005) SPR-MS in functional proteomics, *Brief Funct Genomic Proteomic* 4:39–47.
10. Husi, H., Ward, M. A., Choudhary, J. S., Blackstock , W. P., Grant, S. G. (2000) Proteomic analysis of NMDA receptor-adhesion protein signaling complexes, *Nature Neuroscience* 3:661–69.
11. Abul-Husn, N. S. and Devi, L. A. (2006) Neuroproteomics of the synapse and drug addiction, *Journal of Pharmacology and Experimental Therapeutics* 318:461–68.
12. Li, K. W., Hornshaw, M. P., Van der Schors, R. C. et al. (2004) Proteomics analysis of rat brain postsynaptic density—Implications of the diverse protein functional groups for the integration of synaptic physiology, *Journal of Biological Chemistry* 279:987–1002.
13. Moron, J. A., Abul-Husn, N. S., Rozenfeld, R. et al. (2007) Morphine administration alters the profile of hippocampal postsynaptic density-associated proteins—A proteomics study focusing on endocytic proteins, *Molecular and Cellular Proteomics* 6:29–42.

14. Piccoli, G., Verpelli, C., Tonna, N. et al. (2007) Proteomic analysis of activity-dependent synaptic plasticity in hippocampal neurons, *Journal of Proteome Research* 6:3203–15.
15. Choudhary, J. and Grant, S. G. N. (2004) Proteomics in postgenomic neuroscience: The end of the beginning, *Nature Neuroscience* 7:440–45.
16. Marcus, K., Schmidt, O., Schaefer, H. et al. (2004) Proteomics—Application to the brain, *International Review of Neurobiology* 61:285–311.
17. Garbis, S., Lubec, G. and Fountoulakis, M. (2005) Limitations of current proteomics technologies, *Journal of Chromatography A* 1077:1–18.
18. Hu, L. H., Ye, M. D., Jiang, X. G., Feng, S., Zou, H. F. (2007) Advances in hyphenated analytical techniques for shotgun proteome and peptidome analysis—A review, *Analytica Chimica Acta* 598:193–204.
19. Aebersold, R. and Mann, M. (2003) Mass spectrometry-based proteomics, *Nature* 422:198–207.
20. Le Bihan, T., Duewel, H. S. and Figeys, D. (2003) On-line strong cation exchange mu-HPLC-ESI-MS/MS for protein identification and process optimization, *Journal of the American Society for Mass Spectrometry* 14:719–27.
21. Liu, X. Y., Valentine, S. J., Plasencia, M. D. et al. (2007) Mapping the human plasma proteome by SCX-LC-IMS-MS, *Journal of the American Society for Mass Spectrometry* 18:1249–64.
22. Williams, K., Wu, T., Colangelo, C. and Nairn, A. C. (2004) Recent advances in neuroproteornics and potential application to studies of drug addiction, *Neuropharmacology* 47:148–66.
23. Hoffman, S. A., Joo, W. A., Echan, L. A. and Speicher, D. W. (2007) Higher dimensional (Hi-D) separation strategies dramatically improve the potential for cancer biomarker detection in serum and plasma, *Journal of Chromatography B-Analytical Technologies in the Biomedical and Life Sciences* 849:43–52.
24. Bodzon-Kulakowska, A., Bierczynska-Krzysik, A., Dylag, T. et al. (2007) Methods for samples preparation in proteomic research, *Journal of Chromatography B-Analytical Technologies in the Biomedical and Life Sciences* 849:1–31.
25. Cañas, B., Pineiro, C., Calvo, E., Lopez-Ferrer, D. and Gallardo, J. M. (2007) Trends in sample preparation for classical and second generation proteomics, *Journal of Chromatography A* 1153:235–58.
26. Moron, J. A. and Devi, L. A. (2007) Use of proteomics for the identification of novel drug targets in brain diseases, *Journal of Neurochemistry* 102:306–15.
27. Tannu, N. S., Hemby, S. E. (2006) Methods for proteomics in neuroscience. In *Progress in brain research*, eds. S. E. Hemby and S. Bahn, 41–82 (Amsterdam: Elsevier BV).
28. Banks, R. E., Dunn, M. J., Forbes, M. A. et al. (1999) The potential use of laser capture microdissection to selectively obtain distinct populations of cells for proteomic analysis—Preliminary findings, *Electrophoresis* 20:689–700.
29. Espina, V., Wulfkuhle, J. D., Calver, V. S. et al. (2006) Laser-capture microdissection, *Nature Protocols* 1:586–603.
30. Jain, K. K. (2002) Application of laser capture microdissection to proteomics, *Laser Capture Microscopy and Microdissection* 356:157–67.
31. Mouledous, L., Hunt, S., Harcourt, R. et al. (2003) Navigated laser capture microdissection as an alternative to direct histological staining for proteomic analysis of brain samples, *Proteomics* 3:610–15.
32. Wilson, K. E., Marouga, R., Prime, J. E. et al. (2005) Comparative proteomic analysis using samples obtained with laser microdissection and saturation dye labelling, *Proteomics* 5:3851–58.
33. Pan, S., Shi, M., Jin, J. H. et al. (2007) Proteomics identification of proteins in human cortex using multidimensional separations and MALDI tandem mass spectrometer, *Molecular and Cellular Proteomics* 6:1818–23.

34. Stevens, S. M., Zharikova, A. D. and Prokai, L. (2003) Proteomic analysis of the synaptic plasma membrane fraction isolated from rat forebrain, *Molecular Brain Research* 117:116–28.

35. Taylor, S. W., Fahy, E. and Ghosh, S. S. (2003) Global organellar proteomics, *Trends in Biotechnology* 21:82–88.

36. Chang, M. S., Ji, Q., Zhang, J. and El-Shourbagy, T. A. (2007) Historical review of sample preparation for chromatographic bioanalysis: Pros and cons, *Drug Development Research* 68:107–33.

37. Chen, J. Z., Anderson, M., Misek, D. E., Simeone, D. M. and Lubman, D. M. (2007) Characterization of apolipoprotein and apolipoprotein precursors in pancreatic cancer serum samples via two-dimensional liquid chromatography and mass spectrometry, *Journal of Chromatography A* 1162:117–25.

38. Jiang, L., He, L. and Fountoulakis, M. (2004) Comparison of protein precipitation methods for sample preparation prior to proteomic analysis, *Journal of Chromatography A* 1023:317–20.

39. Huber, L. A., Pfaller, K. and Vietor, I. (2003) Organelle proteomics—Implications for subcellular fractionation in proteomics, *Circulation Research* 92:962–68.

40. Gorg, A., Weiss, W. and Dunn, M. J. (2004) Current two-dimensional electrophoresis technology for proteomics, *Proteomics* 4:3665–85.

41. Rabilloud, T. (2002) Two-dimensional gel electrophoresis in proteomics: Old, old fashioned, but it still climbs up the mountains, *Proteomics* 2:3–10.

42. Wang, J., Gu, Y., Wang, L. H. et al. (2007) HUPOBPP pilot study: A proteomics analysis of the mouse brain of different developmental stages, *Proteomics* 7:4008–15.

43. Levenson, R. M., Anderson, G. M., Cohn, J. A., Blackshear, P. J. (2005) Giant two-dimensional gel electrophoresis: Methodological update and comparison with intermediate-format gel systems, *Electrophoresis* 11:269–79.

44. Gygi, S. P., Corthals, G. L., Zhang, Y., Rochon, Y. and Aebersold, R. (2000) Evaluation of two-dimensional gel electrophoresis-based proteome analysis technology, *Proceedings of the National Academy of Sciences of the United States of America* 97:9390–95.

45. Whitelegge, J. (2005) Tandem mass spectrometry of integral membrane proteins for top-down proteomics, *Trac-Trends in Analytical Chemistry* 24:576–82.

46. Lubec, G., Krapfenbauer, K. and Fountoulakis, M. (2003) Proteomics in brain research: Potentials and limitations, *Progress in Neurobiology* 69:193–211.

47. Cooper, J. W., Gao, J. and Lee, C. S. (2004) Electronic gel protein transfer and identification using matrix-assisted laser desorption/ionization-mass spectrometry, *Electrophoresis* 25:1379–85.

48. Gorg, A., Boguth, G., Kopf, A. et al. (2002) Sample prefractionation with Sephadex isoelectric focusing prior to narrow pH range two-dimensional gels, *Proteomics* 2:1652–57.

49. Wu, S., Tang, X. T., Siems, W. F. and Bruce, J. E. (2005) A hybrid LC-Gel-MS method for proteomics research and its application to protease functional pathway mapping, *Journal of Chromatography B-Analytical Technologies in the Biomedical and Life Sciences* 822:98–111.

50. Righetti, P. G., Castagna, A., Antonucci, F. et al. (2004) Critical survey of quantitative proteomics in two-dimensional electrophoretic approaches, *Journal of Chromatography A* 1051:3–17.

51. Unlu, M., Morgan, M. E. and Minden, J. S. (1997) Difference gel electrophoresis: A single gel method for detecting changes in protein extracts, *Electrophoresis* 18:2071–77.

52. Kakisaka, T., Kondo, T., Okano, T. et al. (2007) Plasma proteomics of pancreatic cancer patients by multi-dimensional liquid chromatography and two-dimensional difference gel electrophoresis (2D-DIGE): Up-regulation of leucine-rich alpha-2-glycoprotein in pancreatic cancer, *Journal of Chromatography B-Analytical Technologies in the Biomedical and Life Sciences* 852:257–67.

53. Heller, M., Michel, P. E., Morier, P. et al. (2005) Two-stage Off-Gel (TM) isoelectric focusing: Protein followed by peptide fractionation and application to proteome analysis of human plasma, *Electrophoresis* 26:1174–88.

54. Michel, P. E., Crettaz, D., Morier, P. et al. (2006) Proteome analysis of human plasma and amniotic fluid by Off-Gel (TM) isoelectric focusing followed by nano-LC-MS/MS, *Electrophoresis* 27:1169–81.

55. Ros, A., Faupel, M., Mees, H. et al. (2002) Protein purification by Off-Gel electrophoresis, *Proteomics* 2:151–56.

56. Rigaut, G., Shevchenko, A., Rutz, B. et al. (1999) A generic protein purification method for protein complex characterization and proteome exploration, *Nature Biotechnology* 17:1030–32.

57. Burre, J., Zimmermann, H. and Volknandt, W. (2007) Immunoisolation and subfractionation of synaptic vesicle proteins, *Analytical Biochemistry* 362:172–81.

58. Maccarrone, G., Milfay, D., Birg, I. et al. (2004) Mining the human cerebrospinal fluid proteome by immunodepletion and shotgun mass spectrometry, *Electrophoresis* 25:2402–12.

59. Pieper, R., Gatlin, C. L., Makusky, A. J. et al. (2003) The human serum proteome: Display of nearly 3700 chromatographically separated protein spots on two-dimensional electrophoresis gels and identification of 325 distinct proteins, *Proteomics* 3:1345–64.

60. Burgess, J. A., Lescuyer, P., Hainard, A. et al. (2006) Identification of brain cell death associated proteins in human post-mortem cerebrospinal fluid, *Journal of Proteome Research* 5:1674–81.

61. Lescuyer, P., Hochstrasser, D. F. and Sanchez, J. C. (2004) Comprehensive proteome analysis by chromatographic protein prefractionation, *Electrophoresis* 25:1125–35.

62. Iida, T., Kamo, M., Uozumi, N., Inui, T. and Imai, K. (2005) Further application of a two-step heparin affinity chromatography method using divalent cations as eluents: Purification and identification of membrane-bound heparin binding proteins from the mitochondrial fraction of HL-60 cells, *Journal of Chromatography B-Analytical Technologies in the Biomedical and Life Sciences* 823:209–12.

63. Bierczynska-Krzysik, A., Kang, S. U., Silberrring, J. and Lubec, G. (2006) Mass spectrometrical identification of brain proteins including highly insoluble and transmembrane proteins, *Neurochemistry International* 49:245–55.

64. Hooker, B. S., Bigelow, D. J. and Lin, C. T. (2007) Methods for mapping of interaction networks involving membrane proteins, *Biochemical and Biophysical Research Communications* 363:457–61.

65. Sanders, P. R., Cantin, G. T., Greenbaum, D. C. et al. (2007) Identification of protein complexes in detergent-resistant membranes of Plasmodium falciparum schizonts, *Molecular and Biochemical Parasitology* 154:148–57.

66. Luque-Garcia, J. L. and Neubert, T. A. (2007) Sample preparation for serum/plasma profiling and biomarker identification by mass spectrometry, *Journal of Chromatography A* 1153:259–76.

67. Monzo, A., Bonn, G. K. and Guttman, A. (2007) Lectin-immobilization strategies for affinity purification and separation of glycoconjugates, *Trac-Trends in Analytical Chemistry* 26:423–32.

68. Wuhrer, M., Catalina, M. I., Deelder, A. M. and Hokke, C. H. (2007) Glycoproteomics based on tandem mass spectrometry of glycopeptides, *Journal of Chromatography B-Analytical Technologies in the Biomedical and Life Sciences* 849:115–28.

69. Itoh, S., Kawasaki, N., Hashii, N. et al. (2006) N-linked oligosaccharide analysis of rat brain Thy-1 by liquid chromatography with graphitized carbon column/ion trap-Fourier transform ion cyclotron resonance mass spectrometry in positive and negative ion modes, *Journal of Chromatography A* 1103:296–306.
70. Li, Y., Liu, Y., Tang, J., Lin, H., Yao. N., Shen, X., Deng, C., Yang, P., Zhang, X. (2007) $Fe_3O_4@Al_2O_3$ magnetic core-shell microspheres for rapid and highly specific capture of phosphopeptides with mass spectrometry analysis, *J. Chromatogr. A* 1172:57–71.
71. Schmidt, S. R., Schweikart, F. and Andersson, M. E. (2007) Current methods for phosphoprotein isolation and enrichment, *Journal of Chromatography B-Analytical Technologies in the Biomedical and Life Sciences* 849:154–62.
72. Issaq, H. J. (2001) The role of separation science in proteomics research, *Electrophoresis* 22:3629–38.
73. Wagner, K., Racaityte, K., Unger, K. K. et al. (2000) Protein mapping by two-dimensional high performance liquid chromatography, *Journal of Chromatography A* 893:293–305.
74. Larmann, J. P., Lemmo, A. V., Moore, A. W. and Jorgenson, J. W. (1993) 2-Dimensional separations of peptides and proteins by comprehensive liquid-chromatography capillary-electrophoresis, *Electrophoresis* 14:439–47.
75. Stroink, T., Ortiz, M. C., Bult, A. et al. (2005) On-line multidimensional liquid chromatography and capillary electrophoresis systems for peptides and proteins, *Journal of Chromatography B-Analytical Technologies in the Biomedical and Life Sciences* 817:49–66.
76. Opiteck, G. J., Ramirez, S. M., Jorgenson, J. W. and Moseley, M. A. (1998) Comprehensive two-dimensional high-performance liquid chromatography for the isolation of overexpressed proteins and proteome mapping, *Analytical Biochemistry* 258:349–61.
77. Carol, J., Gorseling, M. C. J. K., Jong, C. F. d. et al. (2005) Determination of denatured proteins and biotoxins by on-line size-exclusion chromatography–digestion–liquid chromatography–electrospray mass spectrometry, *Analytical Biochemistry* 246:150–57.
78. Ye, M. L., Jiang, X. G., Feng, S., Tian, R. J. and Zou, H. F. (2007) Advances in chromatographic techniques and methods in shotgun proteome analysis, *Trac-Trends in Analytical Chemistry* 26:80–84.
79. Nagele, E., Vollmer, M., Horth, P. and Vad, C. (2004) 2D-HPLC/MS techniques for the identification of proteins in highly complex mixtures, *Expert Review of Proteomics* 1:37–46.
80. Delahunty, C. and Yates, J. R. (2005) Protein identification using 2D-LC-MS/MS, *Methods* 35:248–55.
81. McDonald, W. H., Ohi, R., Miyamoto, D. T., Mitchison, T. J. and Yates, J. R. (2002) Comparison of three directly coupled HPLC MS/MS strategies for identification of proteins from complex mixtures: Single-dimension LC-MS/MS, 2-phase MudPIT, and 3-phase MudPIT, *International Journal of Mass Spectrometry* 219:245–51.
82. Washburn, M. P. and John R. Yates, I. (2000) New methods of proteome analysis: Multidimensional chromatography and mass spectrometry *Trends in Biotechnology* 18 (Suppl 1):27–30.
83. Wolters, D. A., Washburn, M. P. and Yates, J. R. (2001) An automated multidimensional protein identification technology for shotgun proteomics, *Analytical Chemistry* 73:5683–90.
84. Swanson, S. K. and Washburn, M. P. (2005) The continuing evolution of shotgun proteomics, *Drug Discovery Today* 10:719–25.
85. Washburn, M. P., Wolters, D. and Yates, J. R. (2001) Large-scale analysis of the yeast proteome by multidimensional protein identification technology, *Nature Biotechnology* 19:242–47.

86. Millea, K. A., Kass, I. J., Cohen, S. A. et al. (2005) Evaluation of multidimensional (ion-exchange/reversed-phase) protein separations using linear and step gradients in the first dimension, *Journal of Chromatography A* 1079:287–98.
87. Wang, Y., Zhang, J., Liu, C. L., Gu, X. and Zhang, X. M. (2005) Nano-flow multidimensional liquid chromatography with electrospray ionization time-of-flight mass spectrometry for proteome analysis of hepatocellular carcinoma, *Analytica Chimica Acta* 530:227–35.
88. Miller, C. A. and Hoerth, P. (2006) Off-gel electrophoresis vs. SCX fractionation for the HPLC/MS analysis of human serum. Paper presented at the Human Proteome Organization.
89. Issaq, H. J., Chan, K. C., Janini, G. M. and Muschik, G. M. (1999) A simple two-dimensional high performance liquid chromatography high performance capillary electrophoresis set-up for the separation of complex mixtures, *Electrophoresis* 20:1533–37.
90. Issaq, H. J., Chan, K. C., Liu, C. S. and Li, Q. B. (2001) Multidimensional high performance liquid chromatography—Capillary electrophoresis separation of a protein digest: An update, *Electrophoresis* 22:1133–35.
91. Moore, A. W. and Jorgenson, J. W. (1995) Rapid comprehensive 2-dimensional separations of peptides via RPLC optically gated capillary zone electrophoresis, *Analytical Chemistry* 67:3448–55.
92. Moore, A. W. and Jorgenson, J. W. (1995) Comprehensive 3-dimnensional separation of peptides using size-exclusion chromatography reversed-phase liquid-chromatography optically gated capillary zone electrophoresis, *Analytical Chemistry* 67:3456–63.
93. Shi, Y., Xiang, R., Horvath, C. and Wilkins, J. A. (2004) The role of liquid chromatography in proteomics, *Journal of Chromatography A* 1053:27–36.
94. Xiong, L. and Regnier, F. E. (2002) Use of a lectin affinity selector in the search for unusual glycosylation in proteomics, *Journal of Chromatography B-Analytical Technologies in the Biomedical and Life Sciences* 782:405–18.
95. Fenselau, C. (2007) A review of quantitative methods for proteomic studies, *Journal of Chromatography B-Analytical Technologies in the Biomedical and Life Sciences* 855:14–20.
96. Gygi, S. P., Rist, B., Gerber, S. A. et al. (1999) Quantitative analysis of complex protein mixtures using isotope-coded affinity tags, *Nature Biotechnology* 17:994–99.
97. Gerber, S. A., Rush, J., Stemman, O., Kirschner, M. W. and Gygi, S. P. (2003) Absolute quantification of proteins and phosphoproteins from cell lysates by tandem MS, *Proceedings of the National Academy of Sciences of the United States of America* 100:6940–45.
98. Kirkpatrick, D. S., Gerber, S. A. and Gygi, S. P. (2005) The absolute quantification strategy: A general procedure for the quantification of proteins and post-translational modifications, *Methods* 35:265–73.
99. Mayya, V., Rezual, K., Wu, L. F., Fong, M. B. and Han, D. K. (2006) Absolute quantification of multisite phosphorylation by selective reaction monitoring mass spectrometry—Determination of inhibitory phosphorylation status of cyclin-dependent kinases, *Molecular and Cellular Proteomics* 5:1146–57.
100. Cheng, D. M., Hoogenraad, C. C., Rush, J. et al. (2006) Relative and absolute quantification of postsynaptic density proteome isolated from rat forebrain and cerebellum, *Molecular and Cellular Proteomics* 5:1158–70.
101. Ross, P. L., Huang, Y. L. N., Marchese, J. N. et al. (2004) Multiplexed protein quantitation in *Saccharomyces cerevisiae* using amine-reactive isobaric tagging reagents, *Molecular and Cellular Proteomics* 3:1154–69.
102. Ogata, Y., Charlesworth, M. C., Higgins, L. et al. (2007) Differential protein expression in male and female human lumbar cerebrospinal fluid using iTRAQ reagents after abundant protein depletion, *Proteomics* 7:3726–34.

103. Kobeissy, F. H., Ottens, A. K., Zhang, Z. Q. et al. (2006) Novel differential neuroproteomics analysis of traumatic brain injury in rats, *Molecular and Cellular Proteomics* 5:1887–98.
104. Grant, S. G. N. and Blackstock, W. P. (2001) Proteomics in neuroscience: From protein to network, *Journal of Neuroscience* 21:8315–18.
105. Kim, S. I., Voshol, H., van Oostrum, J. et al. (2004) Neuroproteomics: Expression profiling of the brain's proteomes in health and disease, *Neurochemical Research* 29:1317–31.
106. Vercauteren, F. G. G., Bergeron, J. J. M., Vandesande, F., Arckens, L. and Quirion, R. (2004) Proteomic approaches in brain research and neuropharmacology, *European Journal of Pharmacology* 500:385–98.
107. Virginia Espina, J. D. W., Valerie S. Calvert, Amy VanMeter, Weidong Zhou, George Coukos, David H. Geho, Emanuel F. Petricoin III, Lance A. Liotta (2006) Laser-capture microdissection, *Nature Protocols* 1:586–603.
108. Grant, S. G. N. (2003) Systems biology in neuroscience: bridging genes to cognition, *Current Opinion in Neurobiology* 13:577–82.
109. Tyers, M. and Mann, M. (2003) From genomics to proteomics, *Nature* 422:193–97.
110. Kapp, E. A., Schutz, F., Connolly, L. M. et al. (2005) An evaluation, comparison, and accurate benchmarking of several publicly available MS/MS search algorithms: Sensitivity and specificity analysis, *Proteomics* 5:3475–90.
111. Moulder, R., Filen, J. J., Salmi, J. et al. (2005) A comparative evaluation of software for the analysis of liquid chromatography-tandem mass spectrometry data from isotope coded affinity tag experiments, *Proteomics* 5:2748–60.
112. Zhang, W. Z. and Chait, B. T. (2000) Profound: An expert system for protein identification using mass spectrometric peptide mapping information, *Analytical Chemistry* 72:2482–89.
113. Hartler, J., Thallinger, G. G., Stocker, G. et al. (2007) MASPECTRAS: A platform for management and analysis of proteomics HPLC-MS/MS data, *BMC Bioinformatics* 8:197.
114. Sriyam, S., Sinchaikul, S., Tantipaiboonwong, P. et al. (2007) Enhanced detectability in proteome studies, *Journal of Chromatography B-Analytical Technologies in the Biomedical and Life Sciences* 849:91–104.
115. Husi, H., Grant, S.G. (2002) Construction of a protein-protein interaction database (PPID) for synaptic biology. In *Neuroscience databases: A pratical guide*, ed. R. Kötter, 51–62 (Boston/Dordrecht/London: Kluwer Academic Publishers).
116. Kitano, H. (2002) Computational systems biology, *Nature* 420:206–10.

4 2-D Fluorescence Difference Gel Electrophoresis (DIGE) in Neuroproteomics

Roberto Diez, Michael Herbstreith,
Cristina Osorio, and Oscar Alzate

CONTENTS

4.1   INTRODUCTION

The brain is of highest interest in biomedical research and in the pharmaceutical industry due to the appearance of widespread neurological diseases such as Alzheimer's disease, Parkinson's disease, Huntington's disease, multiple sclerosis, and stroke. Research into protein function in the brain and its role in health and disease is advancing swiftly. In the past, research technology allowed the examination of only a few proteins at a time; however, it is rapidly becoming more common to examine simultaneously the expression of hundreds or even thousands of proteins (1). This enables a more holistic view for the comprehension of cellular processes. The importance of harnessing quantitative methodologies for the assessment of differences in protein expression is paramount. Thus, a growing body of work in neuroproteomics using two-dimensional fluorescence difference gel electrophoresis (2D-DIGE) is rapidly emerging, and a timely review of these data is warranted.

Why is it important for neuroscience research to examine thousands of proteins simultaneously? Although certain RNA molecules can act as effector molecules, proteins perform the majority of biological actions in the cell. Identifying the thousands of different proteins in a cell, the modifications to these proteins, along with their expressional level changes under different conditions, and all the protein–protein interactions is revolutionizing biology and medicine. There are many potential applications for proteomics in neuroscience (2). Comparative protein expression profiling and post-translational protein modification profiling are tasks that are best performed with DIGE.

4.2   TWO-DIMENSIONAL GEL ELECTROPHORESIS

Two-dimensional electrophoresis is one of the most commonly used techniques in proteomics. The basic principles of two-dimensional gel electrophoresis (2D-GE) remain the same since its introduction in 1975 (3,4), namely, the initial separation of proteins by isoelectric focusing (IEF, first dimension) followed by an orthogonal separation via sodium-dodecyl sulfate polyacrylamide gel electrophoresis (SDS-PAGE) (5). It has been shown that up to 10,000 protein spots can be separated in one gel allowing high resolution proteomic analysis (6).

IEF separates proteins according to their isoelectric point (pI) (Figure 4.1). The pI of a protein is primarily a function of its amino acid sequence, although post-translational modifications can also contribute to the pI. Proteins are amphoteric molecules, capable of acting either as an acid or a base. The side chains of the amino acids in proteins have acidic or basic buffering groups that are protonated or deprotonated, depending on the pH of the solution in which the protein is present. At a particular pH, the sum of the charges of all the amino acids in a protein equals zero. IEF takes advantage of this property by placing proteins in a pH gradient and applying an electric potential. This produces an electric force by which the protein will migrate toward the anode or cathode depending on its net charge. Eventually, the protein will reach its pI and stop migrating. Initially, the preparation and use of the pH gradients needed for IEF was very difficult and inconsistent.

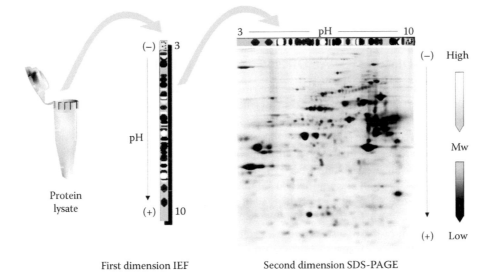

First dimension IEF   Second dimension SDS-PAGE

FIGURE 4.1 The 2D-PAGE concept. For the first dimension, proteins in solution are separated on an IPG strip by applying an electric field. The electric force displaces the proteins
(charges) until they reach their corresponding pI. The strip is loaded onto an SDS-PAGE,
and another electric field is applied. This time the proteins separate depending on their corresponding sizes. Typical values for a screening gel are pI ranges from 3 to 10.

These pH gradients were often in the form of tube gels with carrier ampholytes.
The introduction of chemistries to immobilize the pH gradient into the gel matrix
(7) and the construction of the gel on a solid backing (8,9) were significant steps
for making IEF more widely accessible to researchers and for producing more
consistent results.

The IEF gel or strip is equilibrated in SDS and then placed on top of the SDS-
PAGE gel (Figure 4.1). This equilibration step is necessary to allow the SDS molecules to associate with the proteins and produce the anionic complexes that have
a net negative charge. As a result of the electric force the proteins migrate out of
the IEF gel and into the SDS gel, where they separate according to their molecular weights. Although most applications use denaturing SDS-PAGE, approaches to
separate proteins under non-denaturing conditions have also been developed.

After electrophoresis, the proteins in the gel must be detected. Traditionally, this
is accomplished by the use of a visible stain. Very common visible stains include silver nitrate (10), which is highly sensitive. However, silver staining has complications
with unwanted background formation, lack of quantitative parameters, lack of reproducibility, and lack of mass spectrometry compatibility. Other visible stains such
as zinc imidazol (11) and Coomassie blue (12) have been developed, with the latter
reaching almost universal use. Even though both are more compatible with mass
spectrometry, zinc imidazol has limited quantitation abilities and Coomassie staining is much less sensitive than silver staining. Fluorescent stains have been developed
that attempt to combine simplicity, sensitivity, quantitation, and mass spectrometry

compatibility, such as ruthenium bathophenanthroline disulfonate (Sypro Ruby) (13) and Deep Purple (14,15).

4.3   TWO-DIMENSIONAL DIFFERENCE GEL ELECTROPHORESIS

The problems that have bedeviled the use of conventional 2D-GE for applications like protein expression profiling in neuroscience research are lack of reproducibility and quantitation. Just as much analysis was conducted on the shortfalls of 2D-GE at the turn of the century, a new incremental technology would revitalize the scientific proteomics community and revolutionize the field. Two-dimensional difference gel electrophoresis (2D-DIGE), introduced by Unlu et al. (16), addresses the problems of traditional 2D-GE in an elegant method (Figure 4.2).

DIGE uses direct labeling of proteins with fluorescent dyes (known as CyDyes: Cy2, Cy3, and Cy5) prior to IEF (17–20). CyDyes are spectrally resolvable cyanine dyes carrying an N-hydroxysuccinimidyl ester reactive group that covalently binds the ε-amino groups of lysine residues in proteins. Dye concentrations are kept low, such that approximately one dye molecule is added per protein. The important aspect of the DIGE technology is its ability to label two or more samples with different dyes and separate them on the same gel, eliminating gel-to-gel variability. This makes spot matching and quantitation much simpler and more accurate. Cy2 is used for a normalization pool created from a mixture of all samples in the experiment. This Cy2-labeled pool is run on all gels, allowing spot matching and normalization of signals from different gels. The DIGE approach offers great promise to researchers. The dyes are comparable in sensitivity to silver staining methods, are compatible with mass spectrometry, and offer the best quantitation of any available method.

Sample multiplexing and the use of an internal standard in 2D-DIGE allows the analysis of replicate samples from multiple experimental conditions with unsurpassed statistical confidence for differential display proteomics (21). DIGE experiments can easily accommodate sufficient independent (biological) replicate samples to control for the large interpersonal variation expected from biological samples. The use of multivariate statistical analyses can then be used to assess the global variation in a complex set of independent samples, filtering out the noise from technical variation and normal biological variation, thereby focusing on the underlying differences that can describe various disease or biological states.

4.4   BENEFITS OF DIGE FOR IMAGE ANALYSIS

Once proteins have been visualized, image analysis is required (22). Image analysis can be segregated into spot detection, spot matching, and data analysis. Manually detecting the hundreds to thousands of spots on each traditional 2D-GE gel for the number of gels in an experiment would take several days with countless inconsistencies. For any experiment requiring multiple gels for many different conditions across multiple samples (biological replicates), the different gels must be matched to each other. In other words, a spot at a certain location on one gel must be matched with the same spot on all other gels. In reality, proteins do not migrate to exactly the same point on each gel (IEF or SDS-PAGE). Complex algorithms have been developed that

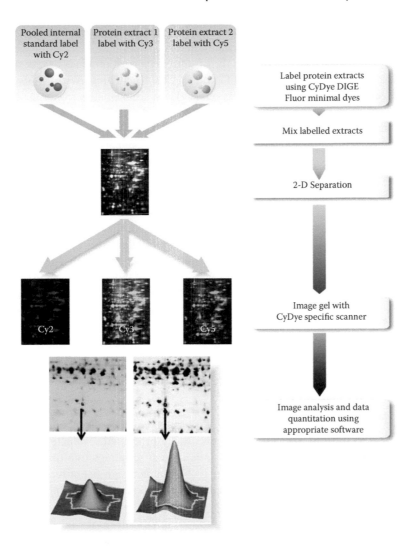

FIGURE 4.2   (See color insert following page 172.) The 2D-DIGE concept. Proteins from different sources (such as a protein lysate from a control sample and a protein lysate from an analytical sample) are covalently labeled with two different CyDyes. Labeled protein lysates are mixed, and separated on a 2D-PAGE. The gels are scanned using an instrument capable of detecting each CyDye independently. The images are analyzed and spots corresponding to same proteins having different expression levels are determined. After 2D-DIGE, protein may be identified by mass spectrometry, or by any other suitable technique.

attempt to match these spots across gels using spot pattern recognition and correcting for shifts in these patterns. The need for computer-based spot matching cannot be underestimated, as matching by hand can take days or even weeks, depending on the sample size. These matching algorithms continue to improve, but the best thing that can be done to aid in spot matching is to develop reproducible laboratory techniques. A common concern with 2-DE is that with a long experimental protocol of protein preparation, IEF, SDS-PAGE, visualization, and data analysis, small errors compound, thus inhibiting meaningful quantitation.

The Cy2-labeled pool used in 2D-DIGE is useful because it provides a consistent spot map on all gels in an experiment, facilitating spot matching. With DIGE the number of gels that need to be matched is reduced and sister images from the same gel will have identical spot patterns. With spots matched and with signal intensities known, comparisons can be made to determine changes in protein expression levels. The DIGE method uses the protein pool to normalize the signal abundances among gels, correcting for any differences in overall signal intensity. This provides a consistent expression measurement across gels.

It is important to keep in mind that while a comprehensive cataloging of protein species within a system could be conclusive, it is also the beginning of much further protein profiling and data analysis. Given the complexity of the neuronal proteome, it is essential to study alterations in protein expression and post-translational modifications within specific brain regions, within specific types of neurons, and within sub-cellular compartments. Awareness of the high level of biological diversity to be expected among both the control and experimental samples and the need for rigorous statistical analysis are important elements to the analysis of protein profiling data.

4.5   VIRTUES OF DIGE

Compared with traditional 2D-GE, 2D-DIGE offers major advantages for studying protein expression changes in biological samples (23):

1. DIGE facilitates the co-separation of two proteomes to be compared, thereby diminishing the number of gels to be processed, evaluated, and interpreted by a factor of two. This saves critical resources like time and manpower. A major advantage of this technique is a significant reduction in inter-gel variability, facilitating spot identification and matching, thus increasing the number of analyzable spots.
2. The pooling of both samples prior to separation diminishes the false quantification of irreproducible losses (e.g., sample entry in the first-dimension gel, transfer from the first-dimension to the second-dimension gel) during the analysis. Any imprecision of the method affects similar proteins in a similar way, compensating methodological errors. Therefore, the reliability of experimental data is significantly increased.
3. The internal pooled standard (17) facilitates the normalization of each spot among all gels, a feature especially useful in a comprehensive study. The internal pooled standard allows the comparison of more than two proteomes without the need to perform pair-wise analysis of all possible combinations of data points.
4. DIGE provides the ability to detect many protein post-translational modifications, such as phosphorylation, ubiquitination, palmitoylation, etc. which often play a key role in modulating protein function and which cannot generally be detected by other protein profiling technologies.

Overall, the major advantages of DIGE are the high sensitivity and linearity of the dyes utilized, its straightforward protocol, as well as its significant reduction of inter-gel variability, which increases the possibility to unambiguously identify biological variability and reduces bias from experimental variation. Moreover, the use of a pooled internal standard, loaded together with the control and experimental samples, increases quantification accuracy and statistical confidence (17).

4.6   SAMPLE PREPARATION

Neuroscientists wishing to implement neuroproteomic approaches for their research will have to surmount difficulties particular to their systems, being the most critical limited sample amounts, heterogeneous cellular compositions in samples, and the fact that many proteins of interest are rare, hydrophobic proteins. There are a number of approaches to reduce the complexity of a protein sample for analysis. Depending on the rationale for a study, analyzing the entire contents of cellular proteins in one experimental run may cause a loss of sensitivity and result in interpretational difficulties. In such cases, it is recommended that the proteome be pre-fractionated in order to remove contaminating material from the sample and to enrich the proteins of interest for further analysis (see Chapter 3 for a complete review of pre-fractionation techniques). The starting material can be fractionated using a variety of approaches including centrifugation (e.g., soluble/insoluble, membrane/cytosolic/nuclear), salt precipitation, liquid chromatographic separation (e.g., ion exchange, affinity, gel filtration), and velocity or equilibrium sedimentation.

Neuroscientists face complications specific to brain and neural tissues. When planning proteomic research, careful attention must be paid to the sample amount available relative to the sample needs of techniques being used. The temptation to use large amounts of brain tissue must be weighed against the need for regional and cellular specificity. All current methods in proteomics, including 2D-GE and DIGE, tend to identify preferentially the most abundant proteins. For neuroscientists, this represents a challenge because many proteins of interest in the brain are expressed at relatively low levels and are hydrophobic proteins. Pre-fractionation is important for extending the dynamic range of the analysis, whether it is enrichment of sub-cellular fractions or enrichment of classes of proteins (e.g., DNA binding proteins or phosphorylated proteins). Because of the deleterious effects that accompany protein overloading of samples, less abundant (and often, more important regulatory) proteins can only be detected with pre-fractionation and/or depletion of highly abundant proteins (see Chapter 3). It is important that any pre-fractionation method be standardized to avoid introduction of potential artifacts and technical variability. Another important sample preparation consideration is that without careful dissection of the tissue being analyzed, significant changes in one cell type or cell population within a given tissue may be diluted by homogenization and mixing with other neighboring unaffected cell types and subsequently disappear beneath the threshold of significance (24). With respect to sample preparation and origin, of notable concern also is postmortem brain tissue (particularly human; see Chapter 2 for a discussion of brain tissue banking and storage), of which little is known about quantitative postmortem changes in the brain protein profile (25). For studies on human brain proteomes it is important to standardize the protocols used for preparation of protein extracts.

Bernocco et al. (26) refined a detergent-based fractionation method which reduces complexity of the protein extracts. The sequential use of detergent-containing buffers on neurons in culture plates yields four extracts enriched in cytosolic, membrane-bound or enclosed, nuclear, and cytoskeletal proteins. Comparison of extracts by DIGE showed a clear difference in protein composition. An extraction efficiency of 85% was calculated for cytosolic proteins in extract 1, 90% for membrane-bound and membrane-enclosed proteins in extract 2, 82% for nuclear proteins in extract 3, and 38% for cytoskeletal and RAFT proteins in extract 4.

One of the distinct advantages of proteomic analysis, not attainable with RNA expression data, is the ability to fractionate the cell's proteins into various subpopulations. Purification of protein from other cellular substances is also necessary. Lipids are particularly abundant in the brain, and along with nucleic acids must be eliminated from the protein sample for good-quality results. The most common methods of protein purification rely on selective precipitation. Acetone, trichloroacetic acid (TCA), and other precipitation methods can be performed, and a number of commercially available kits make this a routine procedure (27,28). In some instances proteins such as IgGs or albumin constitute the vast majority of a protein sample. Selective elimination of these proteins improves detection of less highly expressed proteins (see Chapter 3).

Protein stability and purity are of critical importance to proteomic studies; therefore, it is important to prevent protein degradation and modification(s) during sample preparation. Rapid removal of brain tissue, dissection, and freezing are obvious imperatives for the maintenance of the proteome as close as possible as it was in the animal. Human postmortem studies pose unique challenges, but these can be addressed by careful documentation of postmortem interval, brain pH, and agonal state (29) (and see Chapter 2). Specific proteins have been shown to degrade in a time-dependent manner, highlighting the need for careful selection of controls in human brain postmortem studies (30). Protease and phosphatase inhibitors are commonly used to help prevent degradation and dephosphorylation of proteins during protein preparation (31).

The number of proteomics studies concerning human brain samples has increased in recent years, in particular in the discovery of biomarkers for neurological diseases. The human brain samples are obtained from brain banks, which are interested in providing high quality human nervous tissue. In order to provide brain banks as well as scientists working in the proteomics field with measures for tissue quality, the critical factors after death, the effect of postmortem interval (PMI), and storage temperature on the human brain proteome were investigated (32). This study was focused on the gray matter of the frontal cortex. The PMI was artificially prolonged from the time of autopsy (two hours after death) by storing samples at 47°C or room temperature over 18, 24, and 48 hours. The DIGE experiment revealed the degradation of three proteins: peroxiredoxin-1, stathmin, and glial fibrillary acidic protein. These were further confirmed by Western blot analysis. In the second part of the study, prefrontal cortex samples were compared among three individual donors. 2D-DIGE was selected as the quantitative proteomics technique of choice because delayed PMI was expected to also affect post-translational modifications of proteins, which would be very difficult to monitor by any other method.

4.7   ORTHOGONAL TECHNIQUES

Two-dimensional electrophoresis is also limited in its ability to detect very basic proteins (pI > 10) (33). This assumption has been challenged by a recent study (34). In this instance, 2D-GE was found to be better for small-molecular-weight, hydrophobic, and cysteine-lacking proteins, whereas isotope coded affinity tag (ICAT) was found to be superior in examining high-molecular-weight proteins. Immunoblotting/ Western blotting, enzyme-linked immunoabsorbent assay (ELISA) and immunocytochemistry are all well established methods commonly used in molecular biology laboratories. These methods serve as excellent tools for validation of candidate proteins and further experimentation on targets found by DIGE in combination with mass spectrometry. At the moment, unfortunately, although this validation step should be regarded as an absolute prerequisite, antibodies are not available for every protein and high-throughput confirmation of expression changes from a DIGE experiment is practically and economically not feasible.

There are a wide variety of applications for proteomic technology in neuroscience (2). These applications range from defining the proteome of a particular cell type, identifying changes in brain protein expression under different experimental or disease conditions, profiling protein modifications (e.g., phosphorylation) and mapping protein–protein interactions. With the incredible cellular heterogeneity of the brain, there are a number of defined cell types awaiting proteomic characterization. The possibility of an international collaborative effort, along the lines of the human genome project, to catalog the brain proteome is being discussed to take on this herculean task (35).

Perhaps the proteomic application of greatest interest to neuroscience is protein expression profiling. The number of such studies is still small but growing rapidly. An advantage of profiling by proteomic methods such as DIGE is that it offers an open and unbiased screen. Protein microarrays are closed screens, on the other hand, limited to examining those genes that are deposited on the chip used in the experiment. Techniques like 2D-GE or 2D-DIGE and MS/MS can, theoretically, detect any protein in an unbiased fashion.

4.8   APPLICATIONS IN NEUROPROTEOMICS

Since its inception in 1997, DIGE has been widely adopted in most areas of biology and medicine. DIGE has become an important tool among proteomics technologies for studying the mechanisms of disease, pinpointing new therapeutic targets or finding potential biomarkers (36). Among many other applications, DIGE has been applied to cancer research (37–41), renal physiology (42), plant biology (43–45), and the elucidation of signal transduction pathways (46), just to name a few. In neurology and neuroscience, many applications have been entertained in neurotoxicology and neurometabolism, and used in the determination of specific proteomic aspects of individual brain areas and body fluids in neurodegeneration to identify biomarkers. The concomitant detection of several hundred proteins on a gel provides comprehensive data to elucidate a physiological protein network and its peripheral representatives.

4.8.1 SYNAPTIC PHYSIOLOGY AND STRUCTURE

The auditory forebrain area in songbirds called the caudomedial nidopallium (NCM) plays a key role in auditory learning and song discrimination. The expression of few transcription factors is increased in NCM in response to the neural processes associated with auditory processing of sounds (47). DIGE-based proteomics was used to investigate the NCM of adult songbirds hearing novel songs (47). After one and three hours of stimulation of freely behaving birds with conspecific songs, a significant number of proteins were consistently regulated in NCM. These proteins included metabolic enzymes, cytoskeletal molecules, and proteins involved in neurotransmitter secretion and calcium binding (47) (see Chapter 14 for a detailed discussion of neuroproteomics studies of auditory processing of vocal communications in songbirds).

Neuroproteomics techniques including 2D-DIGE, mass spectrometry, and subcellular fractionation have been used to elucidate the protein constituents of the synapse. The proteomic characterization of the synapses will contribute in the understanding of their essential role in neurotransmission and plasticity. A map of the synapse proteome is emerging, including the proteomes of structures such as pre- and post-synaptic densities.

Burre et al. (48) induced massive exocytosis and analyzed the protein contents of the synaptic vesicle compartment using DIGE-based proteomics, to identify proteins that undergo modifications as a result of synaptic activation. They identified eight proteins that revealed significant changes in abundance following nerve terminal depolarization. Based on these results it was proposed that depolarization of the pre-synaptic compartment induces changes in the abundance of synaptic vesicle proteins and post-translational protein modification.

Synaptic dysfunction is an early event in Alzheimer's disease. In a study by Gillardon et al. (49) synaptosomal fractions from Tg2576 mice over-expressing mutant human APP and from wild-type littermates were analyzed for proteomic changes. Crude synaptosomal fractions from cortical and hippocampal micro-dissected tissue were prepared by differential centrifugation and the proteins were separated by DIGE. Significant alterations were detected in the mitochondrial heat shock protein 70, which is associated with mitochondrial stress response. In addition, numerous changes in the protein subunit composition of the respiratory chain complexes I and III were identified. It is suggested that "early impairment of axonal transport may lead to aberrant axonal amyloid-beta generation and intraneuronal oligomer formation." It is then suggested that amyloid-beta oligomers may then cause mitochondrial dysfunction and synaptic impairment leading to cognitive decline before amyloid plaque deposition (49).

4.8.2   ALZHEIMER'S DISEASE

The amyloid precursor protein (APP) plays a central role in Alzheimer's disease pathology. Hartl et al. (50) analyzed the APP-transgenic mouse model APP23 using DIGE. Cortex and hippocampus of transgenic and wild-type mice at 1, 2, 7, and 15 months of age and cortices of 16-days-old embryos were investigated. A large number of proteins were found to be altered in wild-type mice, which were largely absent in hippocampus of APP23. From these results it was proposed that the absence of developmental proteome alterations along with the down-regulation of proteins related to plasticity suggests the disruption of a normally occurring peak of hippocampal plasticity during adolescence in APP23 mice (50).

In another study, it was found that mortalin (the mitochondrial Hsp70 chaperone) is differentially regulated in the brains of apoE-ε4 targeted replacement (TR) mice, compared with apoE-ε3 TR mice used as controls (51). Similar analysis indicated that this protein is also differentially regulated in Alzheimer's disease patients, and that this protein regulation depends on the *APOE* genotype and the disease state (51).

4.8.3   PHOSPHORYLATION TRENDS

Profiling of protein modifications is as important as determining the proteins present in a cell and their relative expression levels. Many important cellular processes in the nervous system rely on protein modifications as a mechanism to regulate protein function. The most studied protein modifications are phosphorylation cascades in signal transduction. Using DIGE-based proteomics and lambda-phosphatase treatment, a map of phosphorylated proteins in rat cortical neurons was created (52). Losing of a phosphate group makes the pI of a protein more basic by altering its net charge; therefore, the protein migrates to a different position in a 2D-GE map compared to the phosphorylated protein. Because DIGE allows separation of phosphorylated and dephosphorylated protein samples in the same gel, changes in protein patterns after phosphatase treatment are easily detectable. Small differences in migration and small shifts toward more basic pHs are easier to detect with DIGE than with any other technique (52).

4.8.4   BRAIN REGIONS

Jacobs et al. used DIGE to screen for region-specific protein markers (53). By comparing proteome maps of the primary visual and somatosensory areas in mouse brain, they found 22 protein spots with different expression levels in one of the two primary sensory areas. Specific brain-region protein characterization is a promising method for the understanding of cortical networks and their physiological interactions.

4.8.5   BRAIN REGIONS AND DEVELOPMENT

Van den Bergh et al. (54) used DIGE to help elucidate protein expression differences in cat and kitten striate visual cortex, identifying 12 proteins that are differentially expressed. Again using DIGE, the group compared P10–P30 and P30-adult brain protein samples (55). Thirty-four proteins in cat primary visual area-17, whose expression levels from the time of eye opening toward adulthood change with age, were identified. These changes in protein expression levels could be correlated to age-dependent postnatal brain development. Western blot was used to validate some of these results in cat visual cortex (see Chapter 11 for a detailed discussion of neuroproteomics studies on the cat visual cortex). Jin et al. (56) used DIGE to study differential protein expression in immature and mature neurons in culture.

4.8.6 SCHIZOPHRENIA

Comparing human prefrontal cortex from control and schizophrenic individuals using DIGE, 55 proteins with significant differences in expression in white versus grey matter were identified (57). These proteins are known to be associated with metabolism, axonal growth, protein turnover and trafficking, cell signaling, and cytoskeleton structure. In a systematic extension to this study, results were compared with mRNA microarray analysis and measurement of metabolites (58). Many of these proteins were associated with mitochondrial function or with oxidative stress responses, with the changes in protein expression correlating with changes in mRNA expression. Later, the same group used DIGE to analyze liver and red blood cells, providing further evidence for oxidative stress in schizophrenia (59).

4.8.7 PARKINSON'S DISEASE

The protein contents of the striatum of 6 control and 21 MPTP-treated monkeys used as animal models of Parkinson's disease were studied in a DIGE experiment, with or without de novo or long-term L-DOPA administration (60). Several sets of proteins associated with the priming effects of L-DOPA in the striatum in Parkinsonian animals were identified, including proteins involved in energy metabolism and the microtubule cytoskeleton.

4.8.8 DRUG ADDICTION

A study was carried out aimed at understanding the effects of epidemic cocaine use entangled with HIV-1 infection, and the enhancement of HIV-1 replication in normal human astrocytes (61). Twenty-two proteins were identified in normal human astrocytes that were differentially regulated by cocaine, which directly or indirectly play a supportive role in the neuropathogenesis of HIV-1 infection.

The nucleus accumbens has been associated with the reinforcing effects and long-term consequences of cocaine self-administration. To improve the knowledge about cocaine-induced biochemical adaptations in rodent models, DIGE was used to compare changes in cytosolic protein abundance in the nucleus accumbens of rhesus monkeys self-administering cocaine and controls (62). Several protein spots were found to be differentially abundant, of which 18 proteins were identified by mass spectrometry.

In another study, Reynolds et al. (63) used DIGE to analyze the effects of methamphetamine (METH) on HIV-1 infectivity. It was found that METH modulates the expression of several proteins including CXCR3, protein disulfide isomerase, procathepsin B, peroxiredoxin, and galectin-1. This study suggests a mechanism for METH on HIV-1 infectivity.

4.8.9 WEST NILE VIRUS

Dhingra et al. (64) applied DIGE to study the mechanisms of neurodegeneration associated with West Nile virus infections. DIGE was used to characterize protein expression in primary rat neurons and to examine the proteomic profiling to understand the pathogenesis of meningoencephalitis associated with the disease.

4.8.10 OTHER APPLICATIONS

There are many other examples of research projects in which DIGE-based proteomics has been used to identify molecular bases of neurological processes. These examples include protein composition in cerebrospinal fluid of traumatic brain injury patients (65); molecular mechanisms of sleep deprivation and aging (66), Western Pacific amyotrophic lateral sclerosis–Parkinsonism-dementia complex (ALS/PDC) (67); the effects of chronic ethanol treatment on the brain proteome in the long-fin striped strain of zebra fish (68); hypoxia-related protein regulation in medaka fish brain tissue (69); identification of potential biomarkers of developmental neurotoxicity of organohalogen compounds (70); phosphorylation/dephosphorylation cycles (71,72); stress response (51,73); and prion diseases (74).

4.9 SATURATION DYES

There are situations where the available sample is limited, e.g., samples from clinical studies employing cerebrospinal fluid and small but functionally important regions of brain such as pituitary, especially from small animals such as rats, post- and pre-synaptic density preparations, and nerve fibers. In most cases, brain tissue complexity represents a major challenge. In such situations it is common to use laser microdissection (LMD) for obtaining samples suitable for proteomic studies (75). Experimental situations involving such limited amounts of protein may be addressed using other types of fluorescent dyes.

Due to the high lysine content of most proteins, labeling to saturation with the N-hydroxysuccinimide (NHS)-reactive CyDyes described so far in this chapter would require excessive amounts of reagents, and due to the hydrophobic nature of the dyes, it is likely to cause protein precipitation. Cysteine residues are typically less abundant than lysine in mammalian proteins and are therefore more amenable to total labeling. CyDye dyes containing a maleimide group (Cy-maleimides, or Cy-M) that reacts with the thiol groups of cysteine residues are available. These labels do not alter the pI of a labeled protein because they do not have a net charge (19). Each Cy-M dye adds a mass of approximately 680 Da to the protein, but all available cysteine groups are modified. This approach is commonly known as "saturation labeling." Saturation labeling DIGE is more sensitive than traditional stains such as Coomassie, silver, or Sypro Ruby, and as little as 0.1 ng albumin has been detected with Cy5-M dye compared to 1 ng with the Cy5 minimal dye. Saturation labeling also has a greater dynamic range, around 10^3–10^4, which is an order of magnitude greater than for Sypro Ruby and for minimal labeling (76).

Wilson et al. (77) combined LMD with saturation labeling DIGE to identify protein changes in the isolated CA1 pyramidal neuron layer of a transgenic rat carrying a human amyloid precursor protein transgene. Saturation dye labeling proved to be extremely sensitive with a spot map of over 5000 proteins readily produced from 5 µg total protein, with over 100 proteins significantly altered, although identification of those proteins by mass spectrometry represented yet another substantial challenge. Saturation labeling DIGE is useful for applications in which there is limited sample.

4.10   ORGANIZATIONS

The Human Proteome Organization Brain Proteome Project (HUPO-BPP; http://www.hbpp.org/) aims at coordinating neuroproteomics efforts with respect to analysis of development, aging, and evolution in humans and mice and at analyzing normal aging processes as well as neurodegenerative diseases. A major goal of the HUPO-BPP pilot study is the evaluation of different proteomics technologies (78). Frohlich et al. (79) contributed a pilot study with a DIGE analysis of standardized mouse brain samples, consisting of whole brains from mice of three different developmental stages. Five brains per stage were differentially analyzed by DIGE using overlapping pH gradients (pH 4–7 and 6–9). In total, 214 protein spots showing stage-dependent intensity alterations were detected, 56 of which were identified. In another study Focking et al. found 206 protein spots that were differentially expressed among the different stages: 122 spots were highest in intensity in embryonic stage, 26 highest in the juvenile group, and 58 spots highest in the adult stage (80).

4.11   CONCLUSIONS

One of the most fundamental approaches to understanding protein function is to correlate expression level changes as a function of growth conditions, cell cycle stage, disease state, external stimuli, level of expression of other proteins, and other possible variables involving protein regulation and potential protein modifications. Along with the interrogation of DNA microarrays for their ability to indicate relative levels of mRNA expression, protein expression levels, their post-translational modifications, and their interactions should also be interrogated.

Proteomic studies complemented by genomics and more traditional molecular biology techniques are contributing significantly to build complete proteome maps of cells and organelles, under both normal and altered conditions. The valuable information provided in qualitative and quantitative proteome maps will enable further identification of mechanisms of diseases, as well as those underlying drug actions, and may contribute to the development of more effective drug treatments. Understanding the complex changes taking place in biological systems or disease at the molecular level will lead to a better understanding of the underlying mechanisms. DIGE-based proteomics is well suited to describe the molecular anatomy of a system and its changes in levels of protein and their expression pattern, including post-translational modifications.

The studies mentioned in this chapter all indicate that 2D-DIGE-based proteomics provides an alternative approach to explore and understand the molecular basis of complex mechanisms associated with neurodegenerative diseases and neurobiology. They illustrate the potential of DIGE for neuroproteomics to profile differences in the distribution and regulation of thousands of proteins at a time, to study the function of disease markers, and to identify molecular pathways that could lead to novel therapeutic targets.

REFERENCES

1. Patton, W. F. (2002) Detection technologies in proteome analysis, *J Chromatogr B Analyt Technol Biomed Life Sci* 771:3–31.
2. Freeman, W. M. and Hemby, S. E. (2004) Proteomics for protein expression profiling in neuroscience, *Neurochem Res* 29:1065–81.
3. Klose, J. (1975) Protein mapping by combined isoelectric focusing and electrophoresis of mouse tissues. A novel approach to testing for induced point mutations in mammals, *Humangenetik* 26:231–43.
4. O'Farrell, P. H. (1975) High resolution two-dimensional electrophoresis of proteins, *J Biol Chem* 250:4007–21.
5. Gorg, A., Obermaier, C., Boguth, G. et al. (2000) The current state of two-dimensional electrophoresis with immobilized pH gradients, *Electrophoresis* 21:1037–53.
6. Klose, J. and Kobalz, U. (1995) Two-dimensional electrophoresis of proteins: An updated protocol and implications for a functional analysis of the genome, *Electrophoresis* 16:1034–59.
7. Bjellqvist, B., Ek, K., Righetti, P. G. et al. (1982) Isoelectric focusing in immobilized pH gradients: Principle, methodology and some applications, *J Biochem Biophys Methods* 6:317–39.
8. Corbett, J. M., Dunn, M. J., Posch, A. and Gorg, A. (1994) Positional reproducibility of protein spots in two-dimensional polyacrylamide gel electrophoresis using immobilised pH gradient isoelectric focusing in the first dimension: An interlaboratory comparison, *Electrophoresis* 15:1205–11.
9. Gorg, A., Boguth, G., Obermaier, C., Posch, A. and Weiss, W. (1995) Two-dimensional polyacrylamide gel electrophoresis with immobilized pH gradients in the first dimension (IPG-Dalt): The state of the art and the controversy of vertical versus horizontal systems, *Electrophoresis* 16:1079–86.
10. Winkler, C., Denker, K., Wortelkamp, S. and Sickmann, A. (2007) Silver- and Coomassie-staining protocols: Detection limits and compatibility with ESI MS, *Electrophoresis* 28:2095–99.
11. Fernandez-Patron, C., Castellanos-Serra, L., Hardy, E. et al. (1998) Understanding the mechanism of the zinc-ion stains of biomacromolecules in electrophoresis gels: Generalization of the reverse-staining technique, *Electrophoresis* 19:2398–2406.
12. Neuhoff, V., Arold, N., Taube, D. and Ehrhardt, W. (1988) Improved staining of proteins in polyacrylamide gels including isoelectric focusing gels with clear background at nanogram sensitivity using Coomassie Brilliant Blue G-250 and R-250, *Electrophoresis* 9:255–62.
13. Rabilloud, T., Strub, J. M., Luche, S., van Dorsselaer, A. and Lunardi, J. (2001) A comparison between Sypro Ruby and ruthenium II tris (bathophenanthroline disulfonate) as fluorescent stains for protein detection in gels, *Proteomics* 1:699–704.

14. Bell, P. J. and Karuso, P. (2003) Epicocconone, a novel fluorescent compound from the fungus epicoccumnigrum, *J Am Chem Soc* 125:9304–5.
15. Mackintosh, J. A., Choi, H. Y., Bae, S. H. et al. (2003) A fluorescent natural product for ultrasensitive detection of proteins in one-dimensional and two-dimensional gel electrophoresis, *Proteomics* 3:2273–88.
16. Unlu, M., Morgan, M. E. and Minden, J. S. (1997) Difference gel electrophoresis: A single gel method for detecting changes in protein extracts, *Electrophoresis* 18:2071–7.
17. Alban, A., David, S. O., Bjorkesten, L. et al. (2003) A novel experimental design for comparative two-dimensional gel analysis: Two-dimensional difference gel electrophoresis incorporating a pooled internal standard, *Proteomics* 3:36–44.
18. Marouga, R., David, S. and Hawkins, E. (2005) The development of the DIGE system: 2D fluorescence difference gel analysis technology, *Anal Bioanal Chem* 382:669–78.
19. Tonge, R., Shaw, J., Middleton, B. et al. (2001) Validation and development of fluorescence two-dimensional differential gel electrophoresis proteomics technology, *Proteomics* 1:377–96.
20. Westermeier, R. and Scheibe, B. (2008) Difference gel electrophoresis based on lys/cys tagging, *Methods Mol Biol* 424:73–85.
21. Minden, J. (2007) Comparative proteomics and difference gel electrophoresis, *Biotechniques* 43:739, 741, 743 passim.
22. Goldfarb, M. (2007) Computer analysis of two-dimensional gels, *J Biomol Tech* 18:143–46.
23. Karp, N. A., Kreil, D. P. and Lilley, K. S. (2004) Determining a significant change in protein expression with DeCyder during a pair-wise comparison using two-dimensional difference gel electrophoresis, *Proteomics* 4:1421–32.
24. Mirnics, K., Middleton, F. A., Marquez, A., Lewis, D. A. and Levitt, P. (2000) Molecular characterization of schizophrenia viewed by microarray analysis of gene expression in prefrontal cortex, *Neuron* 28:53–67.
25. Voshol, H., Glucksman, M. J. and van Oostrum, J. (2003) Proteomics in the discovery of new therapeutic targets for psychiatric disease, *Curr Mol Med* 3:447–58.
26. Bernocco, S., Fondelli, C., Matteoni, S. et al. (2008) Sequential detergent fractionation of primary neurons for proteomics studies, *Proteomics* 8:930–38.
27. Chan, L. L., Lo, S. C. and Hodgkiss, I. J. (2002) Proteomic study of a model causative agent of harmful red tide, *Prorocentrum triestinum* I: Optimization of sample preparation methodologies for analyzing with two-dimensional electrophoresis, *Proteomics* 2:1169–86.
28. Polson, C., Sarkar, P., Incledon, B., Raguvaran, V. and Grant, R. (2003) Optimization of protein precipitation based upon effectiveness of protein removal and ionization effect in liquid chromatography-tandem mass spectrometry, *J Chromatogr B Analyt Technol Biomed Life Sci* 785:263–75.
29. Hynd, M. R., Lewohl, J. M., Scott, H. L. and Dodd, P. R. (2003) Biochemical and molecular studies using human autopsy brain tissue, *J Neurochem* 85:543–62.
30. Fountoulakis, M., Hardmeier, R., Hoger, H. and Lubec, G. (2001) Postmortem changes in the level of brain proteins, *Exp Neurol* 167:86–94.
31. Olivieri, E., Herbert, B. and Righetti, P. G. (2001) The effect of protease inhibitors on the two-dimensional electrophoresis pattern of red blood cell membranes, *Electrophoresis* 22:560–65.
32. Crecelius, A., Gotz, A., Arzberger, T. et al. (2008) Assessing quantitative post-mortem changes in the gray matter of the human frontal cortex proteome by 2-D DIGE, *Proteomics* 8:1276–91.
33. Bae, S. H., Harris, A. G., Hains, P. G. et al. (2003) Strategies for the enrichment and identification of basic proteins in proteome projects, *Proteomics* 3:569–79.

34. Schmidt, F., Donahoe, S., Hagens, K. et al. (2004) Complementary analysis of the *Mycobacterium tuberculosis* proteome by two-dimensional electrophoresis and isotope-coded affinity tag technology, *Mol Cell Proteomics* 3:24–42.

35. Habeck, P. (2003) Brain proteome project launched, *Nature Medicine* 9:631.

36. Issaq, H. J. and Veenstra, T. D. (2007) The role of electrophoresis in disease biomarker discovery, *Electrophoresis* 28:1980–88.

37. Ciordia, S., de Los Rios, V. and Albar, J. P. (2006) Contributions of advanced proteomics technologies to cancer diagnosis, *Clin Transl Oncol* 8:566–80.

38. Hoffman, S. A., Joo, W. A., Echan, L. A. and Speicher, D. W. (2007) Higher dimensional (Hi-D) separation strategies dramatically improve the potential for cancer biomarker detection in serum and plasma, *J Chromatogr B Analyt Technol Biomed Life Sci* 849:43–52.

39. Maurya, P., Meleady, P., Dowling, P. and Clynes, M. (2007) Proteomic approaches for serum biomarker discovery in cancer, *Anticancer Res* 27:1247–55.

40. Ornstein, D. K. and Petricoin, E. F., 3rd (2004) Proteomics to diagnose human tumors and provide prognostic information, *Oncology (Williston Park)* 18:521–29; discussion 529–32.

41. Somiari, R. I., Somiari, S., Russell, S. and Shriver, C. D. (2005) Proteomics of breast carcinoma, *J Chromatogr B Analyt Technol Biomed Life Sci* 815:215–25.

42. Hoorn, E. J., Hoffert, J. D. and Knepper, M. A. (2006) The application of DIGE-based proteomics to renal physiology, *Nephron Physiol* 104:61–72.

43. Lilley, K. S. and Dupree, P. (2006) Methods of quantitative proteomics and their application to plant organelle characterization, *J Exp Bot* 57:1493–99.

44. Rossignol, M., Peltier, J. B., Mock, H. P. et al. (2006) Plant proteome analysis: A 2004-2006 update, *Proteomics* 6:5529–48.

45. Stroher, E. and Dietz, K. J. (2006) Concepts and approaches towards understanding the cellular redox proteome, *Plant Biol (Stuttg)* 8:407–18.

46. Morandell, S., Stasyk, T., Grosstessner-Hain, K. et al. (2006) Phosphoproteomics strategies for the functional analysis of signal transduction, *Proteomics* 6:4047–56.

47. Pinaud, R., Osorio, C., Alzate, O. and Jarvis, E. D. (2008) Profiling of experience-regulated proteins in the songbird auditory forebrain using quantitative proteomics, *Eur J Neurosci* 27:1409–22.

48. Burre, J., Beckhaus, T., Corvey, C. et al. (2006) Synaptic vesicle proteins under conditions of rest and activation: analysis by 2-D difference gel electrophoresis, *Electrophoresis* 27:3488–96.

49. Gillardon, F., Rist, W., Kussmaul, L. et al. (2007) Proteomic and functional alterations in brain mitochondria from Tg2576 mice occur before amyloid plaque deposition, *Proteomics* 7:605–16.

50. Hartl, D., Rohe, M., Mao, L. et al. (2008) Impairment of adolescent hippocampal plasticity in a mouse model for Alzheimer's disease precedes disease phenotype, *PLoS ONE* 3:e2759.

51. Osorio, C., Sullivan, P. M., He, D. N. et al. (2007) Mortalin is regulated by APOE in hippocampus of AD patients and by human APOE in TR mice, *Neurobiol Aging* 28:1853–62.

52. Raggiaschi, R., Lorenzetto, C., Diodato, E. et al. (2006) Detection of phosphorylation patterns in rat cortical neurons by combining phosphatase treatment and DIGE technology, *Proteomics* 6:748–56.

53. Jacobs, S., Van de Plas, B., Van der Gucht, E. et al. (2008) Identification of new regional marker proteins to map mouse brain by 2-D difference gel electrophoresis screening, *Electrophoresis* 29:1518–24.

54. Van den Bergh, G., Clerens, S., Cnops, L., Vandesande, F. and Arckens, L. (2003) Fluorescent two-dimensional difference gel electrophoresis and mass spectrometry identify age-related protein expression differences for the primary visual cortex of kitten and adult cat, *J Neurochem* 85:193–205.
55. Van den Bergh, G., Clerens, S., Firestein, B. L., Burnat, K. and Arckens, L. (2006) Development and plasticity-related changes in protein expression patterns in cat visual cortex: A fluorescent two-dimensional difference gel electrophoresis approach, *Proteomics* 6:3821–32.
56. Jin, K., Mao, X. O., Cottrell, B. et al. (2004) Proteomic and immunochemical characterization of a role for stathmin in adult neurogenesis, *FASEB J* 18:287–99.
57. Swatton, J. E., Prabakaran, S., Karp, N. A., Lilley, K. S. and Bahn, S. (2004) Protein profiling of human postmortem brain using 2-dimensional fluorescence difference gel electrophoresis (2-D DIGE), *Mol Psychiatry* 9:128–43.
58. Prabakaran, S., Swatton, J. E., Ryan, M. M. et al. (2004) Mitochondrial dysfunction in schizophrenia: Evidence for compromised brain metabolism and oxidative stress, *Mol Psychiatry* 9:684–97, 643.
59. Prabakaran, S., Wengenroth, M., Lockstone, H. E. et al. (2007) 2-D DIGE analysis of liver and red blood cells provides further evidence for oxidative stress in schizophrenia, *J Proteome Res* 6:141–49.
60. Kultima, K., Scholz, B., Alm, H. et al. (2006) Normalization and expression changes in predefined sets of proteins using 2D gel electrophoresis: A proteomic study of L-DOPA induced dyskinesia in an animal model of Parkinson's disease using DIGE, *BMC Bioinformatics* 7:475.
61. Reynolds, J. L., Mahajan, S. D., Bindukumar, B. et al. (2006) Proteomic analysis of the effects of cocaine on the enhancement of HIV-1 replication in normal human astrocytes (NHA), *Brain Res* 1123:226–36.
62. Tannu, N. S., Howell, L. L. and Hemby, S. E. (2008) Integrative proteomic analysis of the nucleus accumbens in rhesus monkeys following cocaine self-administration, *Mol Psychiatry* May 27 [Epub ahead of print].
63. Reynolds, J. L., Mahajan, S. D., Sykes, D. E., Schwartz, S. A. and Nair, M. P. (2007) Proteomic analyses of methamphetamine (METH)-induced differential protein expression by immature dendritic cells (IDC), *Biochim Biophys Acta* 1774:433–42.
64. Dhingra, V., Li, Q., Allison, A. B., Stallknecht, D. E. and Fu, Z. F. (2005) Proteomic profiling and neurodegeneration in West-Nile-virus-infected neurons, *J Biomed Biotechnol* 2005:271–79.
65. Gao, W. M., Chadha, M. S., Berger, R. P. et al. (2007) A gel-based proteomic comparison of human cerebrospinal fluid between inflicted and non-inflicted pediatric traumatic brain injury, *J Neurotrauma* 24:43–53.
66. Pawlyk, A. C., Ferber, M., Shah, A., Pack, A. I. and Naidoo, N. (2007) Proteomic analysis of the effects and interactions of sleep deprivation and aging in mouse cerebral cortex, *J Neurochem* 103:2301–13.
67. Kisby, G. E., Standley, M., Park, T. et al. (2006) Proteomic analysis of the genotoxicant methylazoxymethanol (MAM)-induced changes in the developing cerebellum, *J Proteome Res* 5:2656–65.
68. Damodaran, S., Dlugos, C. A., Wood, T. D. and Rabin, R. A. (2006) Effects of chronic ethanol administration on brain protein levels: A proteomic investigation using 2-D DIGE system, *Eur J Pharmacol* 547:75–82.
69. Oehlers, L. P., Perez, A. N. and Walter, R. B. (2007) Detection of hypoxia-related proteins in medaka (Oryzias latipes) brain tissue by difference gel electrophoresis and de novo sequencing of 4-sulfophenyl isothiocyanate-derivatized peptides by matrix-assisted laser desorption/ionization time-of-flight mass spectrometry, *Comp Biochem Physiol C Toxicol Pharmacol* 145:120–33.

70. Alm, H., Scholz, B., Fischer, C. et al. (2006) Proteomic evaluation of neonatal exposure to 2,2′,4,4′,5-pentabromodiphenyl ether, *Environ Health Perspect* 114:254–59.
71. Cid, C., Garcia-Bonilla, L., Camafeita, E. et al. (2007) Proteomic characterization of protein phosphatase 1 complexes in ischemia reperfusion and ischemic tolerance, *Proteomics* 7:3207–18.
72. Garcia-Bonilla, L., Cid, C., Alcazar, A. et al. (2007) Regulatory proteins of eukaryotic initiation factor 2-alpha subunit (eIF2 alpha) phosphatase, under ischemic reperfusion and tolerance, *J Neurochem* 103:1368–80.
73. Kim, H. G. and Kim, K. L. (2007) Decreased hippocampal cholinergic neurostimulating peptide precursor protein associated with stress exposure in rat brain by proteomic analysis, *J Neurosci Res* 85:2898–2908.
74. Crecelius, A. C., Helmstetter, D., Strangmann, J. et al. (2008) The brain proteome profile is highly conserved between Prnp-/- and Prnp+/+ mice, *Neuroreport* 19:1027–31.
75. Emmert-Buck, M. R., Bonner, R. F., Smith, P. D. et al. (1996) Laser capture microdissection, *Science* 274:998–1001.
76. Shaw, J., Rowlinson, R., Nickson, J. et al. (2003) Evaluation of saturation labelling two-dimensional difference gel electrophoresis fluorescent dyes, *Proteomics* 3:1181–95.
77. Wilson, K. E., Marouga, R., Prime, J. E. et al. (2005) Comparative proteomic analysis using samples obtained with laser microdissection and saturation dye labelling, *Proteomics* 5:3851–58.
78. Stuhler, K., Pfeiffer, K., Joppich, C. et al. (2006) Pilot study of the Human Proteome Organisation Brain Proteome Project: Applying different 2-DE techniques to monitor proteomic changes during murine brain development, *Proteomics* 6:4899–4913.
79. Frohlich, T., Helmstetter, D., Zobawa, M. et al. (2006) Analysis of the HUPO brain proteome reference samples using 2-D DIGE and 2-D LC-MS/MS, *Proteomics* 6:4950–66.
80. Focking, M., Boersema, P. J., O'Donoghue, N. et al. (2006) 2-D DIGE as a quantitative tool for investigating the HUPO Brain Proteome Project mouse series, *Proteomics* 6:4914–31.

5 Mass Spectrometry for Proteomics

Carol E. Parker, Maria R. Warren,
and Viorel Mocanu

CONTENTS

5.1 INTRODUCTION

"Proteomics" is a word coined in 1994 by Marc Wilkins as an alternative to "the protein complement of the genome" (1). Proteomics is still defined in various ways (2), from "the large-scale analysis of the proteome" to "the simultaneous study of all proteins in the cell." In this chapter, we define it as the study of proteins and their interactions. Proteomics is a new field—only 10 years old—and the rapid evolution of this field is due in large part to many improvements in mass spectrometry (MS) that have occurred during the past several years.

Mass spectrometers do one thing—they measure mass. In proteomics, the mass gives information on the protein identity, its chemical modifications, and its structure. Every mass spectrometer has three main components: a source, an analyzer, and a detector. Mass spectrometers measure masses of charged species, so the source must be able to produce ions, the analyzer must be able to separate these ions based on their mass (or, more accurately, mass-to-charge ratio), and the detector must be able to detect charged particles and then amplify the response to give a measurable signal.

In 2002, the award for the Nobel Prize in chemistry was given to two scientists (John Fenn and Koichi Tanaka) responsible for the development of two ionization techniques that have revolutionized biomedical mass spectrometry in general, and proteomics in particular. These two techniques, electrospray (3) and MALDI (matrix-assisted laser desorption ionization) (4) mass spectrometry, were groundbreaking in that they allow the vaporization and ionization (and thus the analysis) of relatively large, non-volatile biomolecules such as proteins and peptides. In addition, simultaneous improvements to and development of mass analyzers and detectors have greatly increased the use of mass spectrometry for biological studies.

5.2   MS SOURCE DESIGN

To really appreciate the dramatic changes in mass spectrometry that have resulted from these new ionization techniques, one has to understand the limitations of mass spectrometry for the analysis of biomolecules before MS was developed. As recently as the late 1970s, mass spectrometry was limited to the analysis of volatile and low-molecular-weight components. One of the most common ionization techniques was electron impact (EI), which uses an "open" low pressure source, in which an electron beam (from a filament) is used to "knock an electron off" a molecule to give a positive ion. This is a fairly high-energy process, often resulting in fragmentation and production of molecular ions (M^+) at low relative abundances.

Another available ionization technique was positive ion chemical ionization (CI), in which a high-pressure (0.5–1.0 torr) reagent gas (usually methane) is ionized via a filament in a closed, higher-pressure source. The protonated reagent ions then transfer a proton to the target analytes, producing a positive ion and providing a more gentle ionization technique with less fragmentation than electron impact. In the late 1970s, negative chemical ionization was developed. This technique, similar to positive chemical ionization, uses methane (0.3–0.5 torr) to "slow" the electrons released from the filament to the point where they can be "captured" by electronegative analytes to form negative ions by electron capture (M^-), or by proton abstraction $(M – H)^-$. Both EI and CI produce singly charged ions, M^+ from electron impact and M^-, $(M–H)^-$, M^+, and $(M+H)^+$, from chemical ionization. These EI and CI sources were typically interfaced with mass analyzers that could detect only up to ~650 Da.

The ionization technique for which John Fenn received the Nobel Prize in 2002 was electrospray ionization (ESI) (3). In this technique, developed around 1985, the sample solution is passed through a needle which is kept at high voltage (~2–5 kV). This results in a spray of small charged droplets containing the analyte. As the solvent evaporates the droplet shrinks until completely desolvated, resulting in a highly-charged species. The extremely high charge density results in a "coulombic explosion" to produce multiply charged, yet stable analyte ions (Figure 5.1). Multiply charged analytes are a desirable feature especially for proteomics, since mass spectrometers separate ions based on their mass-to-charge (m/z) ratios. With multiple

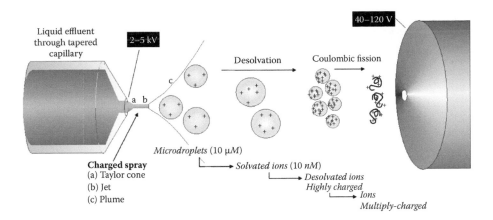

FIGURE 5.1 Electrospray ionization.

or higher charges, higher-molecular-weight compounds (that typically have masses outside the range of the mass spectrometer) could then be analyzed because they have an apparent low-molecular m/z. For example, if you have an instrument with a mass range of 1000 Da, with EI or CI, analysis was limited to analytes with molecular masses of 1000 Da or below. In contrast, if the sample was introduced by ESI, a compound of Mw 4000, with 10 charges, for example, would appear at [(4000 + 10 × 1.008 {the mass of a proton})/10] Da, or approximately 401.0 Da, which is within the detectable mass range.

Figure 5.2A shows the ESI spectrum of a 38 kDa protein obtained by nanoelectrospray ionization. In this high-sensitivity technique, the sample is placed in a glass needle with a 10–20 μm orifice, and is pulled into the vacuum system without the use of any liquid chromatography pumps. The flow rates are very low (a few nanoliters per minute). Because each peak differs from the next by a single charge, a simple calculation can be used to "deconvolute" this spectrum into the average molecular weight of each form of the protein (Figure 5.2B).

The 2002 Nobel Prize was shared between Dr. Fenn and Koichi Tanaka, whose research led to the development of an ionization method called "matrix-assisted laser desorption/ionization" (MALDI) in which a laser and a UV-absorbing chemical compound are used to vaporize and ionize the analyte (Figure 5.3). In MALDI, a laser pulse, typically at a wavelength of 337 nm, is fired at a solid analyte that has been co-crystallized with a chemical matrix, usually a substituted benzoic acid, typically alpha-cyano-4-hydroxybenzoic acid or 2,5-dihydroxybenzoic acid (DHB). This creates a vapor plume of analyte, matrix, and their ions.

MALDI and ESI can be considered as complementary techniques. They deal with the analyte at two different physical states—solid and liquid. As with electrospray, MALDI produces multiply charged analytes, although typically with less charge than with electrospray. Figure 5.4 shows the MALDI spectrum of the same 38 kDa protein shown above (Figure 5.2A) with electrospray. When compared to Figure 5.2A you can see the difference between the ESI spectra and the MALDI spectra in the

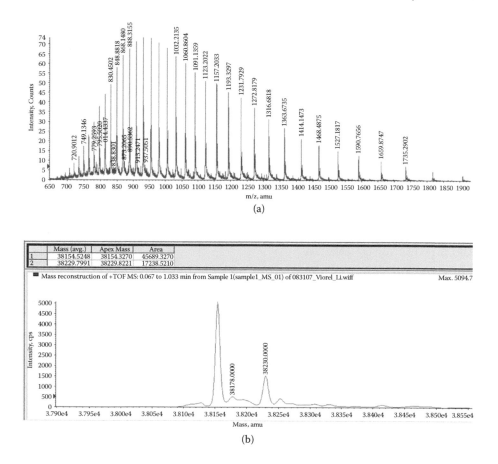

FIGURE 5.2 ESI mass spectrum of a 38 kDa protein (collaborator: Hengming Ke). (a) Spectrum as acquired, showing charge states. (b) Deconvoluted spectrum showing average and apex masses.

number of charges the protein incorporates, the measured m/z, and the improvement in mass resolution using the nanoelectrospray mode.

In terms of detecting intact proteins, electrospray is more limited, with an upper mass limit of 65–70 kDa in a standard commercial instrument, versus the 250 kDa practical mass limit with MALDI. These limits, however, may not be a consequence of the ionization process alone, but may also be due to the mass analyzers and detectors used. With design and software modifications to commercial ESI instruments, a few research groups have detected protein complexes as large as 800 kDa (5). Although MALDI is capable of volatilizing analytes as large as 250 kDa, the output is a "humpogram" rather than sharp peaks, both because of the lower high-mass sensitivity and the lower effective resolution (m/Δm) at very high mass. One disadvantage of electrospray is its sensitivity to salts and detergents, and the fact that it is not as "high throughput" as MALDI, where MS spectra of a series of samples can be obtained in only a few seconds per sample.

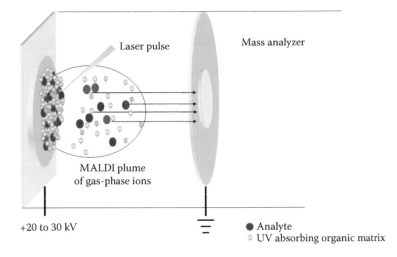

Laser pulse

Mass analyzer

MALDI plume
of gas-phase ions

+20 to 30 kV

● Analyte
◦ UV absorbing organic matrix

FIGURE 5.3   MALDI ionization.

Because what is measured is the mass-to-charge ratio, it is essential to accurately determine the charge on the peptide. Higher resolution instruments make it possible to accurately determine the charge state of the ions. For example, the separation of peaks in the isotope cluster of a peptide with a +1 charge state are 1 Da apart, while those of a +2 ion are 0.5 Da apart (+3 charge state, 0.33 Da apart), and so on (Figure 5.5). This allows the correct determination of the peptide molecular weight and allows more confident identification of an analyte.

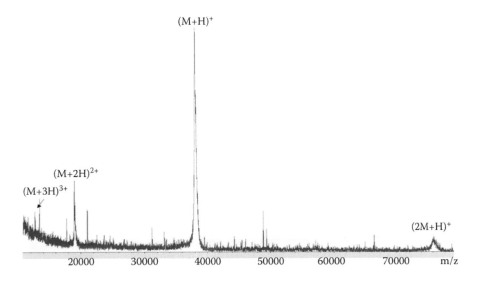

$(M+H)^+$

$(M+2H)^{2+}$

$(M+3H)^{3+}$

$(2M+H)^+$

20000   30000   40000   50000   60000   70000   m/z

FIGURE 5.4   MALDI spectrum of the same 38 kDa protein as in Figure 5.2 (collaborator: Hengming Ke).

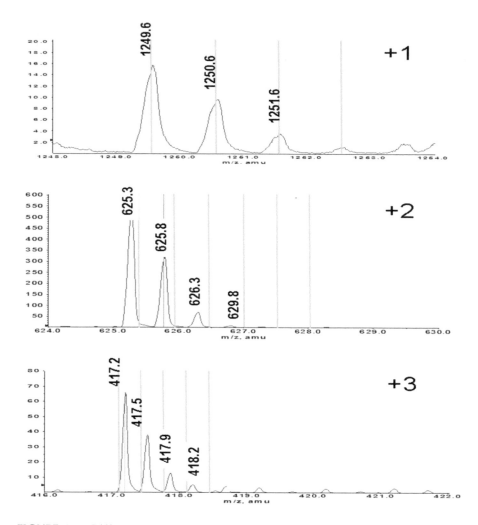

FIGURE 5.5 Different charge states of the same peptide ion (FKDLGEEHFK from BSA), showing different spacing of isotopes.

Not all peptides ionize equally (Figure 5.6A and C), and this ionization efficiency also depends on the ionization mode. On average, it has been found that approximately 25% of peptides ionize exclusively in one mode or the other (Figure 5.7). For projects requiring the highest possible sequence coverage (for example, modification site determination), both techniques are often used.

5.3 LIQUID CHROMATOGRAPHY–MASS SPECTROMETRY CONSIDERATIONS

At the same time as these ionization techniques were being developed, there was growing interest in coupling liquid chromatography (LC) with mass spectrometry

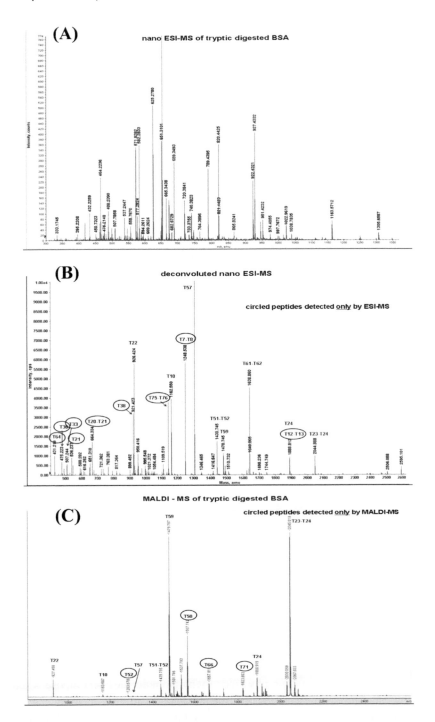

FIGURE 5.6   BSA digest spectrum in (a) ESI; (b) ESI, deconvoluted to +1 charge state; and (c) MALDI modes.

(A)

```
  1 MKWVTFISLL LLFSSAYSRG VFRRDTHKSE IAHRFKDLGE EHFKGLVLIA
 51 FSQYLQQCPF DEHVKLVNEL TEFAKTCVAD ESHAGCEKSL HTLFGDELCK
101 VASLRETYGD MADCCEKQEP ERNECFLSHK DDSPDLPKLK PDPNTLCDEF
151 KADEKKFWGK YLYEIARRHP YFYAPELLYY ANKYNGVFQE CCQAEDKGAC
201 LLPKIETMRE KVLTSSARQR LRCASIQKFG ERALKAWSVA RLSQKFPKAE
251 FVEVTKLVTD LTKVHKECCH GDLLECADDR ADLAKYICDN QDTISSKLKE
301 CCDKPLLEKS HCIAEVEKDA IPENLPPLTA DFAEDKDVCK NYQEAKDAFL
351 GSFLYEYSRR HPEYAVSVLL RLAKEYEATL EECCAKDDPH ACYSTVFDKL
401 KHLVDEPQNL IKQNCDQFEK LGEYGFQNAL IVRYTRKVPQ VSTPTLVEVS
451 RSLGKVGTRC CTKPESERMP CTEDYLSLIL NRLCVLHEKT PVSEKVTKCC
501 TESLVNRRPC FSALTPDETY VPKAFDEKLF TFHADICTLP DTEKQIKKQT
551 ALVELLKHKP KATEEQLKTV MENFVAFVDK CCAADDKEAC FAVEGPKLVV
601 STQTALA
```

41.2 % coverage
(by ESI – MS)

(B)

```
  1 MKWVTFISLL LLFSSAYSRG VFRRDTHKSE IAHRFKDLGE EHFKGLVLIA
 51 FSQYLQQCPF DEHVKLVNEL TEFAKTCVAD ESHAGCEKSL HTLFGDELCK
101 VASLRETYGD MADCCEKQEP ERNECFLSHK DDSPDLPKLK PDPNTLCDEF
151 KADEKKFWGK YLYEIARRHP YFYAPELLYY ANKYNGVFQE CCQAEDKGAC
201 LLPKIETMRE KVLTSSARQR LRCASIQKFG ERALKAWSVA RLSQKFPKAE
251 FVEVTKLVTD LTKVHKECCH GDLLECADDR ADLAKYICDN QDTISSKLKE
301 CCDKPLLEKS HCIAEVEKDA IPENLPPLTA DFAEDKDVCK NYQEAKDAFL
351 GSFLYEYSRR HPEYAVSVLL RLAKEYEATL EECCAKDDPH ACYSTVFDKL
401 KHLVDEPQNL IKQNCDQFEK LGEYGFQNAL IVRYTRKVPQ VSTPTLVEVS
451 RSLGKVGTRC CTKPESERMP CTEDYLSLIL NRLCVLHEKT PVSEKVTKCC
501 TESLVNRRPC FSALTPDETY VPKAFDEKLF TFHADICTLP DTEKQIKKQT
551 ALVELLKHKP KATEEQLKTV MENFVAFVDK CCAADDKEAC FAVEGPKLVV
601 STQTALA
```

41.2 % coverage
(by ESI – MS)

(C)

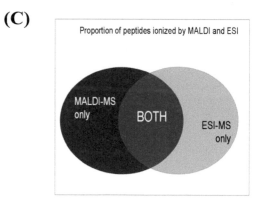

FIGURE 5.7 Selective ionization in (A) ESI (41.2% coverage); (B) MALDI (25.2% coverage)—some peptides can only be detected in one mode or the other, while others can be detected in both ionization modes; (C) Venn diagram of peptide ionization in ESI and MALDI modes.

as an alternative to gas chromatography since many biomolecules decompose when analyzed by gas chromatography. While the advantages of this technique were clear (less decomposition, and the reduction of "suppression effects" by having few types of ions in the source at a time), the challenge of this was enormous because it involves coupling a liquid-based separation technique with a gas phase technique

(6). The first method to accomplish this coupling was the "moving belt" interface (7). This was an on-line technique, and the 1-mL/min flow rate, which was common at the time, meant that significant amounts of solvent had to be removed. This was accomplished by depositing the LC effluent on a belt that moved through various vacuum chambers, and the solvent was evaporated before entering the MS source. Inside the source, the analyte was volatilized by flash evaporation and ionized by electron impact or chemical ionization. One problem with this technique was carryover of one sample to the next due to incomplete cleaning of the belt. Interestingly, one early application of this technique was for the analysis of peptides, but only after derivatization (8).

The other early on-line approaches (ca. 1980) were based on the use of a chemical ionization source. The first technique, called direct liquid injection (DLI), relied on splitting the LC effluent so that only a few microliters per minute entered the source (the LC solvent becomes the CI reagent gas) (9). In this technique, the effluent is sprayed through a tapered capillary or 10-μm pinhole in a diaphragm, producing a fine spray of droplets that can be desolvated in the source. This technique led to the interest in the development of microbore liquid chromatography–mass spectrometry (LC-MS) (10), and eventually capillary LC-MS systems, where the entire LC effluent (200 nL/min to 1 μL/min) can enter the CI source without the need for splitting.

Alternately, heat could be used to desolvate a higher flow rate (1 mL/min) of reversed-phase solvent in a technique called "thermospray," which was perhaps the first easily used LC interface (11). This technique, developed in 1982, used ammonium acetate to provide CI-type ionization, with or without the use of a filament. Although the moving belt interface, the DLI interface, and thermospray are no longer used, these techniques laid the groundwork for the LC-MS methods currently in use today, and provided the impetus for the development of low-flow-rate LC systems which are commonly used in modern LC-MS techniques.

The solution-based technique of ESI was the easiest to couple to mass spectrometry, for the analysis of both proteins and for peptides. MALDI is still most often used for the analysis of single analytes or digests of a single protein. However, LC-MALDI of mixtures would be a very useful technique for mixtures, and although on-line LC-MALDI has been attempted, the more promising approach appears to be off-line LC-MALDI (the basic concept for on-line versus off-line LC methods is discussed in Chapter 3). Various LC-MALDI spotting devices have been developed, using electrodeposition or simply depositing droplets of the eluent on a target. Matrix can be added later, or the matrix solution can be added to the eluent before the droplet is formed. This technique can be thought of as a direct descendant of the "moving belt" interface, but since such low flow rates (usually 200 nL/min to 1 μL/min) are used, and spots are collected every 10–15 seconds, no additional vacuum or heating device is needed. Also, the off-line "decoupling" of the MS from the deposition step greatly simplifies the procedure.

5.4   MASS ANALYZERS

5.4.1   Quadrupole Mass Filters

The revolution in biological mass spectrometry was not only due to improvements in ionization techniques, in particular MALDI and ESI, but also from improvements to existing mass analyzers and the design of new and hybrid instruments. Early mass analyzers, such as the quadrupole mass filter, had a mass range of only ~650 Da, while modern quadrupole instruments can scan up to 4000 Da. Also, while earlier quadrupole mass spectrometers were capable of only unit resolution (which means that they could distinguish ions 1 Da apart), new advances and the coupling of quadrupole and time-of-flight (qQTOF) analyzers made it possible to achieve higher resolution, up to 0.01 Da in these hybrid instruments.

In quadrupole mass analyzers (otherwise known as quadrupole mass filters), a combination of RF and DC voltages on the four parallel rods allows only ions with specific m/z to pass through to the detector (Figure 5.8). The quadrupole can scan a specified m/z range by ramping the voltages to produce MS spectra, or they can be set to allow the passage of only preselected masses (for peptide sequencing [MS/MS], or selected ion monitoring).

5.4.2   Time-of-Flight Analyzers

With time-of-flight (TOF) instruments, the mass-to-charge ratio is indirectly calculated from the length of time it takes for the ion to reach the detector in a field-free vacuum. "Packets" of ions are released into the TOF analyzer at time = 0, and then ions drift toward the detector and generate a signal at t = final. The mass of the

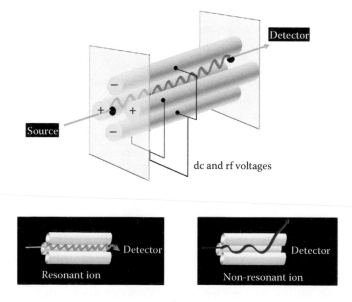

FIGURE 5.8   Diagram of single quadrupole mass analyzer.

analyte is proportional to the flight time (measured in nanoseconds) and heavier ions travel more slowly, taking more time to reach the detector.

The first TOF design was the linear TOF in which the ions traveled a linear path from the opening or orifice on the mass analyzer to the detector (Figure 5.9, top). However, the nature of MALDI ionization produces a plume of ions that may experience slightly different acceleration voltages depending on where they were formed in the "plume" prior to reaching the field-free vacuum. These ions may therefore reach the detector at slightly different times. This translates into broader peaks and reduces resolution. TOF analyzers were greatly improved by two modifications: "delayed extraction" of ions leaving the source, and the addition of a set of reflectron lenses to the standard linear design. The increasing voltages applied to these successive ring electrodes "reflect" ions to a second detector. Since ions with higher kinetic energy penetrate "deeper" into the reflectron, and thus have to travel farther, this helps correct for the variability in ion kinetic energies, resulting in sharper, better resolved peaks (Figure 5.9, bottom).

5.4.3 Ion Traps

During the 1950s, Paul and Steinwedel patented a novel alternation to the quadrupole mass analyzer (12,13). The principles of mass filtering were similar in that RF and DC voltages were used to filter out masses. However the physical design was quite different (Figure 5.10). Instead of using four rods arranged in parallel the new design had only three electrodes. The hyperbolically shaped entrance and exit electrodes allowed selective flow of ions to and from the mass analyzer, and a third electrode between them—the ring electrode—was used to trap ions along this path. The design effectively acts as a three-dimensional quadrupole. Ion traps normally have unit mass resolution.

5.4.4 Fourier Transform MS

A type of mass spectrometer less commonly used for proteomics, but which has the highest mass resolution, is the Fourier transform (FT) MS. These are the most expensive instruments and are available in two types: FT-ICR and the Orbitrap.

FT-ICR (Fourier transform ion cyclotron resonance) MS was first described in the late 1940s by Hipple, Thomas, and Sommer (14,15), but only became practical in the 1970s when multiple ions could be detected (16). The principal components of ICR mass spectrometers are an ICR cell, which typically consists of pairs of plates arranged as a cube or capped cylinder, and a surrounding strong magnetic field (Figure 5.11). ICR-MS requires ion trapping, excitation, detection, and data transformation.

Ions generated in an external source are directed into the ICR cell made of two trapping plates, two excitation plates, and two detector or receiver plates. The ions are trapped in the cell by an electric field applied to the trapping electrodes and by the magnetic field (7–18 tesla). Within the cell the trapped ions travel in a circular orbit (cyclotron motion) with orbital frequencies inversely proportional to the ions' mass-to-charge ratio. An RF sweep is then applied across the excitation plates. Ions

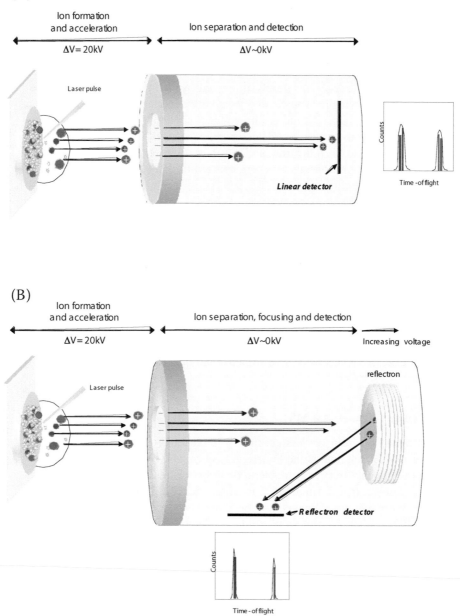

FIGURE 5.9   Diagram of TOF analyzers in (A) linear and (B) reflectron configurations.

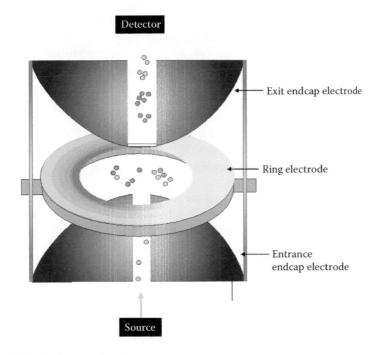

FIGURE 5.10   Diagram of an ion trap.

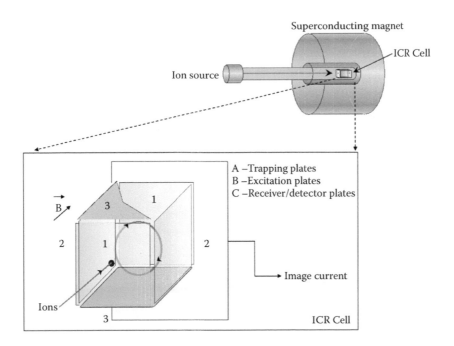

FIGURE 5.11   Diagram of an FT-ICR.

with cyclotron frequencies in resonance with the applied RF are accelerated and increase their orbit radius bringing them closer to detector plates, which then record the image current. The complex image current is deconvoluted to the frequency domain by fast Fourier transformation from which the mass-to-charge ratio is then determined.

By far FT-ICR instruments are the most expensive primarily due to the cost of the superconducting magnet commercially available in field strengths of 7 tesla ($800,000) and 15 tesla for $2.4 million. However, the cost may be justified for certain projects that require the increased sensitivity due to the trapping nature of the technology, the high mass accuracy (up to 0.5 ppm), and the high resolution (100,000 full-width half-mass).

In 2005 Alexander Makarov developed the newest type of mass analyzer to use FT technology, having properties similar to ion traps but with results approaching those of FT-ICR instruments (17). In an Orbitrap, ions enter the trap in a tangent to an inner axial electrode. The voltage of the axial electrode is increased as the ion enters, causing it to spiral. The spiral moves toward the center of the axis then the voltage increase stops and the spiral becomes a ring of ions (Figure 5.12). The rings of ions cycle at different frequencies and the frequencies are converted to ion masses by Fourier transformation of the ion current.

5.4.5 ION-MOBILITY MS

Ion-mobility mass spectrometry is a technology which separates ions based on their collisional cross sections (18). A new commercial instrument based on this technology has recently been introduced by Waters Corporation. A "wave guide" was developed to replace the original RF-only quadrupole-based collision cell, which had been originally developed by ABI. This new wave guide was observed to have the property that it could be used to separate ions based on their shapes instead of to refocus them. This technology is so new that it is difficult to predict its impact on proteomics.

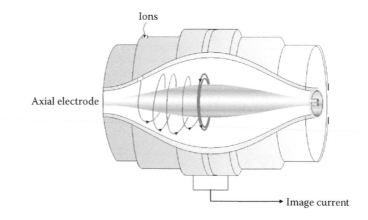

Ions

Axial electrode

Image current

FIGURE 5.12   Diagram of the Orbitrap.

5.5   DETECTOR TECHNOLOGY

The earliest mass spectrometers used photoplate detection, hence the old term "mass spectroscopy," which is no longer appropriate. In the 1920s, multi-stage analyzers were developed, which relied on an electron "cascade" or "avalanche" to provide amplification of the ion signals. Early designs were a kind of "Venetian blind" device with each plate at an increasing voltage compared to the previous level. In the 1980s, this design was replaced with a "continuous dynode," which replaced the plates with a cone-shaped device. Typically, analyte ions strike the detector surface, which releases electrons from the surface. The released electrons in turn strike the surface, emitting more electrons. Each of the newly emitted electrons repeats the process so that a final stream of electrons strikes the cathode (Figure 5.13). For negative ion detection, a "conversion dynode" is placed above the entrance to the multiplier cone. The conversion dynode "converts" the ion polarity of the negative ion. The incoming negative ion strikes the surface of the conversion dynode and produces a positive ion, which then initiates the electron cascade within the electron multiplier. The voltage of the conversion dynode is the same (negative) for both positive and negative ion detection modes. Most TOF instruments now use multichannel plate (MCP) detectors, which are similar in concept to the original electron multipliers but consist of several multipliers in an array (Figure 5.14). These devices are so efficient that they are capable of detecting single ions ("ion counting").

With the FT-based instruments, the ICR and the Orbitrap, the ion current generated from the orbiting ions is detected. In FT-ICR, it is measured by the trapping electrodes. With Orbitrap instruments, split outer electrodes flanking the center of the axial electrode measure the current of the ions oscillating along the axial electrode. With both ICR and Orbitrap instruments, the detected signals are then typically amplified prior to processing by fast Fourier transformation.

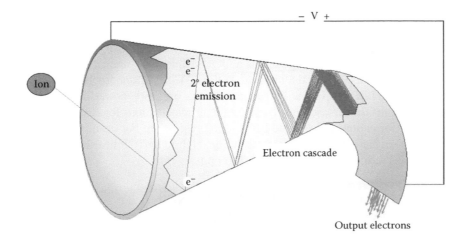

FIGURE 5.13   Continuous dynode detector.

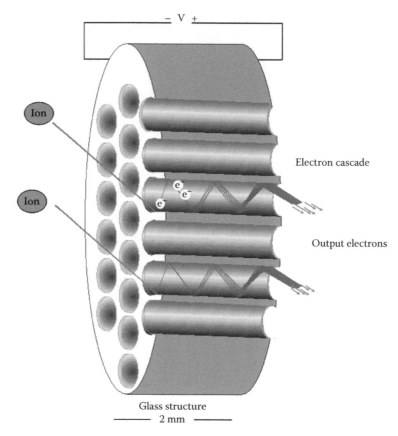

FIGURE 5.14   Diagram of multichannel plate detector.

5.6   HYBRID INSTRUMENTS

Mass spectrometers can be built with a single mass analyzer (single-stage) or can be built with multiple analyzers (multi-stage). The single-stage instruments simply measure the mass of the ion generated by the ion source. Multi-stage instruments can consist of the same mass analyzers such as a QQQ (triple-quadrupole, Figure 5.15) or dissimilar analyzers such as the ion-trap TOF (IT-TOF) or quadrupole-quadru-pole/collision cell-TOF (QqTOF). These instruments are very powerful and can manipulate the ions to increase sensitivity, fragment selected ions for sequence and modification information, and provide better mass resolution. For example, single-stage ion traps normally have unit mass resolution. Hybrid QqTOF instruments normally have mass resolution (m/Δm) of 100–200 ppm.

5.7   COLLISION CELLS: GAS-PHASE SEQUENCING BY MS/MS (TANDEM MASS SPECTROMETRY)

If enough internal energy (20–30 eV) is added to a peptide, it will fragment. We are indeed fortunate that normally this fragmentation occurs along the peptide

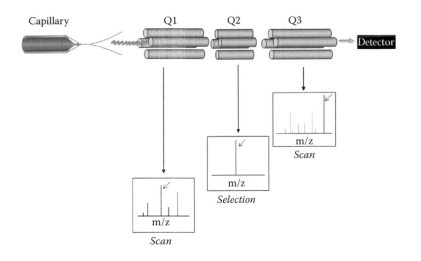

FIGURE 5.15 Diagram of triple quadrupole mass spectrometer.

backbone, although occasionally this fragmentation occurs in the ion source (in-source fragmentation, "skimmer-induced" fragmentation, or "prompt" fragmentation) (19–23). Fragmentation typically occurs in a cell and is induced or activated by colliding the analyte with inert gas molecules such as nitrogen or air. Thus, gas-phase sequencing of peptides is usually done by performing MS/MS—two stages of mass spectrometry surrounding a collision cell. This technique, also called tandem mass spectrometry, involves selection of an ion with the first MS (usually a quadrupole mass filter), fragmenting the peptide ions in a collision cell containing an RF-only quadrupole, and finally detecting the fragments of the selected peptide.

Fragmentation resulting from collision of the analyte with neutral gas is termed collision-induced dissociation (CID). It is a low-energy process, typically resulting in cleavage of the amide bond to form "b" or "y"-type fragment ions, which are most useful for determining the peptide sequence. The fragment ions produced, which contain the N-terminus, are called b ions; those containing the C-terminus are called y-ions (Roepstorff nomenclature) (24) (Figure 5.16).

Fragmentation of a peptide is size dependent since the energy of the collision is absorbed by all the bonds of the peptide. Thus, in practice, if a peptide is larger than 2500 Da, it is difficult to get enough energy into a single bond to cause it to break. For this reason, although MALDI-TOF mass spectrometers are capable of a high mass range, some MALDI TOF/TOF instruments (for example, the 4700 Proteomics Analyzer), which are designed specifically for fragmenting peptides, have effective mass ranges of only 2500 Da.

5.8 NEW FRAGMENTATION TECHNIQUES

Three fragmentation techniques, electron capture dissociation (ECD), infrared multiphoton dissociation (IRMPD), and electron-transfer dissociation (ETD), were

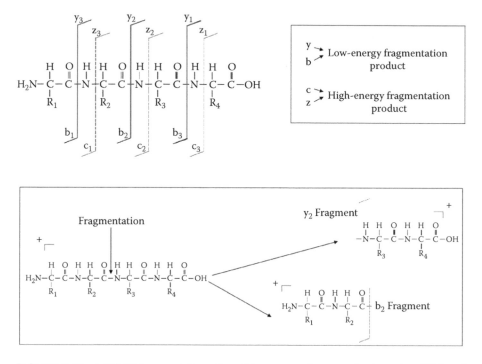

FIGURE 5.16 MS/MS fragmentation of peptides; nomenclature and schematic of "b" and "y" fragmentation.

recently introduced as means of fragmenting larger peptides and even proteins with mass spectrometry. In ECD, low-energy electrons are captured by peptide cations to form odd-electron peptides which then dissociate to fragments primarily of the c- and z-type (25) (Figure 5.17). Unlike CID processes, ECD cleavage does not increase the analyte's internal vibrational energy. Therefore the weakest bonds, such as labile phosphate bonds, are maintained. Similar fragments are formed with ETD, but due to the interaction of the protonated peptide with radical anions rather than with a beam of electrons (26), fragments of b- and y-type are produced by IRMPD when the peptide absorbs energy from IR photons.

ECD and IRMPD are used only on FT instruments. Both of these are used primarily for "top-down" sequencing of proteins, which means that these proteins are fragmented and the molecular weights of the pieces are used to determine the sequence of the protein. This is in contrast to the "bottom-up" approach, which first uses enzymes to cleave the proteins into peptides followed by gas-phase sequencing. ETD is more flexible in that it was first demonstrated in an ion trap but can be used in FT instruments as well.

5.9 GEL-BASED PROTEIN IDENTIFICATION

For the identification of proteins in gel spots or bands, the gel plugs or slices are usually "dehydrated" by soaking them in acetonitrile, and then they are "rehydrated"

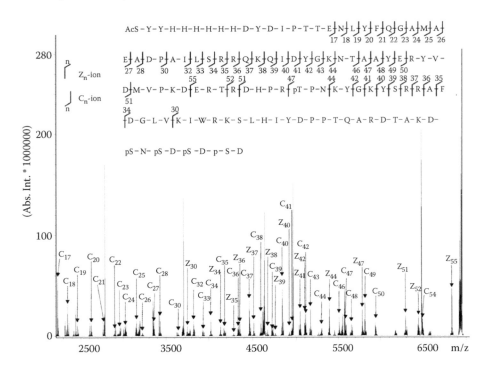

FIGURE 5.17 Top-down fragmentation: MS/MS spectrum of 115 residue protein obtained by ECD on 12T FT-ICR. *Source*: Reprinted from Borchers, C. H., Thapar, R., Petrotchenko, E. V. et al. (2006) Combined top-down and bottom-up proteomics identifies a phosphorylation site in stem-loop-binding proteins that contributes to high-affinity RNA binding, *Proceedings of the National Academy of Sciences of the USA* 103:3094–99, with permission.

with a solution containing trypsin. The proteins are digested into peptides, and the peptides are extracted, lyophilized, and analyzed by MALDI-TOF/TOF. The peptide masses and their partial sequences are compared with the theoretical masses in a database, and the protein is identified. MALDI is the method of choice for these analyses because of the high throughput. Each gel spot ends up as a single MALDI spot on the MALDI target. MALDI is a high-throughput technique, since no separation step is involved, and there is also no possibility of "carryover" of one sample to another.

In our laboratory, ESI is the method of choice for high-sensitivity peptide analysis and high mass-accuracy molecular weight determination. LC-ESI-MS/MS and LC-MALDI-MS/MS are used for modification site determination where high sequence coverage must be obtained. Modification site determination will be discussed in Chapter 6.

5.10 COMPUTERS AND DATABASES

Although the advances in instrumentation described above are clearly critical for the development of MS-based proteomics, proteomics is also dependent on advances in other fields. Genomes had to be sequenced before they could be used to predict protein and peptide sequences. Software packages had to be written to compare the observed peptide molecular weights and sequences with those in the databases. Computers and storage devices need to be developed that can handle the acquisition rates and huge data files (often several gigabytes per analysis) resulting from these analyses. Bioinformatics techniques capable of interpreting the enormous amounts of data generated are also being developed.

5.11 CONCLUSION

Proteomics is a new field, and all of its components are constantly being improved. New mass spectrometers are still being developed, and existing designs continue to be improved by making them more sensitive, and by increasing their mass resolution, mass accuracy, stability, and scan speed. New genomes are being sequenced, and new software packages are being developed to provide protein identifications with lower false-positive and false-negative rates. Computer systems and data storage systems capable of handling terabytes of data—which would have been prohibitively expensive only a few years ago—are becoming common. Thus, this chapter can at best only be an overview of the history and evolution of a rapidly developing field.

REFERENCES

1. Wasinger, V. C., Cordwell, S. J., Cerpa-Poljak, A. et al. (1995) Progress with gene-product mapping of the Mollicutes: *Mycoplasma genitalium. Electrophoresis* 16:1090–4.
2. www.biotechshares.com/glossary.htm.
3. Whitehouse, C. M., Dreyer, R. N., Yamashita, M. and Fenn, J. B. (1985) Electrospray interface for liquid chromatographs and mass spectrometers, *Anal Chem* 57:675–9.
4. Karas, M., Ingendoh, A., Bahr, U. and Hillenkamp, F. (1989) Ultraviolet-laser desorption/ionization mass spectrometry of femtomolar amounts of large proteins, *Biomedical and Environmental Mass Spectrometry* 18:841–3.
5. Sobott, F. and Robinson, C. V. (2004) Characterising electrosprayed biomolecules using tandem-MS-the noncovalent GroEL chaperonin assembly, *International Journal of Mass Spectrometry* 236:25–32.
6. Arpino, P. (1982) On-line liquid chromatography/mass spectrometry? An odd couple! *Anal Chem* 1:154–8.
7. Smith, R. and Johnson, A. (1981) Deposition method for moving ribbon liquid chromatograph-mass spectrometer interfaces, *Anal Chem* 53:739–40.
8. Yu, T. J., Schwartz, H., Giese, R. W., Karger, B. L. and Vouros, P. (1981) Analysis of N-acetyl-N,O,S-permethylated peptides by combined liquid chromatography-mass spectrometry, *Journal of Chromatography* 218:519–33.
9. Henion, J. D. (1978) Drug analysis by continuously monitored liquid chromatography/mass spectrometry with a quadrupole mass spectrometer, *Analytical Chemistry* 50:1687–93.
10. Henion, J. D. (1980) A comparison of direct liquid introduction LC/MS techniques employing microbore and conventional packed columns, *Journal of Chromatographic Science* 18:101–2, 112–15.
11. Blakley, C. R. and Vestal, M. L. (1983) Thermospray interface for liquid chromatography/mass spectrometry, *Analytical Chemistry* 55:750–4.

12. Paul, W. and Steinwedel, H. (1953) A new mass spectrometer without magnetic field, *Zeitschrift fuer Naturforschung* 8a:448–50.
13. Paul, W. and Steinwedel, H. (1956) Separation and indication of ions with different specific charges, *DE 944900 19560628 Patent.*
14. Hipple, J. A. and Thomas, H. A. (1949) A time-of-flight mass spectrometer with varying field, *Physical Review A* 75:1616.
15. Sommer, H., Thomas, H. A. and Hipple, J. A. (1951) Measurement of e/M by cyclotron resonance, *Physical Review* 82:697–702.
16. Comisarow, M. B. and Marshall, A. G. (1975) Resolution-enhanced Fourier transform ion cyclotron resonance spectroscopy, *Journal of Chemical Physics* 62:293–5.
17. Hu, Q., Noll, R. J., Li, H. et al. (2005) The Orbitrap: A new mass spectrometer, *Journal of Mass Spectrometry* 40:430–43.
18. Verbeck, G. F., Ruotolo, B. T., Sawyer, H. A., Gillig, K. J. and Russell, D. H. (2002) A fundamental introduction to ion mobility mass spectrometry applied to the analysis of biomolecules, *J Biomol Tech* 13:56–61.
19. Patterson, S. D. and Katta, V. (1994) Prompt fragmentation of disulfide-linked peptides during matrix-assisted laser desorption/ionization, *Analytical Chemistry* 66:3727–32.
20. Lavine, G. and Allison, J. (1999) Evaluation of bumetanide as a matrix for prompt fragmentation matrix-assisted laser desorption/ionization and demonstration of prompt fragmentation/post-source decay matrix-assisted laser desorption/ionization mass spectrometry, *J Mass Spectrom* 34:741–48.
21. Marzilli, L. A., Golden, T. R., Cotter, R. J. and Woods, A. S. (2000) Peptide sequence information derived by pronase digestion and ammonium sulfate in-source decay matrix-assisted laser desorption/ionization time-of-flight mass spectrometry, *Journal of the American Society for Mass Spectrometry* 11:1000–8.
22. Brown, R. S., Feng, J. and Reiber, D. C. (1997) Further studies of in-source fragmentation of peptides in matrix-assisted laser desorption-ionization, *International Journal of Mass Spectrometry and Ion Processes* 169/170.
23. Raska, C. S., Parker, C. E., Huang, C. et al. (2002) Pseudo-MS3 in a MALDI orthogonal quadrupole-time of flight mass spectrometer, *Journal of the American Society for Mass Spectrometry* 13:1034–41.
24. Roepstorff, P. and Fohlman, J. (1984) Proposal for a common nomenclature for sequence ions in mass spectra of peptides, *Biomed Mass Spectrom* 11:601.
25. Zubarev, R. A., Kelleher, N. L. and McLafferty, F. W. (1998) Electron capture dissociation of multiply charged protein cations. A nonergodic process, *J Am Chem Soc* 120:3265–66.
26. Syka, J. E. P., Coon, J. J., Schroeder, M. J., Shabanowitz, J. and Hunt, D. F. (2004) Peptide and protein sequence analysis by electron transfer dissociation mass spectrometry, *Proceedings of the National Academy of Sciences of the United States of America* 101:9528–33.
27. Borchers, C. H., Thapar, R., Petrotchenko, E. V. et al. (2006) Combined top-down and bottom-up proteomics identifies a phosphorylation site in stem-loop-binding proteins that contributes to high-affinity RNA binding, *Proceedings of the National Academy of Sciences of the USA* 103:3094–99.

6 Mass Spectrometry for Post-Translational Modifications

Carol E. Parker, Viorel Mocanu, Mihaela Mocanu, Nedyalka Dicheva, and Maria R. Warren

CONTENTS

6.1   INTRODUCTION

Post-translational modification of proteins is important for the regulation of cellular processes, including the cellular localization of protein, the regulation of protein function, and protein complex formation. Post-translational modification of proteins is part of what makes proteomics so much more challenging than genomics. Not only does the proteomic "alphabet" contain more letters than the genome (21 common amino acids as opposed to four nucleotides), but these amino acids can also be modified by literally hundreds of modifications that change their molecular weights, the fundamental physical property measured by mass spectrometry. Of these modifications, the most common and naturally occurring are cleavage, acetylation, formylation, methionine oxidation, phosphorylation, ubiquitination, and glycosylation (which is a whole set of modifications rather than a single modification). A more comprehensive list of protein modifications can be found elsewhere (1). In addition, other non–post-translational modifications, such as crosslinking, fluorescent labels, and spin labels, can be used to probe protein structure and function.

6.2   GENERAL CONSIDERATIONS

Regardless of the modification there are some general points to consider. Analysis of post-translational modifications (PTMs) by mass spectrometry can be difficult and the level of difficulty is dependent on (i) the mass shift in the peptide molecular weight, (ii) the overall abundance of the modified peptide, (iii) the stability of the modification during mass spectrometry (MS) and MS/MS analysis, and (iv) the effect of the modification on the peptide's ionization efficiency (and therefore the sensitivity).

The modified peptide must be detected. While it is possible to *identify* a protein (or a family of homologous proteins) from MS/MS spectra of only a few of its peptides, locating the exact site of a modification requires detection and in most cases MS/MS sequencing of the *specific* modified peptide. This can be difficult since not all peptides are equally detected. The "detectability" of a peptide depends on the peptide abundance and its proton affinity, which is a function of its sequence and modifications. This is illustrated in the mass spectrum of a bovine serum albumin (BSA) tryptic digest in Figure 5.6 in the Chapter 5. Although each peptide is present in the same molar abundances, there is high variability in the ion abundances, with some peptides being completely absent.

The purification method is an important factor to consider. Proteomic samples typically are analyzed from polyacrylamide gel with sodium dodecyl sulfate-polyacrylamide gel electrophesis (SDS-PAGE) gels. Although it is possible to get high sequence coverage of highly abundant proteins from a gel, it is more common to have incomplete sequence coverage, partially due to losses during peptide extraction (2). For this reason, we prefer to use non-gel protein purification methods for modification-site determination by electrospray ionization–liquid chromatography (ESI-LC)/MS/MS, although PTM analysis from gels is sometimes the only option, and gel-based separation combined with LC–matrix-assisted laser desorption ionization (MALDI) has been successful for certain projects.

6.3   DATABASE SEARCH CONSIDERATIONS

Peptide identification from a protein or translated genomic database is probability based. This is done by comparing the observed MS/MS spectrum with the theoretical spectra for the predicted proteolytic peptides (see Figure 6.1) of all proteins in the database. If more than a few modifications at a time are considered, the search time increases exponentially, and the probability score is decreased if more modifications are used in order to achieve a "match."

Another factor critical to database searching is that the modification must be specific to only certain amino acids. Non-specific modifications, for example, modifications that can react with *any* amino acid, cannot be searched with some common database search engines, such as Mascot™ (3).

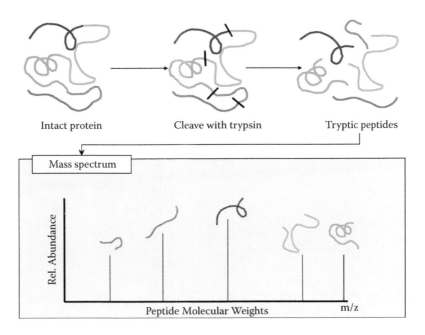

| Intact protein | Cleave with trypsin | Tryptic peptides |

FIGURE 6.1 Schematic of MS spectrum of a tryptic-digested protein.

6.4 MS/MS-BASED MODIFICATION SITE DETERMINATION

With some modifications attachment sites can be determined directly from the MS/MS spectrum. This requires that the modification be stable enough to withstand the energetics of MS and/or MS/MS analysis. Also, the added mass of the modification should not shift the total mass of the peptide outside the mass range suitable for MS/MS sequencing (~800–2500 Da). For stable modifications, not only is the peptide's molecular weight shifted, but all fragment ions containing the modified amino acid are mass-shifted as well due to the modification (Figure 6.2).

The types of molecules that can modify a protein encompass a wide range of compound classes—with different physical and chemical properties. It is difficult to discuss all of even the most common modifications thoroughly so we focus here on a few that represent increasing levels of difficulty in mass spectrometric analysis. This mass shift is characteristic of the particular type of post-translational modification (e.g., acetylation, 42 Da; phosphorylation, 80 Da, etc.).

6.5 ACETYLATION AND FORMYLATION

N-terminal acetylation is a common modification, and is one of the groups that block Edman sequencing of a protein. Acetylation is also common on the amino groups of lysine and arginine, and is one of the easiest post-translational modification sites to identify due to the 42 Da mass shift (resulting from the replacement of one of the hydrogens from the amine group with $COCH_3$) on the modified amino acid. Formylation is similar to acetylation (but with a smaller, 28 Da mass shift). N-formylation is a

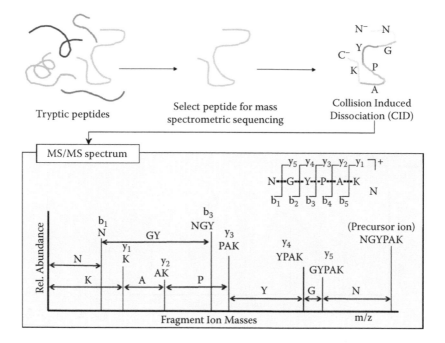

FIGURE 6.2 Schematic of MS/MS fragmentation, showing shift in m/z due to peptide modification. All fragments containing this modification (m) show an increase in m/z corresponding to the mass of the modification.

common modification in bacterial proteins. Formylation from formic acid can also occur as an artifact during protein purification if formic acid is used.

The 42-Da mass shift due to acetylation is shown in Figure 6.3. In the corresponding MS/MS spectra, all of the fragment ions containing the acetyl group are also shifted by 42 Da. In this example, all of the "b" ions in the acetylated peptide are shifted in mass (b'), while the "y" ion series shows no change in mass. This indicates that the acetylation is on the N terminus.

6.6 PHOSPHORYLATION

Phosphorylation is an important regulatory process involved in cell signaling and cancer. Phosphorylation site analysis presents numerous analytical challenges. First, phosphopeptides are present in low abundance, making them more susceptible to suppression by other components in the source. Second, the presence of the phosphoryl group decreases the sensitivity of the peptide—the more phosphoryl groups, the lower the sensitivity. Although this loss of sensitivity has been reported by many research groups (4–6) in both MALDI and ESI, new work by Steen et al. (7) claims that this is "mythology," and the reason that phosphopeptides are so difficult to detect is simply due to their low copy number. Whatever the reason, the proposed approach to phosphopeptide analysis is the same—enrich the sample in phosphopeptides, and, as much as possible, prevent loss of the phosphoryl group during ionization.

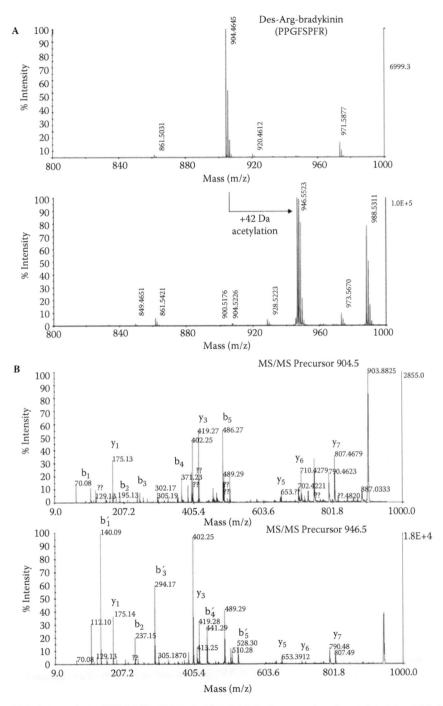

FIGURE 6.3   A: MALDI-MS of Des-Arg bradykinin (*top trace*) and acetylated bradykinin (*bottom trace*). B: MALDI-MS/MS spectra of Des-Arg bradykinin (*top trace*) and acetylated bradykinin (*bottom trace*).

To answer this question, we collected MS spectra from a set of three peptides from the kinase domain of the insulin receptor, which had zero, one, or three phosphotyrosines. As can be seen from the spectra in Figure 6.4, which were obtained from equimolar solutions with the same number of MALDI shots or the same number of nanospray spectra obtained from each sample, the MALDI spectra clearly show a decrease in the signal intensity with MALDI ionization as a function of increasing phosphorylation. There is also a decrease in sensitivity of these peptides with increasing phosphorylation in electrospray, although it is not as pronounced (Figure 6.4).

Because the phosphoryl group is somewhat labile, some modified peptides will lose their phosphoryl groups during the ionization process. Phosphotyrosine can lose HPO_3, giving the unmodified amino acid; phosphoserine and phosphothreonine often undergo loss of 80 Da and/or H_3PO_4 (98 Da), although we have observed the loss of 98 Da from phosphotyrosine-containing peptides as well. To reduce these decomposition reactions during MALDI-MS/MS analysis of phosphopeptides, a "cold" matrix, 2,5-dihydroxybenzoic, is usually used for phosphopeptide analysis, because less energy is transmitted to the analyte than when alpha-cyano-4-hydroxycinnamic acid ("alpha-cyano"), a "hot" matrix, is used (Figure 6.5).

Because of the low abundance of most phosphopeptides, and the requirement for high sequence coverage, we normally use LC/MS/MS techniques to separate the peptides before analysis and to reduce suppression effects. In phosphorylation site determination by LC/MS, ion signals that are 80-Da higher than the calculated mass of a peptide indicate the addition of a phosphoryl group. MassLynx (Q-tof) and Analyst (Q-trap) software allow several different types of scanning techniques, including data-dependent triggering of MS/MS spectra and "neutral loss" scanning, which uses the loss of 98 Da (corresponding to the elements of H_3PO_4 from phosphoserine- or phosphothreonine-containing peptides) to trigger MS/MS analysis. MS/MS analysis can then reveal where the phosphorylation/dephosphorylation sites are located on the peptides (Figure 6.6).

Analytical "challenges" make phosphorylation site determination a "project" rather than a routine analysis. Our current strategy is to first perform LC/MS/MS with the ThermoFisher Orbitrap because of its high sensitivity. If the LC/MS/MS of the entire digest mixture is unsuccessful in finding the phosphorylated peptides, the second step is to use a variety of affinity purification techniques to try to increase the concentration of the phosphopeptide in the sample. These affinity techniques include the use of IMAC beads (8), anti-phosphotyrosine beads (9), or Perkin-Elmer's "Phostrap" titanium oxide beads, or zirconium oxide beads, followed by direct MALDI analysis of the captured phosphopeptides, or elution followed by nanoelectrospray ionization. In "direct MALDI analysis," the beads are placed directly on the MALDI target, rather than eluting the trapped peptides from the beads prior to analysis (9).

Although there are new enrichment techniques developed each year, phosphorylation is still a challenging area of research. Unfortunately, so far, there is no "magic bullet," and there seem to be advantages and disadvantages of each new method—some enrichment techniques seem to work better for multiply phosphorylated peptides, some for singly phosphorylated peptides, etc., with different degrees

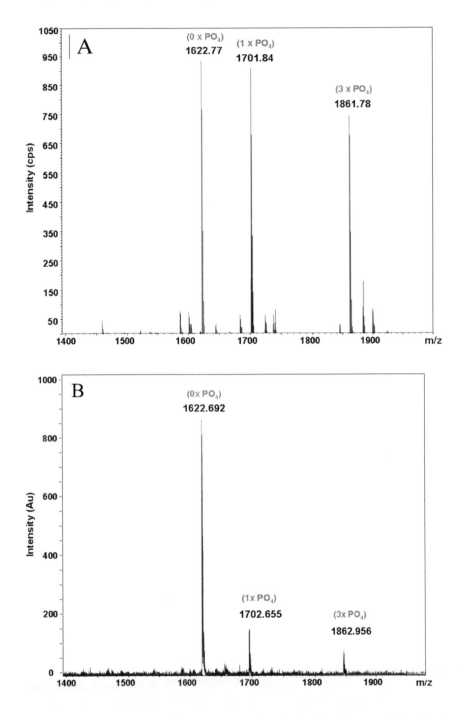

FIGURE 6.4 Sensitivity as a function of the number of phosphoryl groups. A: Positive ion electrospray (deconvoluted spectrum). B: Positive ion MALDI-MS (DHB matrix).

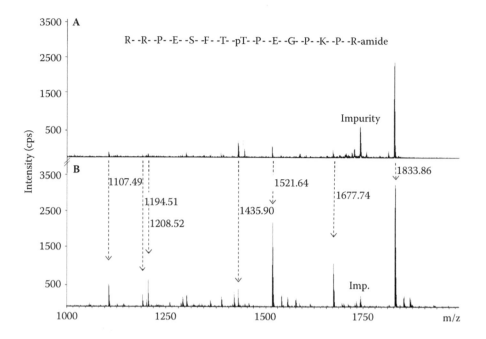

FIGURE 6.5 Comparison of peak intensities in the MS spectra of a phosphopeptide mixture analyzed by MALDI-MS using (A) alpha-cyano-4-hydroxycinnamic acid and (B) dihydroxybenzoic acid (DHB) as the matrix showing the overall increase in signal intensities with DHB.

of non-specific binding of non-phosphorylated peptides. Often the same methods that trap the phosphoryl group also will trap acidic peptides.

Another strategy is to compare the spectra of phosphorylated peptides before and after treatment with phosphatase (10). Peptides which previously contained phosphoryl groups will show a decrease in molecular weight of 80 or 98 Da. Often, peptides will "appear" only after phosphatase treatment—this is especially true for multiply phosphorylated peptides, either because of their low initial sensitivities or because of their strong binding to the affinity media (Figure 6.7).

In conclusion, often the most critical factor in studying phosphorylation is getting enough of the purified protein in order to have enough of the low-abundance phosphoprotein. Even then, phosphorylation site determination is very dependent on the particular peptide sequence.

6.7   TYROSINE SULFATION

Sulfation of tyrosine residues is a fairly common modification, but is even more difficult to determine than phosphorylation [for a recent excellent review of tyrosine sulfation determination, see Monigatti et al. (11)]. Most of the work done on tyrosine sulfation has been through the use of radioactive ^{35}S, analogous to the way ^{32}P is used in phosphorylation studies. The sulfate modification is even more labile to ionization than phosphorylation—easily losing SO_3. MALDI-MS of sulfotyrosine-containing

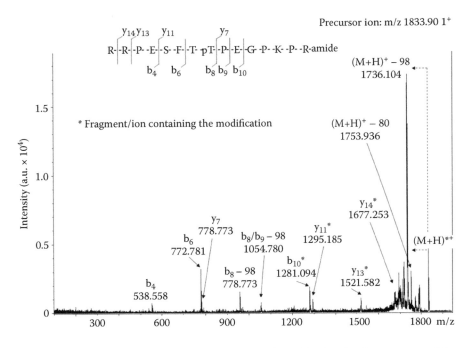

FIGURE 6.6 MS/MS spectrum of a phosphorylated peptide showing fragments resulting from loss of 80 and 98 Da corresponding to a loss of HPO_3 and H_3PO_4.

peptides can provide molecular weight information—although only in the negative ion mode. Molecular weight information can be provided by both positive and negative ion ESI or by negative ion MALDI (Figure 6.8), but localization of the sulforyl group can be difficult because this group is so labile (11). The neutral loss of SO_3,

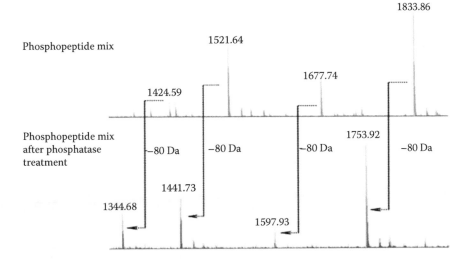

FIGURE 6.7 Peptide mass spectra before and after phosphatase treatment.

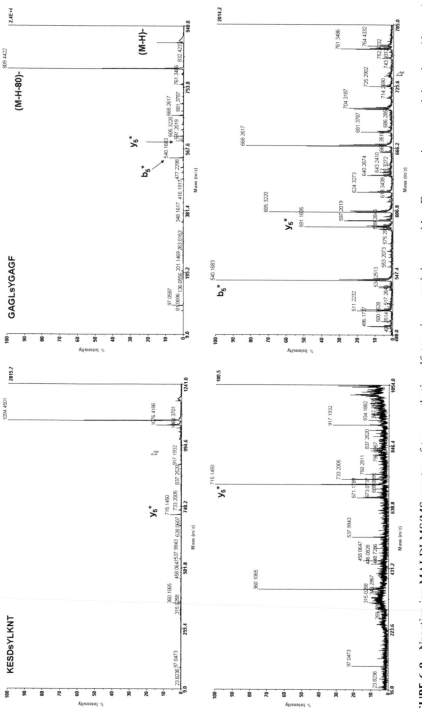

FIGURE 6.8 Negative ion MALDI MS/MS spectra of two synthetic sulfotyrosine-containing peptides. Fragment ions containing the sulfotyrosine residue are marked with an asterisk (*). (*Upper boxes*: full spectra; *lower boxes*: expanded region). Collaborator: David Klapper.

even with low collision energies, leaves the peptide in its original unmodified state, making the localization of the modification difficult (11).

Two model tyrosine-sulforylated peptides (synthesized by the University of North Carolina Peptide Synthesis Facility as C-terminal amides) were examined with a series of MALDI matrices in the positive and negative ion modes. As expected, the negative ion mode produced "cleaner" spectra (less sensitivity to contaminants) with more abundant molecular ions compared to ions resulting by loss of the sulforyl group. With DHB as the matrix, one of the synthetic peptides (KESDsYLKNT) produced an abundant $(M+H)^+$ ion, even in the positive ion mode, but the positive ion MS/MS spectra in all matrices showed only loss of 80 Da. When "cool" matrices, such as DHB and THAP, were used in the negative ion mode, the sulforyl groups in both peptides *could* be localized, as is shown in Figure 6.8. Newer "gentle" ionization techniques such as ETD, IRMPD, and ECD (described in Chapter 5) may also be useful for the MS/MS analysis of sulfotyrosine-containing peptides (11,12).

Another analytical challenge is that the mass shift due to sulforylation (a shift of 80 Da due to the addition of SO_3) is nearly isobaric with that due to phosphorylation on all but the highest resolution mass spectrometers (monoisotopic masses: SO_3 79.9568; HPO_3 79.9663). The Leary group has developed a strategy using alkaline phosphatase to remove phosphorylated peptides, leaving only the sulforylated peptides (13). This research group has also pioneered the use of anti-sulfotyrosine antibodies for selective enrichment of sulfotyrosine-containing peptides (14).

6.8   BIOTINYLATION

Because of the strong interaction between biotin and avidin (and the even stronger interaction with streptavidin), biotinylation is often used to aid in protein purification. Although we have done "on-target" analysis of biotinylated peptides (9), the avidin–biotin interaction is so strong that in some protein purification studies, we usually perform a tryptic digestion on the affinity-bound protein. The biotinylated peptide may remain on the bead, but the protein can be identified from the other tryptic peptides. If the goal of the study is the localization of the biotinylation site, it may be removed from the avidin beads by an acidic solution containing acetonitrile and formic acid, or by "competing" it off with biotin (Figure 6.9).

6.9   NITROSYLATION

Nitric oxide (NO) can modify the redox state of cysteine, and NO-modified cysteine can be considered an intermediate between reduced cysteine (Cys-S-H) and oxidized cysteine (Cys-S-S-Cys). Thus NO can affect protein folding, regulating protein activity by inducing the formation of disulfide bonds (15). S-nitrosylation of cysteine is usually not studied directly, as are other modifications. Instead, the NO moiety is replaced with a His tag (15) or a biotin tag (16) to facilitate purification and detection (Figure 6.10). Using these techniques, S-nitrosylation of cysteines has been found to be important for signal transduction, including initiation of apoptosis (17). It is now thought that S-nitrosylation may be as important as phosphorylation for signaling.

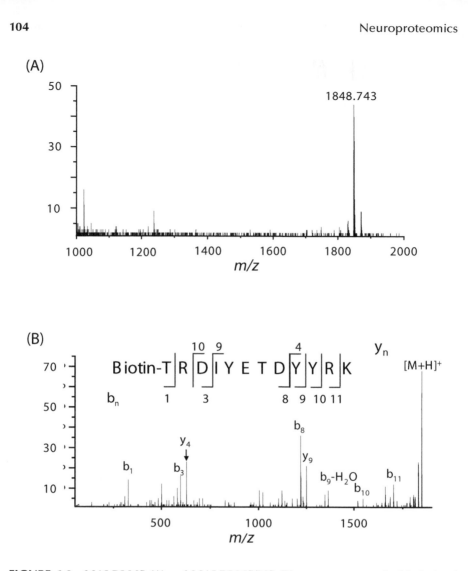

FIGURE 6.9 MALDI-MS (A) and MALDI-MS/MS (B) mass spectrum of a biotinylated peptide. *Source*: Reprinted from Raska, C. S., Parker, C. E., Sunnarborg, S. W. et al. (2003) Rapid and sensitive identification of epitope-containing peptides by direct matrix-assisted laser desorption/ionization tandem mass spectrometry of peptides affinity-bound to antibody beads, *J Am Soc Mass Spectrom* 14:1076–85, with permission.

6.10 UBIQUITINATION

Ubiquitin is an 8800 Da protein that attaches to lysine residues in the target protein, and may target the protein for degradation. Ubiquitination plays a role in cell division, cell death, and signal transduction, and has been found to be an important factor in synaptic remodeling (18). From an analytical perspective, digestion of a ubiquitinated protein with trypsin cleaves the ubiquitin chain, leaving a residual GG tag on the modified lysine (complete cleavage) or an LRGG tag (incomplete cleavage), as in Figure 6.11. The resulting mass shifts can be used to identify the modified lysine

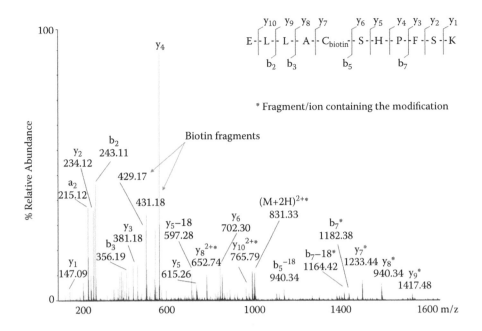

FIGURE 6.10   MS/MS spectrum of a nitrosylated peptide after a biotin switch and the characteristic biotin fragment ions at m/z 429 and 430. *Source*: Reprinted from Whalen, E. J., Foster, M. W., Matsumoto, A., Ozawa, K., Violin, J. D., Que, L. G., Nelson, C. D., Benhar, M., Keys, J. R., Rockman, H. A., Koch, W. J., Daaka, Y., Lefkowitz, R. J., and Stamler, J. S. (2007) Regulation of alpha-adrenergic receptor signaling by S-nitrosylation of G-protein-coupled receptor kinase 2, *Cell (Cambridge, MA, United States)* 129 (3), 511–512, with permission.

residue. Because of the transient nature of ubiquitinated proteins, ubiquitination is very difficult to detect and identify unless you have large amounts of material. For a more complete discussion of the mass spectrometric determination of ubiquitination, we refer the readers to reference (19) and the references cited therein.

6.11   COLLISIONAL-INDUCED DISSOCIATION– CLEAVABLE MODIFICATIONS

Most of the above examples are modifications that remain intact and remain on the amino acid while the peptide backbone fragments, or are labile and fall off completely during ionization or during collisional-induced dissociation (CID). Figure 6.12 shows an example where the modification and the backbone *both* fragment under MS/MS conditions, making interpretation of the spectrum much more difficult.

6.12   GLYCOSYLATION

Of all of the post-translational modifications, glycosylation is one of the most challenging because of the variability in the attached glycans, and the isobaric nature of many of these glycans. Often, the glycans are removed by glycosidases and their

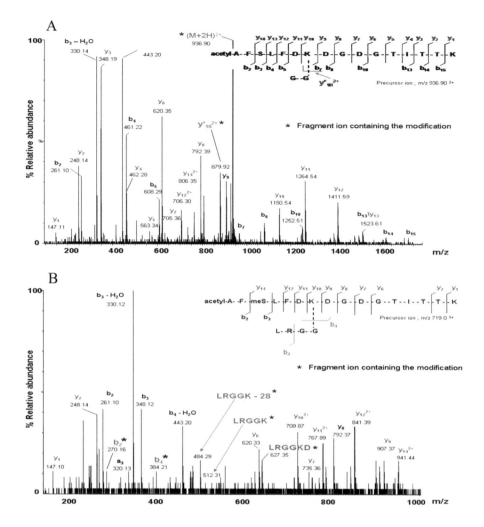

FIGURE 6.11   MS/MS fragmentation of a model (A) GG- and (B) LRGG-tagged peptide.

branch structure is determined separately from the protein. For proteomics, the opposite approach is often used—the glycans (which can shift the peptide molecular weights out of the range for peptide sequencing) are enzymatically removed, and the deglycosylated protein is studied separately.

Glycosylation is an extreme example of a case where the modification undergoes fragmentation. Adding to the analytical challenge is that glycosylation is not merely a single modification, but is often a set of modifications in which a single glycosylated site on a peptide may have a number of glycan isoforms with different chain lengths and branches attached. Because of this heterogeneity, there is no single mass shift associated with glycosylation, unlike other modifications. Glycosylation is variable, with different arrangements of glycans and branch structures (high mannose,

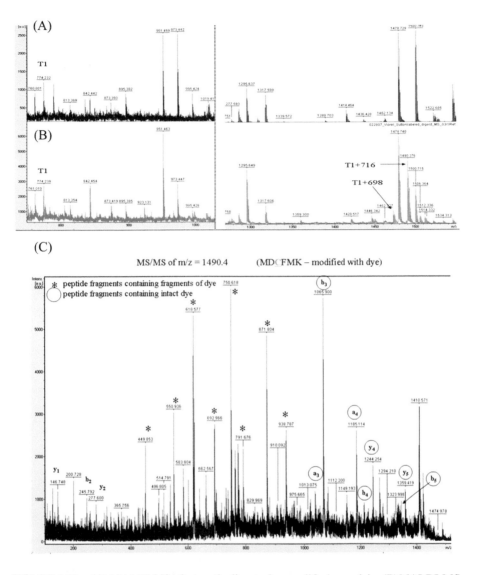

FIGURE 6.12 (A) MALDI-MS of a tryptic digest of unmodified synuclein. (B) MALDI-MS of a tryptic digest of dye-modified synuclein. (C) MS/MS of dye-modified tryptic peptide. Collaborators: Gary Pielak and Rebecca Ruf, unpublished results.

hybrid, complex) (Figure 6.13). In addition, not only the branch structure but the extent of glycosylation at a particular site can vary as well.

What adds to the analytical challenge is that intact glycans also are high-molecular weight modifications (as much as several thousand Da), which can shift the molecular weights of the peptides out of the mass region where they can be sequenced. Also, because of the heterogeneity of glycan structures, these modified peptides can appear as "humps" on the baseline rather than as discrete peaks. Often, glycosylated peptides

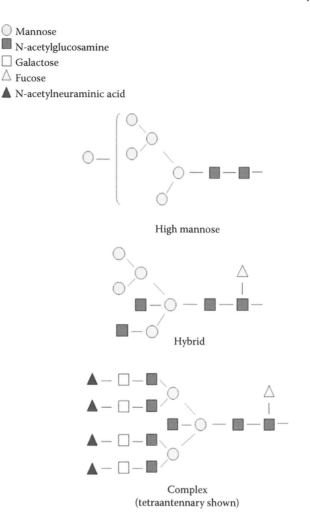

FIGURE 6.13   Types of N-glycan structures (sialic acid residues are not shown).

simply do not appear in the peptide digest spectra at all. Glycosylated peptides are also difficult to extract from PAGE gels, a standard purification technique used in proteomics. The consensus sequence for N-glycosylation is NxS or NxT (where x ≠ P), so potential glycosylation sites can be predicted from the sequence. There is no corresponding consensus sequence for O-glycosylation.

    There are three basic approaches to the study of protein glycosylation. The first is to leave the glycosylation on the protein. This approach can define the site of attachment and determine the type of glycosylation at each site (i.e., high mannose, hybrid, or complex), even though the exact branch structure cannot be determined. The analytical approach to determine the class of N-linked glycosylation is shown in Figure 6.14 (20). In this approach, the terminal sialic acid groups are first removed by neuramidase. After determination of the type of branch structure by MALDI-MS, the

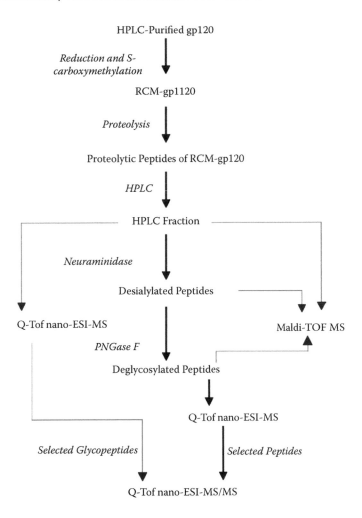

FIGURE 6.14   Analytical scheme for the analysis of the type of glycan attached to particular NxS or NxT sites. *Source*: Reprinted from Zhu, X., Borchers, C., Bienstock, R. J. and Tomer, K. B. (2000) Mass spectrometric characterization of the glycosylation pattern of HIV-gp120 expressed in CHO cells, *Biochemistry* 39:11194–204, with permission.

N-linked glycans can be enzymatically removed from glycopeptides with PNGaseF, and the amino acid sequence can be determined by MALDI-MS/MS or by nanoESI. The PNGase cleaves the glycan from the asparagine, releasing the glycan and converting the asparagine to aspartic acid, resulting in a shift of +1 Da from the Mw of the native peptide.

The type of glycan structure on a particular peptide (site-specific determination of glycosylation) can be inferred from the masses observed during MS/MS analysis. A series of fragment ions 162 Da apart in the MS/MS spectra indicates the presence of mannose moieties. Similarly, loss of an N-acetyl glucosamine residue ("GlcNac") leads to a mass difference of 203.08 Da, while the loss of

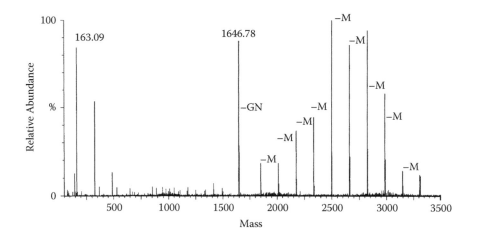

FIGURE 6.15   ESI-MS/MS spectrum of a glycosylated peptide of at (M+3H)+3 = 1103.4 Da, showing losses of mannose and GlcNac moeties. MaxEnt software was used to transform the multiply-charged peptides to their corresponding +1 forms. *Source*: Reprinted from Zhu, X., Borchers, C., Bienstock, R. J. and Tomer, K. B. (2000) Mass spectrometric characterization of the glycosylation pattern of HIV-gp120 expressed in CHO cells, *Biochemistry* 39:11194–204, with permission.

a fucose causes a Mw shift of –146 Da, loss of a sialic acid (–291 Da), and loss of GlcNac-Gal (–365 Da). Obviously, this can get complicated, especially since more than a single type of glycan or branch structure can be present on a particular peptide, so that these patterns of losses overlap each other. For a fairly simple example of a peptide that contains a high-mannose N-linked glycan, see Figure 6.15.

The second approach to the study of protein glycosylation is to remove the glycans, and to determine their structures separately. This results in an average picture of the glycosylation on the protein, but not the details of where each type of the glycosylation was located on the protein. An example of this type of study is shown in Figure 6.16.

Several mass spectrometrists have specialized in the area of determination of the glycan structure on particular peptides—notably the research groups of Orlando, Harvey, and Morelle—and the reader is referred to their research and review articles for a more detailed description of the analysis of glycoproteins (21–28).

6.13   CONCLUSION

In this book chapter we have described mass spectrometric approaches to the study of protein modification. This book chapter is intended only as a broad overview of the use of mass spectrometry for the study of post-translational modifications— for more detailed studies, the reader is referred to review articles and current literature on specific modification types. New developments in instrumentation (as discussed in the Chapter 5) will have a profound effect on this rapidly changing area of research.

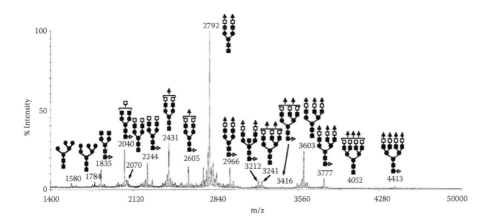

FIGURE 6.16 MALDI-MS of permethylated human serum N-glycans, analyzed after removing them from the proteins. *Source*: Reprinted from Morelle, W., Canis, K., Chirat, F., Faid, V. A. and Michalski, J.-C. (2006) The use of mass spectrometry for the proteomic analysis of glycosylation, *Proteomics* 6:3993–4015, with permission.

ACKNOWLEDGMENTS

Many of the examples shown here were acquired at the UNC-Duke Proteomics Center, which was partially funded by an anonymous gift in honor of Michael Hooker. We would also like to acknowledge support from the Specialized Cooperative Center for Reproductive and Infertility Research (2 U54 HD35041-11), and from the UNC Lineberger Cancer Center (5 P30 CA16806-31).

REFERENCES

1. Krishna, R. G. and Wold, F. (1997) Identification of common post-translational modifications. In *Protein structure: A practical approach*, ed. T. E. Creighton, 91–116 (Oxford: Oxford University Press).
2. Larsen, M. R., Sorenson, G. L., Fey, S. J., Larsen, P. M. and Roepstorff, P. (2001) Phospho-proteomics: Evaluation of the use of enzymatic de-phosphorylation and differential mass spectrometric peptide mass mapping for site specific phosphorylation assignment in proteins separated by gel electrophoresis, *Proteomics* 1:223–8.
3. Mascot (www.matrixscience.com).
4. Bonewald, L. F., Bibbs, L., Kates, S. A. et al. (1999) Study on the synthesis and characterization of peptides containing phosphorylated tyrosine, *Journal of Peptide Research* 53:161–9.
5. Janek, K., Wenschuh, H., Bienert, M. and Krause, E. (2001) Phosphopeptide analysis by positive and negative ion matrix-assisted laser desorption/ionization mass spectrometry, *Rapid Communications in Mass Spectrometry* 15:1593–9.
6. Lee, K. A., Craven, K. B., Niemi, G. A. and Hurley, J. B. (2002) Mass spectrometric analysis of the kinetics of in vivo rhodopsin phosphorylation, *Protein Science: A Publication of the Protein Society* 11:862–74.

7. Steen, H., Jebanathirajah, J. A., Rush, J., Morrice, N. and Kirschner, M. W. (2006) Phosphorylation analysis by mass spectrometry. Myths, facts, and the consequences for qualitative and quantitative measurements, *Molecular and Cellular Proteomics* 5:172–81.

8. Raska, C. S., Parker, C. E., Dominski, Z. et al. (2002) Direct MALDI-MS/MS of phosphopeptides affinity-bound to immobilized metal ion affinity chromatography beads, *Anal Chem* 74:3429–33.

9. Raska, C. S., Parker, C. E., Sunnarborg, S. W. et al. (2003) Rapid and sensitive identification of epitope-containing peptides by direct matrix-assisted laser desorption/ionization tandem mass spectrometry of peptides affinity-bound to antibody beads, *J Am Soc Mass Spectrom* 14:1076–85.

10. Torres, M. P., Thapar, R., Marzluff, W. F. and Borchers, C. H. (2005) Phosphatase-directed phosphorylation-site determination: a synthesis of methods for the detection and identification of phosphopeptides, *Journal of Proteome Research* 4:1628–35.

11. Monigatti, F., Hekking, B. and Steen, H. (2006) Protein sulfation analysis—A primer, *Biochimica et Biophysica Acta, Proteins and Proteomics* 1764:1904–13.

12. Mikesh, L. M., Ueberheide, B., Chi, A. et al. (2006) The utility of ETD mass spectrometry in proteomic analysis *Biochimica et Biophysica Acta* 1764:1811–22.

13. Yu, Y., Hoffhines, A. J., Moore, K. L. and Leary, J. A. (2007) Determination of the sites of tyrosine O-sulfation in peptides and proteins, *Nature Methods* 4:583–88.

14. Hoffhines, A. J., Damoc, E., Bridges, K. G., Leary, J. A. and Moore, K. L. (2006) Detection and purification of tyrosine-sulfated proteins using a novel anti-sulfotyrosine monoclonal antibody, *Journal of Biological Chemistry* 281:37877–87.

15. Camerini, S., Polci, M. L., Restuccia, U. et al. (2007) A novel approach to identify proteins modified by nitric oxide: The HIS-TAG switch method, *Journal of Proteome Research* 6:3224–31.

16. Forrester, M. T., Foster, M. W. and Stamler, J. S. (2007) Assessment and application of the biotin switch technique for examining protein S-nitrosylation under conditions of pharmacologically induced oxidative stress, *Journal of Biological Chemistry* 282:13977–83.

17. Jaffrey, S. R., Erdjument-Bromage, H., Ferris, C. D., Tempst, P. and Snyder, S. H. (2001) Protein S-nitrosylation: A physiological signal for neuronal nitric oxide, *Nat Cell Biol* 3:193–7.

18. Ehlers, M. D. (2003) Activity level controls postsynaptic composition and signaling via the ubiquitin-proteasome system, *Nat Neurosci* 6:231–42.

19. Parker, C. E., Mocanu, V., Warren, M. R., Greer, S. F. and Borchers, C. H. (2005) Mass spectrometric determination of protein ubiquitination. In *Ubiquitin-proteosome protocols*, eds. W. C. Patterson and D. M. Cyr, 153–73 (Totowa, NJ: Humana Press).

20. Zhu, X., Borchers, C., Bienstock, R. J. and Tomer, K. B. (2000) Mass spectrometric characterization of the glycosylation pattern of HIV-gp120 expressed in CHO cells, *Biochemistry* 39:11194–204.

21. Atwood, J. A., Cheng, L., Alvarez-Manilla, G. et al. (2008) Quantitation by isobaric labeling: Applications to glycomics, *Journal of Proteome Research* 7:367–74.

22. Colangelo, J. and Orlando, R. (1999) On-target exoglycosidase digestions/MALDI-MS for determining the primary structures of carbohydrate chains, *Analytical Chemistry* 71:1479–82.

23. Harvey, D. J. (2005) Proteomic analysis of glycosylation: Structural determination of N- and O-linked glycans by mass spectrometry, *Expert Review of Proteomics* 2:87–101.

24. Harvey, D. J. (2005) Structural determination of N-linked glycans by matrix-assisted laser desorption/ionization and electrospray ionization mass spectrometry, *Proteomics* 5:1774–86.

25. Harvey, D. J., Royle, L., Radcliffe, C. M., Rudd, P. M. and Dwek, R. A. (2008) Structural and quantitative analysis of N-linked glycans by matrix-assisted laser desorption ionization and negative ion nanospray mass spectrometry, *Analytical Biochemistry* 376:44–60.

26. Morelle, W., Canis, K., Chirat, F., Faid, V. A. and Michalski, J.-C. (2006) The use of mass spectrometry for the proteomic analysis of glycosylation, *Proteomics* 6:3993–4015.

27. Morelle, W. and Michalski, J.-C. (2007) Analysis of protein glycosylation by mass spectrometry, *Nature Protocols* 2:1585–1602.

28. Yang, Y. and Orlando, R. (1996) Identifying the glycosylation sites and site-specific carbohydrate heterogeneity of glycoproteins by matrix-assisted laser desorption/ionization mass spectrometry, *Rapid Communications in Mass Spectrometry* 10:932–36.

7 MALDI Imaging and Profiling Mass Spectrometry in Neuroproteomics

Malin Andersson, Per Andren, and Richard M. Caprioli

CONTENTS

7.1 SUMMARY

This chapter describes the basic technology and application of matrix-assisted laser desorption ionization (MALDI) imaging mass spectrometry (MS) in the field of neurobiology. Examples of applications that search for biomarkers of diseases of the central nervous system for both diagnostic and prognostic purposes are discussed.

Newly developed strategies and analytical methods are discussed for the study of drug pharmacokinetics, neuropeptide metabolism, and treatment-induced or disease-specific effects on the proteome.

7.2   INTRODUCTION

The clinical diagnosis of many major neurodegenerative disorders remains unsatisfactory and consequently there is a need for biomarkers for both diagnostic and prognostic purposes. Moreover, as protocols and methods for high-throughput genomics and proteomics improve, these approaches will provide significant insight into the pathogenesis of many neurodegenerative diseases. The possibility of identifying novel biomarkers, those related to pathogenesis in particular, also promises to help achieve early detection and ultimately to help lead the way toward individualized treatment. Mass spectrometry is a molecular analysis technology that is vital to this biomarker discovery for proteomics.

7.3   MS TECHNOLOGY

MALDI MS is able to detect and measure levels of neuropeptides and proteins and their distributions directly in tissue sections. The discovery and description of MALDI MS was first reported in 1987 (1). Briefly, a laser beam (i.e., a nitrogen laser at 337 or Nd:Yag laser at 355 nm) irradiates a solid sample that is a co-crystal of sample molecules and molecules from an organic matrix material. Absorption of energy from the laser by the matrix causes an ablation of the irradiated surface of the sample, with the desorption and formation (for the most part) of singly protonated molecular species ($[M+H]^+$). The desorbed ions are given an acceleration of 20–25 kV, traverse a time-of-flight (TOF) analyzer, and their mass-to-charge (m/z) is determined. Modern instruments have delayed extraction to achieve high sensitivity, a reflectron to record high-resolution mass spectra at low m/z values (<5000), and post-source decay or MS/MS capabilities to provide structure elucidation (e.g., TOF/TOF or quadrupole TOF [QTOF] for peptide sequencing; for a detailed discussion of MS see Chapters 5 and 6).

The MALDI matrices generally employed for the analysis of peptides and proteins are sinapinic acid (3, 5-dimethoxy-4-hydroxycinnamic acid), CHCA (alpha-cyano-4-hydroxycinnamic acid), and DHBA (2, 5-dihydroxybenzoic acid). Matrix can be applied directly on tissue sections by a spray or spot deposition process to achieve a uniform coating so that the spatial relationships of compounds remain intact.

One of the most common mass analyzers used in combination with MALDI is the TOF mass analyzer, which is well suited for pulsed ion sources. The TOF mass analyzer measures the time it takes for an ion to travel a specific distance in a field-free flight tube to the detector. Typically 30–100 or more laser shots are averaged in order to obtain a spectrum. The sensitivity of commercially available MALDI instruments is in the attomole to low femtomole range, with a mass accuracy of 10–50 ppm using delayed extraction and a reflector to refocus the molecular ions in the mass analyzer. MALDI-TOF instruments are able to detect ions over 200 kDa, although sensitivity falls off rapidly with increased mass range. Other mass analyzers have also been coupled with a MALDI including quadrupole, ion trap, Fourier transform ion cyclotron resonance (FT-ICR), orbitrap, and QTOF (quadrupole TOF MS/MS instrument), as well as other analyzers (see Chapter 5 for a description of different types of MS instruments). A potentially useful instrument for imaging mass spectrometry studies is the FT-ICR mass spectrometer because it is able to measure the molecular weight of peptides and other low molecular weight compounds with an accuracy of 0.1–1 ppm.

We have utilized a MALDI-TOF mass spectrometer for the direct analysis of tissue for spatial detection of molecules using a laser spot size on target of 30–80 microns (2,3). This technology is termed "MALDI imaging mass spectrometry" and will be abbreviated "IMS" for the purposes of this chapter. The versatility of mass spectrometry-based proteomic approaches makes IMS an ideal analytical tool for many applications, including pharmacokinetic studies of centrally acting drugs (4–6), lipidomics (7), neuropeptide metabolism (8,9), and discovery-based proteomics (10–13).

7.3.1 SAMPLE PREPARATION AND ANALYSIS

A thin section (typically 10–12 μm thick) of tissue is cut from a fresh frozen sample and thaw-mounted directly onto a MALDI target plate or compatible glass microscope slide (14). The section is covered with matrix dissolved in acidified organic solution (1%–2% trifluoroacetic acid in either 50% methanol or 50% acetonitrile in water) using a robotic spotter or spray coating device. Spotting robots such as the acoustic picoliter droplet ejector (Portrait 630; LabCyte) and a piezoelectric based chemical printer (ChIP-1000; Shimadzu) have been used with good results, including high sensitivity and highly reproducible spectra (15–17). Spray coating can be performed in several ways, for example, by hand using a spray nebulizer or with a commercially robotic device such as the ImagePrep (Bruker Daltonics) or TM nozzle sprayer (Leap Technology).

Irradiation of a spot on a tissue or biopsy section with 50–200 or more repetitive laser shots gives rise to a mass spectrum containing signals from molecules that were contained in that spot. Depending on the size of the sample and the analytical task at hand, many thousands of such mass spectra can be obtained from a single section through the use of a movable stage that allows repositioning of the area to be irradiated within milliseconds. In the *profiling mode*, the analytical task usually involves comparison of spectra from a relatively small number of spots on a sample either taken at random or targeted to specific areas in the tissue through histology or other means as shown in Figure 7.1a (44). Typically, a biopsy from a relatively homogeneous area of a tissue section would be interrogated in this way. In the *imaging mode*, a raster of the surface of the sample by the laser beam is accomplished by incremental movement of the sample stage to acquire thousands of spectra in an ordered array. These basic processes are illustrated in Figure 7.1b. The laser-ablated spots can be made adjacent (on 30–50 micron centers) for high-resolution imaging or at any other desired spacing. Often, a survey image is first obtained at 200–300 micron centers prior to a high-resolution image of a specific area of interest. The mass spectrum of each laser spot typically contains 500–1000 or more recorded

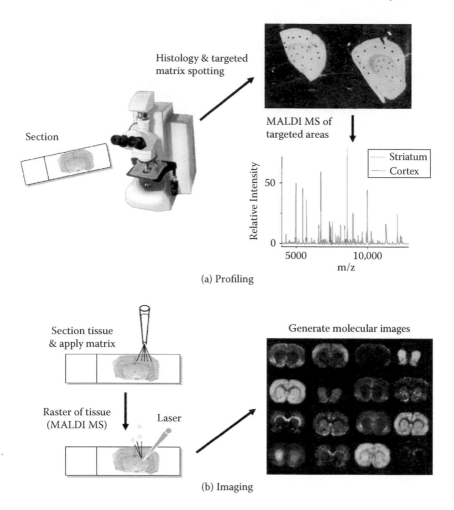

(a) Profiling

(b) Imaging

FIGURE 7.1 (See color insert following page 172.) (a) Region-specific mass spectra can be obtained through histology-directed tissue profiling MS, in this example, targeting the cortex (*blue*) and striatum (*red*) in a rat brain tissue section. Two average mass spectra display different molecular signatures in these regions. (b) MALDI imaging mass spectrometry. A thin tissue section is covered with MALDI matrix and irradiated by a laser in a raster across the tissue section, producing an ordered array of spectra. Each signal or peak in the spectra can be visualized by constructing a molecular density map, displaying the spatial distribution of peptides and proteins.

ion species, with accurate mass assignments (± 1 Da in 10,000 up to about 30,000 Das) for most of these. Computer-generated images at a given molecular weight can be obtained by plotting the intensity value of the chosen molecular species in the ordered array of laser spots (or pixels). Special software has been implemented in our lab in order to speed and automate the imaging process. Spot-to-spot cycle times can be 1 sec or less, depending on the choice of user-set parameters and sample quality desired, enabling a 1000 pixel image to be acquired in less than 20 minutes. Even so, this is not a limiting speed and total acquisition times can be improved with the use of high-speed lasers and data transfer technology.

The lower limit of imaging resolution is determined by (i) the matrix spot size and spot-to-spot distance in the case of robotic spotters, and (ii) by the size of the laser spot, currently approximately 30–50 µm, in the case of spray coating. Other important factors include tissue quality, the sample preparation protocols employed, section size, and user-determined time for analysis and data file size. In order to maximize reproducibility and sensitivity, in terms of peak height measurements and the number of molecular species detected, it is usually necessary to establish the optimum matrix-to-analyte ratio.

Each IMS image file contains multiple data dimensions: x and y coordinates for spatial localization, many hundreds of mass dimensions in terms of mass-to-charge (m/z), and the corresponding signal intensity at each m/z value. Using commercially available software or freeware such as the imaging software tool BioMap (Novartis), individual peptide or protein ion images can be visualized. Comparison of the relative intensities from normalized spectra then can be used to determine the relative amounts of a given compound throughout a section or series of samples.

7.3.2 DATA PROCESSING

Data pre-processing is generally necessary and includes baseline removal, noise reduction, and normalization. These are extremely important steps and validation of the procedures and algorithms used is highly recommended. In a recent study, several normalization algorithms were evaluated for their ability to reduce spectrum-to-spectrum variation (18). One of the most robust algorithms tested is based on the total ion current, i.e., the sum of all ion counts in a spectrum from the mass range 2–20 kDa. By rescaling each spectrum to its total ion current, unevenness in matrix application and crystallization is normalized, leading to increased signal-to-noise and more biologically relevant information (Figure 7.2). Preliminary data from our laboratories suggest this normalization strategy is also beneficial for the image analysis of low-molecular-weight species and peptides in the mass range of 400–3000 Da.

Data reduction for the purpose of comprehensive statistical analyses is an important step in biomarker discovery or exploratory proteomics. For MALDI IMS data, one generally accepted strategy is to apply a peak-picking algorithm for extraction of information regarding protein or peptide species present. This reduces the spectrum to a peak list containing one m/z value with its associated intensity per peak. Further data reduction can be done by averaging multiple spectra from a single organ or brain region that then is representative of that sub-region, although this must be done with great care so as not to obscure real spatial differences. In the end, the user must determine the trade-off of data file size, the speed of the image acquisition, and the resolution required since these parameters are inter-dependent.

Myelin Basic Protein (MBP)   Thymosin β4   PEP-19

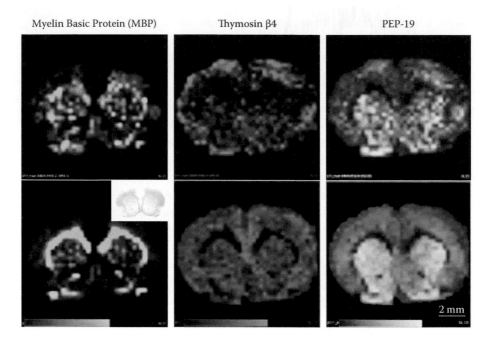

FIGURE 7.2 (See color insert following page 172.) Baseline subtraction, noise removal, and total ion current normalization greatly improves image information. The three examples of myelin basic protein, thymosin beta-4, and PEP-19 were derived from a single IMS data file. Insert in the lower left panel shows the corresponding MBP immunohistochemistry.

7.4 MALDI IMAGING MASS SPECTROMETRY OF THE BRAIN

IMS allows determination of the tissue distribution of hundreds of peptides and proteins virtually at the same time. Since all of the images produced are inherently registered to each other, comparing multiple protein distributions from IMS images is straightforward (Figure 7.3). This is in contrast to dual- or multiple-antigen immunohistochemistry where each antibody must be optimized for each protein.

An example of a mass spectrum that can be obtained directly from a tissue specimen in a profile experiment is shown in Figure 7.4 for a portion of the spectrum obtained from a human glioma biopsy. The complete spectrum is typically recorded over a mass-to-charge range of 2000–50,000, with most signals present in the lower half of this range. The vast majority of the signals in this spectrum are derived from proteins, verified through on-tissue protease digestion and subsequent sequence analysis of the resulting peptides (17). High-molecular-weight proteins (>200,000) can also be recorded directly from tissue, although this requires additional time for adjusting acquisition parameters to optimize sensitivity for this range.

We have imaged a coronal section of mouse brain in which glioma GL261 cancer cells had previously been injected in vivo in one hemisphere of the brain and subsequently a tumor had developed. After 15 days post-injection, the animal was sacrificed and brain sections were cut at 12-μm thickness and mounted on a target plate. A tumor of several millimeters had developed in the brain left lateral ventricle.

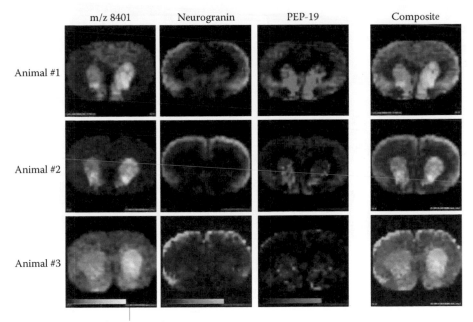

FIGURE 7.3 (See color insert following page 172.) Rat brain sections at the level of the striatum were analyzed by MALDI IMS, and the distribution of three proteins were visualized using three primary colors: green, red, and blue. The spatial (lateral) resolution of these images is 300 μm.

Evidence of metastatic migration of the tumor was seen in the right lateral ventricle. Significant differences in the profiles were observed, e.g., the signals at m/z 4965, 6719, 8565, and 12,134 were found consistently expressed at relatively higher levels in normal tissue, whereas m/z 9737, 9910, 11,641, and 12,372 were found consistently expressed at relatively higher levels in the tumor. A low-resolution image analysis of a section from this brain is shown in Figure 7.5. The perimeter of the tumor has been outlined in the optical image (panel a) for ease of viewing of the figure. Image analysis was performed with a resolution of 110 μm, averaging 20 laser shots per spectrum. Eight different protein expression maps across the section are shown in Figure 7.5b–i, taken from hundreds of such maps that could be generated from a single raster acquisition.

7.5 IMAGING MASS SPECTROMETRY OF NEUROPEPTIDES

One of the most challenging aspects of neurochemistry is the detection of endogenous neuropeptides due to their low in vivo concentrations ranging from pico- to sub-femtomolar levels (19,20). Recently, we have a developed a new, sensitive, and highly reproducible protocol for IMS of peptides and neuropeptides in rat brain sections. Utilizing a matrix spotter, approximately 1000 monoisotopic molecular species were observed and imaged in one section of a rat ventral midbrain (16). Furthermore,

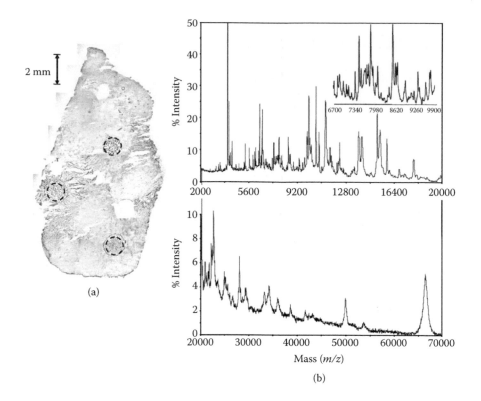

FIGURE 7.4 (a) A human glioma specimen was mounted on a glass slide and a mass spectrum was obtained directly from the section in a profiling experiment. (b) The mass spectrum is typically recorded over a mass-to-charge range of 2000–100,000, with most signals present in the lower half of this range. The inset is an expanded display to demonstrate the molecular resolution and complexity of the dataset.

we have been able to identify neuropeptides such as substance P (SP) in the substantia nigra pars reticulata by MS/MS analysis directly off the rat brain tissue sections using a MALDI QTOF mass spectrometer. The capability of determining a neuropeptide sequence directly in a discrete brain region of a section enables the study of the physiological and disease-related metabolic processing of neuropeptides. For example, disease-state alterations in the metabolic processing of dynorphin A occurs in an animal model of Parkinson's disease (PD) (21) that is well in line with findings of disturbed opioid transmission in both animal models and patients with Parkinson's disease (22,23). Indeed, we have observed both full-length SP as well as the metabolites SP(1-7) and SP(1-9) in the rat substantia nigra pars reticulata (16).

7.6   MALDI PROFILING AND IMAGING OF EXPERIMENTAL NEURODEGENERATIVE DISORDERS

Animal models have been used extensively to better understand the etiology and pathophysiology of human neurodegenerative diseases. Ideally, they should reproduce

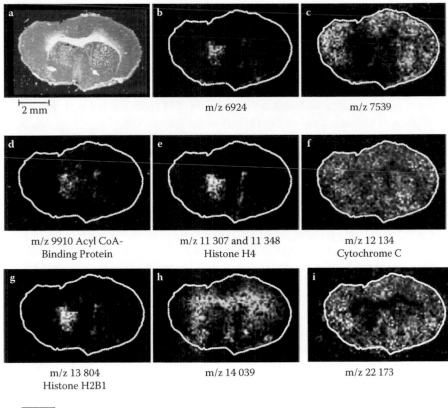

FIGURE 7.5   (See color insert following page 172.) (a) A coronal section of mouse brain with a glioma tumor in one hemisphere; for ease of viewing the tumor is outlined with a dotted line. IMS analysis was carried out at 110-μm resolution and (b–i) display several molecular images, including histones H4 and H2B1, which are typically present in actively growing tumors.

the clinical manifestation of the disease and mimic the selective neuronal loss in the brain. These models can also be used to test therapeutic approaches for treating functional disturbances observed in the disease of interest.

Profiling and imaging MS technology has been applied to animal models of neurodegenerative disorders, particularly to compare peptide and protein patterns in different animal models of PD (13,24,25) and Alzheimer's disease (AD) (3,26). Clinically, PD is characterized by motor features including resting tremor, bradykinesia, and rigidity, and AD by loss of mental functions (dementia). In these affected brains, peptide and protein patterns are complex and new molecular tools, such as MALDI IMS, are needed to identify not only the changes in these patterns but also to identify specific proteins involved in neurodegeneration.

Profiling MALDI MS has been used to generate peptide and protein profiles of brain tissue sections obtained from experimental PD (the unilaterally 6-hydroxy-dopamine lesioned rodent model). Initially, a number of protein expression profile differences were found in the dopamine denervated side of the brain in specific brain regions when compared to the corresponding intact side, for example, calmodulin, cytochrome c, and cytochrome c oxidase (13). This work was the first study utilizing MALDI MS protein profiling on tissue sections to screen for protein expression differences in an animal model of PD.

In addition to the loss of dopamine-producing neurons in substantia nigra, idiopathic PD is characterized by the presence of Lewy bodies. These are abnormal cytoplasmic protein inclusions found in the midbrain of PD patients and are regarded as a pathological characteristic of PD (27). A major component of Lewy bodies is ubiquitin (8.5 kD, 76 residue protein) (28). The ubiquitin pathway degrades intracellular proteins with a ubiquitin tag that marks proteins for proteolysis. The polyubiquinated proteins are enzymatically degraded to peptides, and the ubiquitin moieties are released intact (29). The high levels of ubiquitin and ubiquinated proteins in Lewy bodies suggest that protein degradation is impaired in PD. However, Lewy body formation has not been demonstrated in the acutely induced neurotoxic experimental models of PD (30). Using profiling MALDI MS on brain tissue sections from the unilaterally 6-OHDA lesioned rat model of PD, increased levels of free ubiquitin specifically in the dorsal striatum of the dopamine depleted hemisphere were detected (24). No similar changes were found in the intact hemisphere or in the ventral striatum of the dopamine depleted hemisphere. In addition, this study emphasizes the utility of MALDI MS for direct analysis of proteins and peptides on biological tissue sections. Most antibodies that have been used in immunocytochemical studies to assess the presence of ubiquitin in PD-associated Lewy bodies are directed toward a protein-bound form of ubiquitin, because free monomeric ubiquitin is not immunogenic in most mammalian species used to produce antibodies.

The 12 kDa neuroimmunophilin protein FKBP-12 is abundant in the brain and acts as a receptor for the immunosuppressant drug FK506 (31). The drug belongs to a class of neuroimmunophilin compounds that elicit both neuroprotection and neuroregeneration in neurotoxic chemical models of PD (32). Further, FK506 has also been shown to enhance nerve regeneration in a variety of experimental situations including nerve crush and transection, and spinal cord injury (33–35). In the presence of the drug FK506, FKBP-12 binds to a subunit of the calcium-dependent phosphatase calcineurin by blocking the access to the catalytic site of the subunit (36). Increased levels of FKBP-12 in the dorsal and middle part of the dopamine-depleted striatum were detected using the 6-OHDA model of PD (37). The finding was further confirmed by in situ hybridization, Western blotting, two-dimensional gel electrophoresis, and liquid chromatography electrospray tandem mass spectrometry. Another calcium-binding protein was effectively imaged from brain in the MPTP mouse model of PD (25). PEP-19 (Purkinje cell protein 4; PCP 4) is a 6.7 kDa protein that belongs to a family of proteins involved in calmodulin-dependent signal transduction (38). IMS showed that PEP-19 protein was predominantly localized to the striatum of the brain tissue cross sections and that PEP-19 levels were significantly reduced by 30% after MPTP administration (25) (Figure 7.6).

A pathological feature of AD is the presence of amyloid deposits in plaques and in blood vessel walls. These accumulations are mainly composed of amyloid beta (Aβ) peptides of 4–5 kDa and are all derived from the amyloid precursor protein (APP) (39). In an animal model of AD, a transgenic mouse model over-expressing

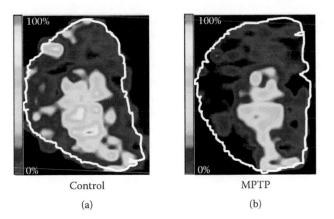

Control   MPTP

(a)   (b)

FIGURE 7.6   (See color insert following page 172.) MALDI IMS analysis of brain tissue sections of an experimental model of PD. The relative ion density of PEP-19 from one control (a) and one MPTP-administered animal (b) displays a reduction of PEP-19 expression in the striatum (lateral resolution 280 µm). About 400 distinct mass signals were detected in the mass range of 2–30 kDa in this area of the tissue. *Source*: Modified with permission from Sköld, K., Svensson, M., Nilsson, A. et al. (2006) Decreased striatal levels of PEP-19 following MPTP lesion in the mouse, *J Proteome Res* 5:262–9.

human APP23, the spatial distribution of different Aβ peptides was studied using IMS (26,40). In this animal model of AD, an increased production of Aβ and Aβ peptide fragments has been shown to be induced by the mutation of the amyloid precursor protein gene, and A β deposits develop in cortical and hippocampal structures upon aging. IMS detected the Aβ peptides (1–37, 1–38, 1–39, 1–40, and 1–42) in the brain tissue sections located in the parietal and the occipital cortical lobe and in the hippocampus region. By normalizing the distributions to a standard protein insulin it was shown that Aβ (1–40) and Aβ (1–42) were the most abundant amyloid peptides (26,40).

7.7   THREE-DIMENSIONAL IMAGING MASS SPECTROMETRY OF THE BRAIN

The ability to spatially resolve different proteins while simultaneously acquiring quantitative information concerning their relative abundance is particularly important in heterogeneous tissues such as the brain. Analyses of three-dimensional (3D) brain data by different imaging modalities such as computer tomography, magnetic resonance imaging (MRI), and positron emission tomography (PET) have directly led to major advances in understanding both normal and pathological function. Most in vivo imaging methods provide only structural information, although PET can localize and quantitatively assess levels of certain receptors and transporters. Three-dimensional reconstructions of histological material, such as immunohistochemically stained tissue sections, can provide excellent anatomical resolution, but quantitation of such material is difficult and the use of antibodies to reveal the proteins of interest limits the approach to revealing few proteins in a given section and precludes any unbiased approach to monitoring proteins. By contrast, MALDI IMS detects hundreds of molecular species and recent developments have provided a novel basis of interrogation of protein localization in 3D (16,41,42). In one study, a series of ventral midbrain sections was collected and scanned (16). After MS acquisition, the matrix was washed off and the sections were stained using a histological stain (cresyl violet). Three-dimensional reconstructions of the brain region were made from the histologically stained sections, which then served as a template for inserting the registered MALDI IMS data into the volume reconstruction. This allowed the mapping of several known peptides and proteins, such as substance P and PEP-19, but also unknown proteins could be visualized simultaneously with a known protein (in this case PEP-19) and information gathered on the relative localization of the unknown species in the ventral midbrain (Figure 7.7).

The structural heterogeneity of many tissues, such as brain, requires an approach like this that permits relatively high anatomical resolution coupled with quantitative analysis of proteins and the ability to do so in an unbiased fashion.

7.8 IMAGING WITH TANDEM MASS SPECTROMETRY

An extension of the early work using MALDI-TOF technology is the use of tandem mass spectrometry (MS^2 or MS/MS), whereby both the molecular species and also

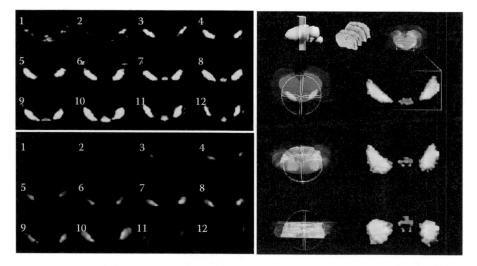

FIGURE 7.7 (See color insert following page 172.) Assessment of distribution in 3D of PEP-19 (*green, top left panel*) and an unknown 7.4 kDa protein (*red, bottom left panel*) in the rat ventral midbrain. Tissue sections 12-μm thick were collected at 200-μm spacing throughout the rat ventral midbrain, and a series of sections covering the anterior-posterior length of the substantia nigra were imaged using MALDI IMS. After data acquisition, the matrix was washed off and sections were stained with cresyl violet (*right panel, top row*). Three-dimensional reconstructions of histologically stained sections provided the volume coordinates (*right panel, left column*) to insert the two MALDI IMS images into the same volume reconstruction (*right panel, right column*). Co-localization is suggested by the appearance of the yellow color.

a key structural variety of the molecule are simultaneously monitored to give added validity to molecular specificity, particularly at low molecular weights. This involves an instrument containing two consecutive mass analyzers in conjunction with a collision-activated dissociation (CAD) process (see Chapter 5 for a discussion on fundamentals and instrumentation of mass spectrometry). A parent molecular ion is selected by the first mass analyzer and transmitted to a collision cell where it is fragmented by collisions with a neutral gas, such as helium or nitrogen. The fragments are then analyzed by the second mass analyzer. From the fragmentation pattern, the identity of the precursor ion can be determined. The use of MS/MS for image analysis of low-molecular weight compounds provides extremely high specificity and typically reduces levels of chemical noise, thus enhancing the signal-to-noise ratio and the limits of detection (6,43). The MS/MS imaging analysis is practically limited to a discrete number of parent ions that can be recorded in a given experiment. For example, Khatib-Shahidi et al. (4) demonstrated the distribution of clozapine and three of its metabolites at various time points after administration. Clozapine was clearly detected in the central nervous system, whereas the metabolites were mostly present at the sites of conversion: the liver and kidney.

7.9   QUANTITATION

The reproducibility and ability of the technology to provide relative quantitation have been published and cited in several reviews (44,45). Protocols have been developed for preparation of the section and matrix application that provide a high level of reproducibility. For example, in one study of glioma biopsies designed to test the reproducibility of the spectral patterns on a given biopsy from areas of similar histology, 687 individual spectra were obtained from 89 biopsies taken from 58 patients. Histologically, the samples were classified as 18 non-tumor, 25 grade 2, 17 grade 3, and 38 grade 4 biopsies. The statistical result showed 7% variability in the MS signal variation within a morphologically similar region of a given biopsy (46). A second issue is the question of reproducibility of a given signal from different parts of a tissue section where there are morphologically dissimilar areas. Although we have not performed a statistical analysis, at least in the case of a brain section, we do not observe significant variations for proteins known to be evenly distributed in tissue. Thus, for a coronal brain section, the image intensity of such a protein is quite similar throughout the section. However, MALDI MS of sections from different organs display significant variations in the spectral signature, reflecting in part the inherent complexity of the molecular compositions of the different cell types present.

7.10   IDENTIFICATION OF PROTEINS

Since the measurement of the molecular weight of a protein does not specifically identify the protein, additional studies are necessary and this can involve two approaches. First, protein extracts from tissue samples can be fractionated by high performance liquid chromatography (HPLC) (see Chapter 3), one-dimensional polyacrylamide gel electrophoresis with sodium dodecyl sulfate (1D SDS-PAGE), or solution phase isoelectric focusing, and fractions having the protein of interest, as determined by MALDI MS, are then subjected to tryptic digestion and analysis by LC-MS/MS (as described in Chapter 5). Peptides and their corresponding proteins of origin are identified from MS-MS spectra with a protein database search program that correlates uninterpreted MS-MS spectra with theoretical spectra from database sequences. Confirmation of protein identities are based on the apparent molecular weight of the MS-MS identified proteins compared to pattern-specific signals detected in the MALDI profiles.

For MALDI IMS analysis, targeted proteins can be identified directly from tissue through on-tissue proteolysis. Trypsin is deposited using a robotic spotter in a Cartesian array across the tissue section. Multiple passes are required to keep the trypsin in solution and allow sufficient time for the enzymatic reaction to take place. Subsequently, matrix solution (e.g., DHB in 50% methanol, 0.3% TFA) is placed directly on top of the trypsin digests and these spots are analyzed by MALDI-TOF. The data set can be used to map tryptic peptides with similar distribution patterns using cluster analysis algorithms for first-stage peptide mapping analysis. Second, MS/MS analysis can be performed directly off tissue, which has the advantage of allowing the tryptic peptide parent mass and fragment masses to be mapped to a specific localization in the tissue (16,17). MS/MS analysis results in a list of tryptic masses coupled with a positive identification of a single protein. These tryptic masses can then be visualized in a section, mapping to the individual digest spots. Taken together, this in situ peptide analysis approach provides high power of molecular identification including spatial specificity. Another advantage of this approach is the possibility to map high-molecular-weight proteins that are otherwise not readily detectable in their intact form.

A second identification approach uses the combination of LC-MS-MS analyses and stable isotope tags. In its simplest form, equal amounts of protein extracts from two samples to be compared are chemically tagged with light and heavy (unlabeled and deuterium or ^{13}C-labeled) reagents. They are combined and digested and the tagged peptides are analyzed by LC-MS-MS. Peptides derived from the two samples are distinguished by pairs of signals in full scan MS separated by the mass difference of the light and heavy isotope tags. Pairs of signals whose intensities deviate from unity represent proteins that were differentially present in the original two samples. MS-MS spectra acquired in the same LC-MS-MS analyses allow unambiguous identification of the differentially expressed proteins. Two commercial products are currently available. One commonly used procedure utilizes thiol-reactive isotope coded affinity tag (ICAT) reagents (47), and isobaric tags for relative and absolute quantification (iTRAQ), another acid-cleavable reagent that offers more efficient recovery of tagged peptides and produces high quality MS-MS spectra for identification (48). N-terminal isotope tagging of tryptic peptides enables identification of proteins that differ in posttranslational modifications rather than protein expression level per se (49,50). A major advantage of the tagging reagents is that they provide a quantitative comparison of the proteins of interest.

7.11   INTEGRATING HISTOLOGY AND IMAGING MASS SPECTROMETRY

An important aspect of IMS is the ability to identify histological features of interest within the tissue sections and then obtain ion images of that same area. A protocol has been developed that allows the histology and MS analyses to be performed on the same section (44), avoiding significant problems of comparing the histology and mass spectra taken from different sections. A conductive glass slide is used to mount tissue sections, allowing visualization of tissue sections with transmitted light and also serving as the target plate for IMS measurements. Typically, a pathologist studies a photomicrograph of the stained section and marks the areas of interest. In this way, the number of spectra are reduced to include only the tissue regions of interest, and a high number of specimens can be examined in a short time. Several tissue staining procedures compatible with IMS can be used to permit a pathologist to perform routine histology (hematoxylin and eosin [H&E] stained tissue produces poor mass spectral patterns). Figure 7.4 presents the analysis of a grade IV human glioma section after staining with methylene blue, providing a strong nuclear as well as faint cytoplasmic staining. The spot size on the tissue analyzed can vary from 80 μm up to 1 mm or larger. The resulting spectrum displays signals in the m/z range from 2000 to 70,000. Several other dyes may also be used, including toluidine blue, cresyl violet, and nuclear fast red (14).

Histology-directed profiling is especially important in the research of invasive tumors where connective tissues and normal intact cells often surround discrete clusters of cancer cells. Cornett et al. (44) demonstrated how this approach could discern, by unsupervised cluster analysis, different grades of breast tumor stages. Earlier forms of this strategy involved manually depositing matrix and subsequent manual acquisition of MS, which was successfully used for the classification of various glioma tumor grades and for the development of prognostic markers of short-term versus long-term survival of glioma patients (46).

7.12   CELL ISOLATION BY LASER CAPTURE MICRODISSECTION AND ANALYSIS BY IMS

Laser capture microdissection (LCM) is a technology that permits the isolation of single cells or single populations of cells from thin tissue sections, typically 5–10 μm thick, mounted on a glass slide (51). Using IMS, protein signatures can be obtained from 50–300 captured cells from a single cell type within a heterogeneous sample (52). For the Pixcell LCM instrument, a narrow laser beam (7.5–30 μm in diameter) irradiates a heat-sensitive polymer film on a cap that is in contact with a tissue section. The heated polymer adheres to the targeted cell(s) and these cells are subsequently removed from the section when the cap is lifted. We have established protocols to analyze cancer cells by MALDI MS after their LCM capture isolation from a thin tissue section (53).

7.13   LIMITATIONS OF TECHNOLOGY

IMS has several technical limitations, some of which are inherent in the laser desorption/ionization process and others in the performance of instrumentation available today, but advances are under way through which significant improvements will be made. Certainly one of the inherent limitations is the "ion suppression" effect that occurs in the MALDI ionization process. Basically, proteins that can act as better bases to capture protons can show preferential ionization yields. Although careful preparation protocols can minimize this effect, it can be significant. In that light, while one can estimate relative intensities for a given protein in several samples, it does not necessarily apply for two different proteins in a spectrum. Another limitation is the fact that only a small window of many hundreds of proteins are measured from a single spot on a sample, whereas many times this number exist in the tissue at moderate levels. Also, using current protocols, soluble proteins are favored since generally the preparation does not involve the use of detergents since these significantly compromise sensitivity. To this end we have synthesized several new families of cleavable detergents that, for example, when matrix is added to give low pH, the detergent hydrolyzes and does not interfere with the analysis (54). One of these detergents is equivalent to SDS in its detergent properties.

A practical limitation of IMS is the laser spot size and the trade-off between image resolution and sensitivity; the smaller the spot size the less material there is to desorb. Smaller spot sizes have been achieved and we have developed a 1–5 μm spot diameter on target, but its effectiveness in tissue analysis is yet to be established. Overall, the sensitivity of IMS on tissue is currently estimated to be in the high attomole range, but it is difficult to measure this accurately. Thus, proteins that are expressed at low copy numbers per cell will not be readily analyzed with current instrumentation.

Although the spatial resolution of MALDI IMS is relatively high, at this point in its development it is insufficient to accurately differentiate subcellular regions in single mammalian cells (such as axons and dendrites). Future advances in MALDI IMS suggest that it will be possible to localize peptides and proteins in much smaller areas (55). Ultimately, combining mapping of a given protein at high spatial resolution coupled with markers of cellular components (e.g., MAP2 for dendrites) should permit unambiguous determination of molecular entities in subcellular structures.

7.14   PERSPECTIVES

Neuropeptides are activated from preprohormones through specific proteases and subjected to posttranslational modifications (PTMs) prior to release. In general, over 200 posttranslational modifications have been reported to date, including glycosylation, C-terminal amidation, N-terminal acetylation, phosphorylation, and sulfation (56). Mass-spectrometry-based approaches are well suited for the detection and identification of previously unknown PTMs. Many PTMs depend on cell- and site-specific enzymes, which require a high spatial resolution in order to be detected. MALDI IMS is able to detect femtomolar levels of neuropeptides in tissue areas as small as 50 microns. Future work will aim to correlate PTMs and protein changes in specific brain areas using MALDI IMS, and promises to elucidate biochemical processes in both normal and pathological states.

Analysis of brain tissue sections using IMS offers many advantages over traditional analytical techniques such as Western immunoblotting, in situ hybridization histochemistry, and immunohistochemistry, which have to rely on known chemistry and well-characterized peptide and protein species. In contrast, IMS has the capability of detecting virtually any compound present in a tissue section, as long as it can be ionized and desorbed into a gaseous phase in the MALDI process. In addition, while proteolytic cleavage and PTMs of proteins and neuropeptides are common mechanisms for the fine-tuning regulation of biological activity, these events may be undetectable using immunology-based methods that target specific antigenic epitopes, but can be readily determined by MS-based methods, including IMS.

Many classical technologies and approaches for protein analyses, such as two-dimensional difference gel electrophoresis (2D-DIGE), are not commonly used for neuropeptide analyses for several reasons. This is because relatively low concentrations of neuropeptides make it difficult to obtain enough material from small regions or subregions of the target brain structure, and the mass range of neuropeptides is usually below that of the effective limits of gel-based approaches (i.e., ~5–10 kDa) (see Chapter 4 for a discussion of 2D-DIGE). In addition, the throughput for multiple samples is low. IMS provides strength in these areas because of its high sensitivity in the mass range up to 20 kDa and its extraordinary high-throughput capabilities. All in all, IMS is ideally suited for unbiased spatial localization and relative quantitation of proteins in brain and other heterogeneous tissues.

ACKNOWLEDGMENTS

The authors thank D. S. Cornett for excellent technical assistance and for the custom imaging preprocessing software, and Ariel Deutch for his consultation and constructive comments, and also acknowledge funding from NIH grant 2RO1 GM58008-09, and the Department of Defense grant W81XWH-05-1-0179.

REFERENCES

1. Karas, M., Bachmann, D., Bahr, U. and Hillenkamp, F. (1987) Matrix-assisted ultraviolet-laser desorption of nonvolatile compounds, *International Journal of Mass Spectrometry and Ion Processes* 78:53–68.
2. Caprioli, R. M., Farmer, T. B. and Gile, J. (1997) Molecular imaging of biological samples: Localization of peptides and proteins using MALDI-TOF MS, *Anal Chem* 69:4751–60.
3. Stoeckli, M., Chaurand, P., Hallahan, D. E. and Caprioli, R. M. (2001) Imaging mass spectrometry: A new technology for the analysis of protein expression in mammalian tissues, *Nat Med* 7:493–6.
4. Khatib-Shahidi, S., Andersson, M., Herman, J. L., Gillespie, T. A. and Caprioli, R. M. (2006) Direct molecular analysis of whole-body animal tissue sections by imaging MALDI mass spectrometry, *Anal Chem* 78:6448–56.

5. Reyzer, M. L. and Caprioli, R. M. (2005) MALDI mass spectrometry for direct tissue analysis: A new tool for biomarker discovery, *J Proteome Res* 4:1138–42.

6. Reyzer, M. L., Hsieh, Y., Ng, K., Korfmacher, W. A. and Caprioli, R. M. (2003) Direct analysis of drug candidates in tissue by matrix-assisted laser desorption/ionization mass spectrometry, *J Mass Spectrom* 38:1081–92.

7. McLean, J. A., Ridenour, W. B. and Caprioli, R. M. (2007) Profiling and imaging of tissues by imaging ion mobility-mass spectrometry, *J Mass Spectrom* 42:1099–1105.

8. Andren, P. E. and Caprioli, R. M. (1999) Determination of extracellular release of neurotensin in discrete rat brain regions utilizing in vivo microdialysis/electrospray mass spectrometry, *Brain Res* 845:123–9.

9. Nydahl, K. S., Pierson, J., Nyberg, F., Caprioli, R. M. and Andren, P. E. (2003) In vivo processing of LVV-hemorphin-7 in rat brain and blood utilizing microdialysis combined with electrospray mass spectrometry, *Rapid Commun Mass Spectrom* 17:838–44.

10. Kruse, R. and Sweedler, J. V. (2003) Spatial profiling invertebrate ganglia using MALDI MS, *J Am Soc Mass Spectrom* 14:752–9.

11. Lemaire, R., Menguellet, S. A., Stauber, J. et al. (2007) Specific MALDI imaging and profiling for biomarker hunting and validation: Fragment of the 11S proteasome activator complex, Reg alpha fragment, is a new potential ovary cancer biomarker, *J Proteome Res* 6:4127–34.

12. Meistermann, H., Norris, J. L., Aerni, H. R. et al. (2006) Biomarker discovery by imaging mass spectrometry: Transthyretin is a biomarker for gentamicin-induced nephrotoxicity in rat, *Mol Cell Proteomics* 5:1876–86.

13. Pierson, J., Norris, J. L., Aerni, H. R. et al. (2004) Molecular profiling of experimental Parkinson's disease: Direct analysis of peptides and proteins on brain tissue sections by MALDI mass spectrometry, *J Proteome Res* 3:289–95.

14. Schwartz, S. A., Reyzer, M. L. and Caprioli, R. M. (2003) Direct tissue analysis using matrix-assisted laser desorption/ionization mass spectrometry: Practical aspects of sample preparation, *J Mass Spectrom* 38:699–708.

15. Aerni, H. R., Cornett, D. S. and Caprioli, R. M. (2006) Automated acoustic matrix deposition for MALDI sample preparation, *Anal Chem* 78:827–34.

16. Andersson, M., Groseclose, M. R., Deutch, A. Y. and Caprioli, R. M. (2008) Imaging mass spectrometry of proteins and peptides: 3D volume reconstruction, *Nat Methods* 5:101–8.

17. Groseclose, M. R., Andersson, M., Hardesty, W. M. and Caprioli, R. M. (2007) Identification of proteins directly from tissue: In situ tryptic digestions coupled with imaging mass spectrometry, *J Mass Spectrom* 42:254–62.

18. Norris, J. L., Cornett, D. S., Mobley, J. A. et al. (2007) Processing MALDI mass spectra to improve mass spectral direct tissue analysis, *Int J Mass Spectrom* 260:212–21.

19. Kendrick, K. M. (1990) Microdialysis measurement of in vivo neuropeptide release, *J Neurosci Methods* 34:35–46.

20. Strand, F. L. (2003) Neuropeptides: General characteristics and neuropharmaceutical potential in treating CNS disorders, *Prog Drug Res* 61:1–37.

21. Klintenberg, R. and Andren, P. E. (2005) Altered extracellular striatal in vivo biotransformation of the opioid neuropeptide dynorphin A(1-17) in the unilateral 6-OHDA rat model of Parkinson's disease, *J Mass Spectrom* 40:261–70.

22. Johansson, P. A., Andersson, M., Andersson, K. E. and Cenci, M. A. (2001) Alterations in cortical and basal ganglia levels of opioid receptor binding in a rat model of l-DOPA-induced dyskinesia, *Neurobiol Dis* 8:220–39.

23. Nisbet, A. P., Foster, O. J., Kingsbury, A. et al. (1995) Preproenkephalin and preprotachykinin messenger RNA expression in normal human basal ganglia and in Parkinson's disease, *Neuroscience* 66:361–76.

24. Pierson, J., Svenningsson, P., Caprioli, R. M. and Andren, P. E. (2005) Increased levels of ubiquitin in the 6-OHDA-lesioned striatum of rats, *J Proteome Res* 4:223–6.
25. Sköld, K., Svensson, M., Nilsson, A. et al. (2006) Decreased striatal levels of PEP-19 following MPTP lesion in the mouse, *J Proteome Res* 5:262–9.
26. Stoeckli, M., Knochenmuss, R., McCombie, G. et al. (2006) MALDI MS imaging of amyloid, *Methods Enzymol* 412:94–106.
27. Gibb, W. R. (1989) Neuropathology in movement disorders, *J Neurol Neurosurg Psychiatry* Suppl:55–67.
28. Fornai, F., Lenzi, P., Gesi, M. et al. (2003) Recent knowledge on molecular components of Lewy bodies discloses future therapeutic strategies in Parkinson's disease, *Curr Drug Targets CNS Neurol Disord* 2:149–52.
29. Snyder, H. and Wolozin, B. (2004) Pathological proteins in Parkinson's disease: Focus on the proteasome, *J Mol Neurosci* 24:425–42.
30. Melrose, H. L., Lincoln, S. J., Tyndall, G. M. and Farrer, M. J. (2006) Parkinson's disease: A rethink of rodent models, *Exp Brain Res* 173:196–204.
31. Liu, J., Farmer, J. D., Jr., Lane, W. S. et al. (1991) Calcineurin is a common target of cyclophilin-cyclosporin A and FKBP-FK506 complexes, *Cell* 66:807–15.
32. Costantini, L. C., Chaturvedi, P., Armistead, D. M. et al. (1998) A novel immunophilin ligand: Distinct branching effects on dopaminergic neurons in culture and neurotrophic actions after oral administration in an animal model of Parkinson's disease, *Neurobiol Dis* 5:97–106.
33. Bavetta, S., Hamlyn, P. J., Burnstock, G., Lieberman, A. R. and Anderson, P. N. (1999) The effects of FK506 on dorsal column axons following spinal cord injury in adult rats: neuroprotection and local regeneration, *Exp Neurol* 158:382–93.
34. Gold, B. G., Katoh, K. and Storm-Dickerson, T. (1995) The immunosuppressant FK506 increases the rate of axonal regeneration in rat sciatic nerve, *J Neurosci* 15:7509–16.
35. Madsen, J. R., MacDonald, P., Irwin, N. et al. (1998) Tacrolimus (FK506) increases neuronal expression of GAP-43 and improves functional recovery after spinal cord injury in rats, *Exp Neurol* 154:673–83.
36. Hemenway, C. S. and Heitman, J. (1999) Calcineurin. Structure, function, and inhibition, *Cell Biochem Biophys* 30:115–51.
37. Nilsson, A., Skold, K., Sjogren, B. et al. (2007) Increased striatal mRNA and protein levels of the immunophilin FKBP-12 in experimental Parkinson's disease and identification of FKBP-12-binding proteins, *J Proteome Res* 6:3952–61.
38. Slemmon, J. R., Feng, B. and Erhardt, J. A. (2000) Small proteins that modulate calmodulin-dependent signal transduction: Effects of PEP-19, neuromodulin, and neurogranin on enzyme activation and cellular homeostasis, *Mol Neurobiol* 22:99–113.
39. Davison, A. N. (1987) Pathophysiology of ageing brain, *Gerontology* 33:129–35.
40. Rohner, T. C., Staab, D. and Stoeckli, M. (2005) MALDI mass spectrometric imaging of biological tissue sections, *Mech Ageing Dev* 126:177–85.
41. Crecelius, A. C., Cornett, D. S., Caprioli, R. M. et al. (2005) Three-dimensional visualization of protein expression in mouse brain structures using imaging mass spectrometry, *J Am Soc Mass Spectrom* 16:1093–9.
42. Sinha, T. K., Khatib-Shahidi, S., Yankeelov, T. E. et al. (2008) Integrating spatially resolved three-dimensional MALDI IMS with in vivo magnetic resonance imaging, *Nat Methods* 5:57–9.
43. Reyzer, M. L. and Caprioli, R. M. (2007) MALDI-MS-based imaging of small molecules and proteins in tissues, *Curr Opin Chem Biol* 11:29–35.
44. Cornett, D. S., Mobley, J. A., Dias, E. C. et al. (2006) A novel histology-directed strategy for MALDI-MS tissue profiling that improves throughput and cellular specificity in human breast cancer, *Mol Cell Proteomics* 5:1975–83.

45. Garden, R. W. and Sweedler, J. V. (2000) Heterogeneity within MALDI samples as revealed by mass spectrometric imaging, *Anal Chem* 72:30–6.

46. Schwartz, S. A., Weil, R. J., Thompson, R. C. et al. (2005) Proteomic-based prognosis of brain tumor patients using direct-tissue matrix-assisted laser desorption ionization mass spectrometry, *Cancer Res* 65:7674–81.

47. Gygi, S. P., Rist, B., Gerber, S. A. et al. (1999) Quantitative analysis of complex protein mixtures using isotope-coded affinity tags, *Nat Biotechnol* 17:994–9.

48. Ross, P. L., Huang, Y. N., Marchese, J. N. et al. (2004) Multiplexed protein quantitation in *Saccharomyces cerevisiae* using amine-reactive isobaric tagging reagents, *Mol Cell Proteomics* 3:1154–69.

49. Pflieger, D., Junger, M. A., Muller, M. et al. (2008) Quantitative proteomic analysis of protein complexes: Concurrent identification of interactors and their state of phosphory-lation, *Mol Cell Proteomics* 7:326–46.

50. Wiese, S., Reidegeld, K. A., Meyer, H. E. and Warscheid, B. (2007) Protein labeling by iTRAQ: A new tool for quantitative mass spectrometry in proteome research, *Proteomics* 7:340–50.

51. Emmert-Buck, M. R., Bonner, R. F., Smith, P. D. et al. (1996) Laser capture microdis-section, *Science* 274:998–1001.

52. Xu, B. J., Caprioli, R. M., Sanders, M. E. and Jensen, R. A. (2002) Direct analysis of laser capture microdissected cells by MALDI mass spectrometry, *J Am Soc Mass Spectrom* 13:1292–7.

53. Xu, B. J., Shyr, Y., Liang, X. et al. (2005) Proteomic patterns and prediction of glomeru-losclerosis and its mechanisms, *J Am Soc Nephrol* 16:2967–75.

54. Norris, J. L., Porter, N. A. and Caprioli, R. M. (2003) Mass spectrometry of intracellular and membrane proteins using cleavable detergents, *Anal Chem* 75:6642–7.

55. Chaurand, P., Schriver, K. E. and Caprioli, R. M. (2007) Instrument design and charac-terization for high resolution MALDI-MS imaging of tissue sections, *J Mass Spectrom* 42:476–89.

56. Mann, M. and Jensen, O. N. (2003) Proteomic analysis of post-translational modifica-tions, *Nat Biotechnol* 21:255–61.

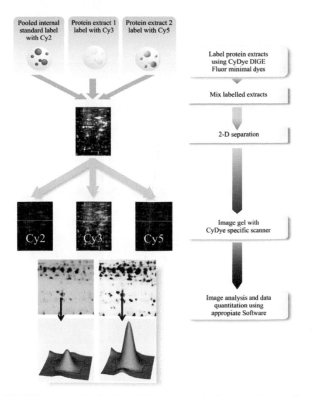

FIGURE 4.2 The 2D-DIGE concept. Proteins from different sources (such as a protein lysate from a control sample and a protein lysate from an analytical sample) are covalently labeled with two different CyDyes. Labeled protein lysates are mixed, and separated on a 2D-PAGE. The gels are scanned using an instrument capable of detecting each CyDye independently. The images are analyzed and spots corresponding to same proteins having different expression levels are determined. After 2D-DIGE, protein may be identified by mass spectrometry, or by any other suitable technique.

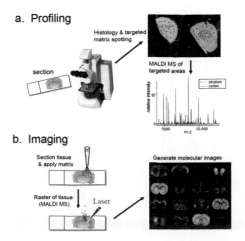

FIGURE 7.1 (a) MALDI imaging mass spectrometry. A thin tissue section is covered with MALDI matrix and irradiated by a laser in a raster across the tissue section, producing an ordered array of spectra. Each signal or peak in the spectra can be visualized by constructing a molecular density map, displaying the spatial distribution of peptides and proteins. (b) Region-specific mass spectra can be obtained through histology-directed tissue profiling MS, in this example, targeting the cortex (*blue*) and striatum (*red*) in a rat brain tissue section. Two average mass spectra display different molecular signatures in these regions.

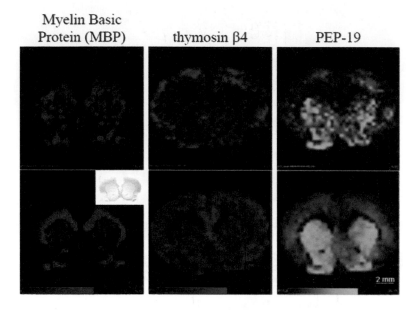

Myelin Basic Protein (MBP)   **thymosin β4**   **PEP-19**

2 mm

FIGURE 7.2 Baseline subtraction, noise removal, and total ion current normalization greatly improves image information. The three examples of myelin basic protein, thymosin beta-4, and PEP-19 were derived from a single IMS data file.

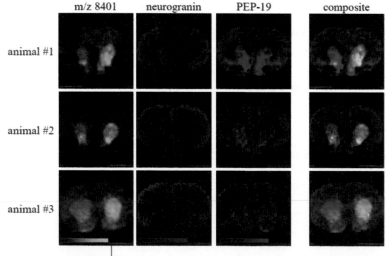

m/z 8401   neurogranin   PEP-19   composite

animal #1

animal #2

animal #3

parkinsonian side

FIGURE 7.3 Rat brain sections at the level of the striatum were analyzed by MALDI IMS, and the distribution of three proteins were visualized using three primary colors: green, red, and blue. The spatial (lateral) resolution of these images is 300 μm.

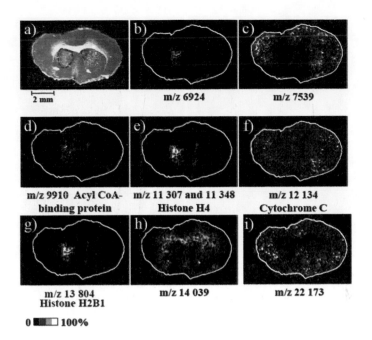

m/z 6924

m/z 7539

**m/z 9910 Acyl CoA-
binding protein**

**m/z 11 307 and 11 348
Histone H4**

**m/z 12 134
Cytochrome C**

**m/z 13 804
Histone H2B1**

m/z 14 039

m/z 22 173

0 ▮▮▮▯▯ 100%

FIGURE 7.5 (a) A coronal section of mouse brain with a glioma tumor in one hemisphere; for ease of viewing the tumor is outlined with a dotted line. IMS analysis was carried out at 110-μm resolution and (b–i) display several molecular images, including histones H4 and H2B1, which are typically present in actively growing tumors.

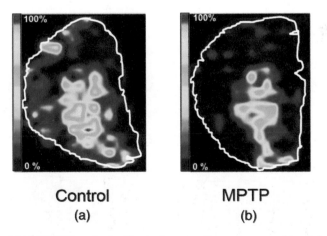

Control
(a)

MPTP
(b)

FIGURE 7.6 MALDI IMS analysis of brain tissue sections of an experimental model of PD. The relative ion density of PEP-19 from one control (a) and one MPTP-administered animal (b) displays a reduction of PEP-19 expression in the striatum (lateral resolution 280 μm). About 400 distinct mass signals were detected in the mass range of 2–30 kDa in this area of the tissue. *Source*: Modified with permission from Sköld, K., Svensson, M., Nilsson, A. et al. (2006) Decreased striatal levels of PEP-19 following MPTP lesion in the mouse, *J Proteome Res* 5:262–9.

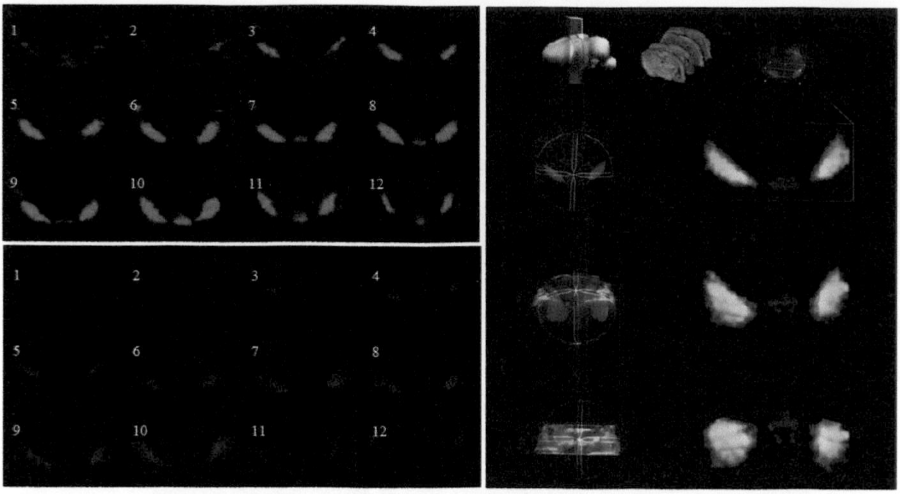

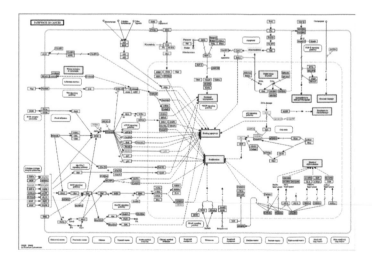

FIGURE 7.7 Assessment of distribution in 3D of PEP-19 (*green, top left panel*) and an unknown 7.4 kDa protein (*red, bottom left panel*) in the rat ventral midbrain. Tissue sections 12-μm thick were collected at 200-μm spacing throughout the rat ventral midbrain, and a series of sections covering the anterior-posterior length of the substantia nigra were imaged using MALDI IMS. After data acquisition, the matrix was washed off and sections were stained with cresyl violet (*right panel, top row*). Three-dimensional reconstructions of histologically stained sections provided the volume coordinates (*right panel, left column*) to insert the two MALDI IMS images into the same volume reconstruction (*right panel, right column*). Co-localization is suggested by the appearance of the yellow color.

FIGURE 9.3 Glioma on the global map of cancer pathways. The global map is a manually combined map of 14 existing cancer pathway maps in KEGG. Proteins are represented by boxes with the default coloring of green and those appearing in glioma are colored blue. Boxes marked with red are causative gene products for this cancer.

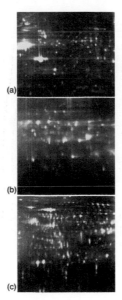

FIGURE 11.2 False-colored two-dimensional difference gel electrophoresis (2D-DIGE) images of the compara-tive analysis of area 17 protein expression patterns between kitten and adult cat. False-colored overlay images of a 2D-DIGE experiment with adult cat, propyl-Cy3-labeled visual area 17 proteins colored in red, and methyl-Cy5-labeled kitten proteins colored in green. Yellow spots contain proteins that have equal expression levels in the two samples, red spots containing proteins with a higher expression in adult cat visual area 17, and green spots with proteins more abundantly expressed in 30-day-old kitten primary visual cortex. Numbers indicate the spots with statistically significant fluorescence levels, further identified as (a) CRMP2, CRMP4; (b) Dyn I; and (c) Syt I using mass spectrometry. Panels (a) and (b) present protein samples separated according to pI and molecular weight over a pH range of pH 4 to 7 or 5.5–6.7. Panel (c) illustrates proteins separated according to hydrophobicity (HPLC pre-fractionation), pI, and molecular weight.

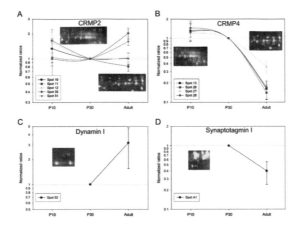

FIGURE 11.3 Detailed visualization of developmental expression differences for CRMP2, CRMP4, DynI, and Syt I as revealed by 2D-DIGE. Mean relative normalized spot volume ratios of postnatal day 10 (P10) or adult levels to postnatal day 30 (P30) protein levels of the protein spots corresponding to CRMP2 (A; five isoforms), CRMP4 (B; four isoforms), dynamin I (C) and synaptotagmin I (D). Error bars indicate standard deviation, while P30 fluorescence level ratios are arbitrarily fixed at 1. 2D-DIGE false-colored overlay images of the protein spots are shown in the insets, comparing P10 (*red*) and P30 (*green*) in left insets and P30 (*green*) and adult (*red*) in right insets. Different CRMP4 isoform spots show strikingly similar expression profiles; CRMP2 isoforms follow a different developmental route of expression. Spot numbers match with those in Figure 11.1.

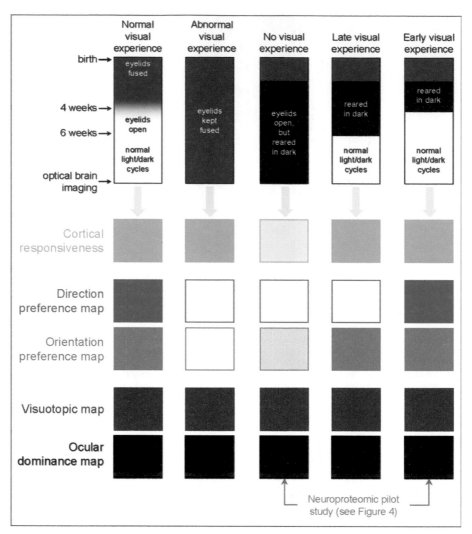

FIGURE 12.3 Illustration of experimental protocol for generating physiological phenotypes that vary in the development of functional maps in ferret visual cortex. Normal ferrets (*far left column*) are characterized by cortical maps of direction and orientation preference, visuotopy, and ocular dominance. The former two maps are "computational," representing the cortical distribution of orientation and direction preference computed by thalamocortical and intracortical circuits, and the latter two maps are "topological," representing the near neighbor relations established in the retina and conveyed to the visual cortex by input from the thalamic lateral geniculate nucleus. Rearing ferrets with abnormal visual experience (*second column from left*) prevents the development of the maps of direction and orientation preference (see Figure 12.2). Rearing animals without visual experience (*middle column*) reduces cortical responsiveness by about half and prevents the development of the map of direction preference, while the map of orientation preference is present but weak (saturation of color in corresponding blocks indicates status of map development). Providing late visual experience after early dark-rearing through the second week after eye-opening (*second column from right*) restores all visual cortical maps, except for the map of direction preference. The normal expression of all functional maps in ferrets that were dark-reared only until eye-opening (*far right column*) highlights the importance of early visual experience, especially for the development of the map of direction preference [see (57)].

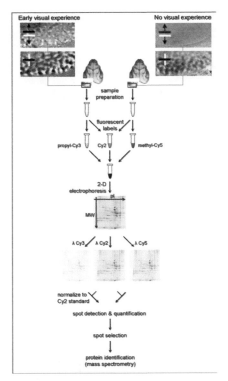

FIGURE 12.4 Schematic overview of 2D-DIGE/MS analysis of visual cortical tissue from juvenile ferrets with early visual experience (and cortical maps of direction preference) and without visual experience (lacking cortical maps of direction preference). See Chapters 3, 4, 5, and 6 for further methodological details.

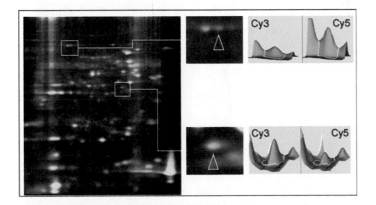

FIGURE 12.5 2D-DIGE analysis of visual cortical samples from normal and visually deprived ferrets. (*Left*) Combined image of 2-D gel generated by the Typhoon 9410 Variable Mode Imager; normal tissue is labeled with Cy3 (*green*) and deprived tissue is labeled with Cy5 (*red*). High magnification insets highlight spots (*arrowheads*) that are expressed more in the dark-reared visual cortex (*upper panel*) or the control cortex (*lower panel*); these two spots were analyzed quantitatively with DeCyder software (*right surface plots*).

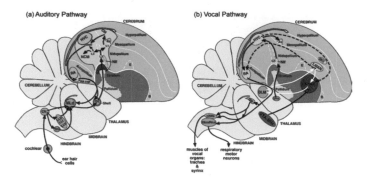

(a) Auditory Pathway   (b) Vocal Pathway

FIGURE 13.1 Diagram of auditory (a) and vocal (b) pathways of the songbird brain. Only the most prominent or most studied projections are indicated. For the auditory pathway, NCM actually lies in a parasagittal plane medial to that depicted and reciprocal connections between pallial areas are not indicated. For the vocal pathway, black arrows show connections of the nuclei (in *dark grey*) of the posterior vocal pathway, white arrows show connections of the nuclei (in *white*) of the anterior forebrain pathway, and dashed lines indicate connections between the two pathways. Only the lateral part of the anterior vocal pathway is shown, and the connection from Uva to HVC is not depicted. *Source*: Modified from Jarvis, E. D., Gunturkun, O., Bruce, L. et al. (2005) Avian brains and a new understanding of vertebrate brain evolution, *Nat Rev Neurosci* 6:151–9.

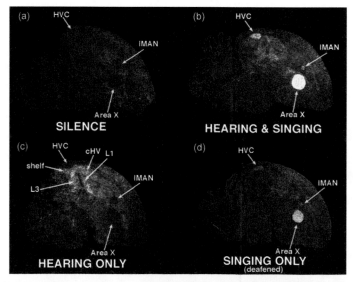

FIGURE 13.2 ZENK mRNA expression in canary brain. Shown are dark field views of cresyl violet stained (*red*) sagittal brain sections reacted in in situ hybridization experiments with a 35S-zenk riboprobe (*white grains*). One can see a subset of structures (*yellow arrows*), as outlined in Figure 13.1, which showed singing or hearing song-induced zenk expression. (a) Bird exposed to silence and that did not sing; (b) bird that sang 61 songs during a 30-minute period; (c) bird in an adjacent cage that did not sing during the same 30-minute period; and (d) deafened bird that sang 30 songs during 30 minutes. *Source*: Figure modified from Jarvis, E. D. and Nottebohm, F. (1997) Motor-driven gene expression, *Proc Natl Acad Sci U S A* 94:4097–4102.

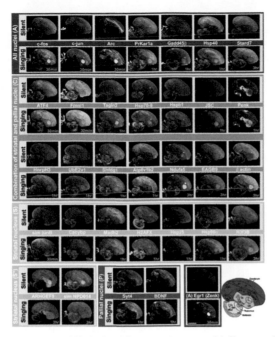

FIGURE 13.4 In situ hybridizations of 33 singing-driven genes in song nuclei. Shown are inverse images of auto-radiographs; white is mRNA expression. Images are ordered from top to bottom according to four overall expression patterns and from left to right in temporal order of peak expression (time shown in *lower right corner*). Zenk expression is shown to the left of the brain diagram (*bottom right*) for anatomical reference. *Abbreviations*: A, arcopallium; Av, avalanche; DM, dorsal medial nucleus; LX, lateral AreaX of the striatum; MO, oval nucleus of the mesopallium; N, nidopallium; NIf, interfacial nucleus of the nidopallium; P, pallidum; RA robust nucleus of the arcopallium; St, striatum. (Scale bar, 2 mm.). Gene abbreviations and further description in Figure 13.5. *Source*: Figure from Wada, K., Howard, J. T., McConnell, P. et al. (2006) A molecular neuroethological approach for identifying and characterizing a cascade of behaviorally regulated genes, *Proc Natl Acad Sci U S A* 103:15212–7, with permission.

FIGURE 13.5 Summary of singing-driven genes in songbird song nuclei. (a) Table of 36 singing-regulated genes discovered to date. Shown are the inferred cellular locations, molecular functions, and biological processes based on ontology definitions of homologous genes in other species. The list is organized according to cellular location (nucleus-to-extracellular space), the number of song nuclei showing regulated expression, and peak time (0.5–3 hours) of expression. Sim, similar to the named gene at 60%–74% protein identity. (b and c) Pie-chart quantifications of cellular location (b) and molecular function (c) of all genes (except the one gene [#36] of unknown function). (d) Percentage of genes regulated by singing in each song nucleus. Numbers in parentheses ( ) show number of genes for each nucleus. Only 34 genes are included in the calculation, as two of the remaining genes were not examined in all song nuclei. n.d., not determined.

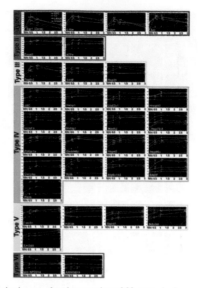

FIGURE 13.6 Time course of singing-regulated expression of 33 genes in four song nuclei. The genes are categorized into one of six types of temporal expression patterns by Wada et al. (2006) based on expression peak time (0.5, 1, or 3 hours) and expression profile. Expression levels were measured as pixel density on x-ray film minus background on the film. The * indicates significant differences at one or more time points relative to silent controls at the 0 hour (0 h) (ANOVA by post hoc probable least-squares difference test. *, $P < 0.05$; **, $P < 0.01$). *Source*: Figure from Wada, K., Howard, J. T., McConnell, P. et al. (2006) A molecular neuroethological approach for identifying and characterizing a cascade of behaviorally regulated genes, *Proc Natl Acad Sci U S A* 103:15212–7.

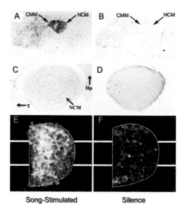

Song-Stimulated   Silence

FIGURE 14.3 Induction of zenk mRNA and protein levels in the songbird forebrain following exposure to song. A: In situ hybridization autoradiogram of a section of an adult male zebra finch exposed to 45 minutes of playbacks of conspecific songs. B: Unstimulated control. These sections correspond to parasagittal plane 250 μm lateral to the medial surface of the brain. C–D: ZENK expression, revealed by immunocytochemistry, in the caudomedial telencephalon. Notice the presence of numerous ZENK-labeled cell nuclei after song stimulation (C) in the caudomedial nidopallium (NCM) of a zebra finch from the hearing-only group. D: Unstimulated control. E: Density/intensity map of ZENK expression in the NCM of a canary resulting from presentation of a playback of conspecific songs. F: Unstimulated control. Colors and brightness correspond to relative density and intensity of labeling of ZENK positive cells. Anatomical abbreviations not mentioned above or in text: Hp, hippocampus; T, telencephalon. Sources: Adapted from Mello, C. V. and Pinaud, R. (2006) Immediate early gene regulation in the auditory system. In Immediate early genes in sensory processing, cognitive performance and neurological disorders, eds. Pinaud, R. and Tremere, L. A., 35–56 (New York: Springer-Verlag); Mello, C. V., Vicario, D. S. and Clayton, D. F. (1992) Song presentation induces gene expression in the songbird forebrain, *Proc Natl Acad Sci USA* 89:6818–22; and Mello, C. V. and Ribeiro, S. (1998) ZENK protein regulation by song in the brain of songbirds, *J Comp Neurol* 393:426–38.

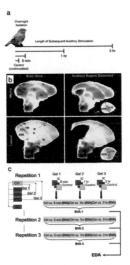

FIGURE 14.4 (a) Experimental approach used to investigate experience-regulated proteins in the songbird NCM. All animals were isolated overnight in soundproof chambers and randomly included into one of four experimental groups. Control animals were sacrificed after isolation without stimulation. The remaining animals were stimulated with playbacks of conspecific songs for 5 minutes, 1 hour, or 3 hours. (b) Representative brain sections were obtained with a vibratome for dissection of NCM. The red dotted lines indicate the approximate locations where sections were performed to isolate NCM. Note the darker color of the thalamo-recipient field L2 in these preparations, which significantly facilitates the isolation of NCM. (c) Representation of the 2D-DIGE experiments conducted by Pinaud et al. (94). Gels were run for individual comparisons across experimental groups (control vs. 5 minutes; control vs. 1 hour; and control vs. 3 hours). Each experiment was repeated three times with different animals. Intra-gel (DIA), inter-gel (BVA), as well as across-repetition (EDA) comparisons were conducted in a quantitative manner in order to identify statistically significant differences across these variables.

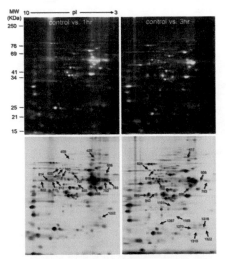

FIGURE 14.5 Top: Representative 2D-DIGE gels illustrating fractionated proteins from NCM in a comparison for the 1-hour (control vs. 1 hour; *left*) and 3-hour (control vs. 3 hours; *right*) groups. Control samples were labeled with Cy3 (*green*) while experimental samples were labeled with Cy5 (*red*). Internal control samples were labeled with Cy2 (blue, not shown). Bottom: Coomassie blue-stained gels illustrating differentially regulated spots (*arrows*) in the 1-hour (*left*) and 3-hour (*right*) conditions, as revealed by quantitative and statistical analyses with DeCyder software. All differentially regulated spots underwent protein fingerprinting by mass-spectrometry. *Source*: Figure from Pinaud, R., Osorio, C., Alzate, O. and Jarvis, E. D. (2008) Profiling of experience-regulated proteins in the songbird auditory forebrain using quantitative proteomics, *Eur J Neurosci* 27:1409–22.

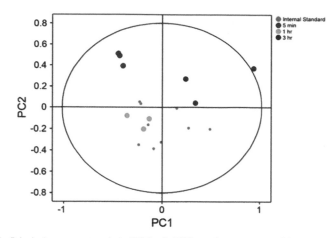

FIGURE 14.6 Principal component analysis (PCA) of ~2500 protein spots per condition reveals that the NCM's proteome is significantly different across experimental conditions. Independent gels for each of the conditions are not significantly different from each other. The circle represents the 95% confidence interval. Values that cluster in a quadrant are significantly different from values falling within other quadrants. See Chapter 4 description of internal standard (IS). *Source*: Figure from Pinaud, R., Osorio, C., Alzate, O. and Jarvis, E. D. (2008) Profiling of experience-regulated proteins in the songbird auditory forebrain using quantitative proteomics, *Eur J Neurosci* 27:1409–22.

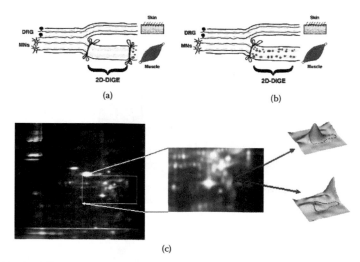

FIGURE 15.7 2D-DIGE results of individual terminal nerve branches of the rat femoral nerve. (a) and (b) Diagram of the two terminal nerve branches of the rat femoral nerve, showing the sensory cutaneous branch (continuing as the saphenous nerve and providing local cutaneous innervation) and the muscle branch to the quadriceps muscle. Some animals received two ligations: one just distal to the parent femoral nerve bifurcation and one just proximal to the quadriceps muscle (as shown in a). Other animals received a single ligation just distal to the parent femoral nerve bifurcation (as shown in B). Individual muscle branch samples were harvested 7 days post-ligation. Muscle-derived proteins (indicated by red dots in b) include trophomorphic factors. The translucent tube-like structures within the nerve represent Schwann cell tubes that remain following nerve degeneration. DRG = dorsal root ganglion, MNs = motor neurons. (c) Nerve samples were labeled with CyDyes, combined, and run on the same 2D-DIGE gel. The combined image of the gel is shown (all three CyDyes), with an enlarged view to the right. Red spots are proteins expressed more highly in single-ligated nerve samples. Spots that appear white are expressed equally in all three samples. Graphical images of spot intensities from the single-ligated nerve sample (*upper image*) and double-ligated nerve sample (*lower image*) illustrate an example of expressions differences for a candidate trophomorphic mediator. Spots showing large expression differences were subsequently selected for protein identification.

8 Protein Interaction Networks

Alexei Vazquez

CONTENTS

8.1 INTRODUCTION

Protein interactions are an important layer of connectivity between cell components and cell processes. They allow the formation of protein complexes and mediate post-translational protein modifications. Disruption of protein–protein interactions may result in disruption of the cell component or process to which they contribute, compromising the cell viability or even leading to cell death. The study of protein interactions and their function has recently jumped to an organism scale, following the development of high-throughput yeast–two-hybrid screens mapping protein–protein interactions (1–7) and affinity purification followed by mass spectrometry (MS) detecting protein complexes (8,9) (see Chapters 5, 6, and 7 for a thorough discussion of mass spectrometry and MS-based tissue imaging).

At the cellular level, protein interactions are represented by the *protein interaction network*, which is the union of all proteins and the interactions among them. The protein interaction network involves asking questions similar to those asked for specific protein interactions: how to map these interactions, which properties characterize them, how are these proteins connected to cell processes and functions, and what is the evolution of the protein network? These questions are now investigated at different levels, from specific protein–protein interactions, passing through protein complexes, signaling cascades, and other protein interaction network motifs, ending in the whole protein interaction network.

The role of the protein interaction network is beyond the network itself and it materializes through its interaction with other cellular networks. For example, the cell response to a change in environmental conditions requires the coordination between the signaling (a subnet of the protein interaction network), and the gene regulatory and metabolic networks. Consequently the study of protein interaction networks should be put in the context of the cell function. It is only in this way that the protein interaction network description can be transformed into better understanding and prediction of novel features and behavior. In this review I cover these different aspects of protein interaction networks. It is not intended to be an exhaustive review but is complemented by cited literature, including reviews of more specific topics.

8.2   NETWORK PROPERTIES

A graph is composed of a set of *nodes* and a set of *edges*, with *links* or connections among them. In the case of the protein interaction graph, the nodes represent proteins and the links represent the existence of protein–protein interactions. In addition to the existence of protein–protein interactions we consider some measure of *interaction strength*. The protein interaction graph together with this measure of strength constitutes the protein interaction network. Protein–protein interactions do not imply any particular direction; therefore, the protein interaction network is said to be undirected. There are, however, applications where directionality is important, e.g., transmission of information in signaling cascades. In the latter case the network is said to be directed. Furthermore, several proteins interact with themselves and in this case we speak of self-interactions or loops.

One of the first observations about protein interaction networks was the variability of *protein degree* (10,11), which is defined as the number of interacting partners of a protein. The protein interaction networks of most organisms are characterized by a wide degree of variation, from a majority of nodes participating on none or a few interactions to a few (yet significant) nodes participating in up to hundreds of interactions (2–7). For example, the gene products of the human genes frataxin (FXN), parkin (PARK2), huntingtin (HTT), and ataxin 1 (ATXN1) interact with about 3, 29, 60, and 159 other proteins, respectively (12).

The wide degree of variation can be characterized by the *degree of distribution*, defined as the probability p_k that a node has degree k (Figure 8.1A). It has been proposed that the degree of distribution follows a power-law $p_k = Ak^{-\gamma}$, where A is a constant and γ is a positive exponent (10) (Figure 8.1A). Besides being broad, this power law form implies that the degree of distribution has the same shape at degree k than at some rescaled degree ak, up to a constant factor ($p_{ak} = A'k^{-\gamma}$, where $A' = Aa^{-\gamma}$). Therefore, power laws are said to be homogeneous or "scale-free" laws, in the sense they exhibit the same shape at all scales. The scale-free property of the degree distribution motivated the introduction of the term *scale-free networks* (13), although strictly speaking we should say networks with a scale-free degree distribution. It is also worth mentioning that some studies report that not all protein interaction network datasets exhibit a power law degree distribution (14).

Beyond the precise form of the degree distribution, the degree heterogeneity is a characteristic feature of currently available maps of protein interaction networks. More important, several magnitudes exhibit a strong correlation with the degree. The first association came with the report of a positive correlation between the degree of a protein and its essentiality for cell viability (10). Further study on this direction indicates that conservation as well as essentiality are strongly correlated with the

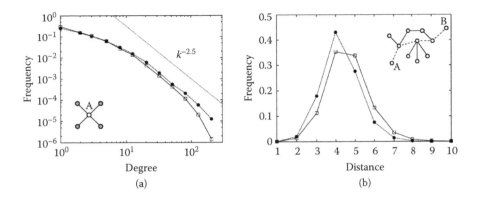

FIGURE 8.1   a: The *degree* of a *node* is given by the number of *links* between that node and other nodes in the network. For example, the inset shows a node A with degree four. The main panel shows the degree distribution of the human protein interaction network, as obtained from different sources indicated by the different symbols (described below). A power law decay with exponent 2.5 is shown as a reference (*dotted line*). b: The distance, or shortest past distance, between two nodes is given by the smallest number of links that need to be crossed to go from one node to the another, provided that there is at least one path between them. For example, the inset shows two nodes A and B at distance four. The main panel shows the distance distribution of the human protein interaction network, as obtained from different sources indicated by the different symbols. The empty squares are obtained from the union of literature curated (LC) data, and two recent high-throughput yeast–two-hybrid screens (5,6), as has been reported in reference (17). The solid circles were obtained using data reported in the Human Protein Reference Database (HPRD) (12).

degree (15). In the context of protein function inference algorithms based on protein interaction networks, it has been shown that the success rate increases with increasing the unclassified protein degree (16). Finally, virus proteins preferentially target host proteins with higher degree (17).

Protein interaction networks also exhibit the so-called *small-world* property (18), meaning that the average distance between two proteins in the protein interaction network is small (Figure 8.1B). The notion of small world does not require the existence of a high clustering coefficient (defined below), a confusion that appeared following the introduction of the small-world model (18). The degree of heterogeneity and the small-world property neither imply nor exclude each other. More important, the fact that protein interaction networks exhibit both a broad degree distribution and the small-world property results into several important consequences. Networks with these two properties are characterized, under general conditions, by a resilience against random perturbations (19–21), as first noticed by direct simulated experiments (10). This means that genetic mutations disrupting a protein or a protein–protein interaction do not affect the integrity of the network, the latter being measured by the existence of a large cluster of interconnected proteins (*giant component*), a manifestation of the small-world property. On the other hand, because of these two properties, protein interaction networks are fragile to disruption of high degree proteins, the *hubs* of the protein interaction network (19,20). Indeed, removal of only a few hubs is sufficient to fragment the giant component into small sub-networks (10), potentially disrupting the communication between different functional modules in the cell.

Although less known among the protein interaction network research community, the degree heterogeneity together with the small-world property carry as a consequence the fast and robust spread of information along the network. In the context of protein interaction networks, this would imply that the spread of information from one protein to almost all other proteins is not disrupted by random transmission failures between pairs of interacting proteins (22), and it takes place in a time scale of the order of the transmission time between interacting partners (23). The study of signaling cascades will certainly benefit from further study in this direction.

Modularity is the third characteristic feature of protein interaction networks (24–26), manifested through the existence of protein groups that are highly connected among them (the modules) yet with lesser connections between modules. The first indication of modularity came from the observation that protein interaction networks are characterized by a high average *clustering coefficient* (27), defined as the fraction of interacting partners of a protein that interact among themselves. This observation has been corroborated using other measures of modularity (24–26,28). Direct and comparative methods have been proposed to identify the network modules, also under the name of motifs or communities. Direct approaches aim to identify modules following their definition, i.e., protein groups which are highly connected among them yet with lesser connections with protein outside the group (25). Comparative approaches consist, instead, on identifying network modules, or motifs, that are overrepresented in the real network relative to a random network reference with the same degree of distribution (24,29).

The observation of all these network properties is conditioned by the quality and coverage of the available protein interaction network (30,31). For example, it has been shown that a broad degree distribution can result from a low coverage sampling of interactions in a network with a narrow degree distribution (31). However, this hypothesis has the caveat that, to observe a broad distribution up to degrees of the order of hundreds (Figure 8.1A), the original network should have an average degree of about a hundred interactions. Despite potential sampling biases, in the next section it becomes clear that the degree heterogeneity, and small-world and modularity properties are the "natural expectation" rather than the exception.

8.3   NETWORK EVOLUTION

Preferential attachment has been proposed as a general mechanism for network evolution (32). In essence it postulates that nodes acquire new interactions at a rate proportional to their current degree. The outcome is a network with a power law degree distribution, with an exponent that depends on the details of the model (33,34). In the context of protein interaction networks, the empirical data indicates that the number of interactions a protein gains during its evolution is proportional to its connectivity (35), implicating preferential attachment in the protein interaction network evolution. Yet, what are the molecular mechanisms behind the observed preferential attachment and other protein interaction network properties?

Current models of protein interaction network evolution are built on two well-accepted mechanisms of gene evolution, namely, *duplication* and *divergence*. There is substantial evidence of gene duplication from a single gene to the whole genome (36,37). The duplication of a gene results into two identical copies of that gene. Later their sequences diverge following mutations and selective pressures. In the classical model of gene duplication–divergence the evolution is thought to be asymmetrical (38), one copy retaining the functions of the ancestor while the other is either lost or develops a new functionality. Symmetric models of gene evolution have been proposed as well. In this case both copies diverge from the ancestor, resulting in a stochastic partition of the ancestor's functions and the development of new functionalities as well (39).

These models of gene evolution have been translated to the evolution of protein interactions (11,27,40,41) (Figure 8.2). Duplication of a protein (more precisely the gene encoding it) results in two protein duplicates, together with the interactions between the duplicates and the interacting partners of the ancestor protein. Following divergence some of these interactions may be lost. As for the divergence of gene duplicates, we can envision asymmetric (Figure 8.2A–C) and symmetric (Figure 8.2D–F) modes of protein interaction network evolution. In the asymmetric mode only one copy is allowed to lose interactions (40), while in the symmetric mode both copies are allowed to lose their interactions (27). In practice we would expect both modes to operate.

To understand the impact of these evolution mechanisms on the degree of distribution we should analyze both the dynamics of the protein duplicates degree and of the interacting partners. Because the chance that a duplication takes place among the interacting partners of a protein is proportional to the degree of that protein, duplication and divergence imply preferential attachment for the interacting partners of duplicate proteins (27,40). In contrast, the degree dynamics of the duplicate itself does not exhibit preferential attachment. When both contributions are put together we obtain a broad degree distribution that does not follow, however, a power law (27).

The networks obtained using these evolution rules exhibit the small-world property as well (42). Hence, duplication and divergence implies two of the key features observed in protein interaction networks: the degree heterogeneity and the small-world property. The evolution of self-interactions provides the additional ingredient necessary to understand the origin of a high clustering coefficient (27), modularity (28), and protein complex formation (43). Indeed, a model based on duplication–divergence and taking into account the existence of self-interactions provides a fit significantly better than a model with duplication divergence alone (44). A mathematical derivation of these results can be found in reference (45) and more rigorously in reference (46).

Taken together these results indicate that the properties of currently mapped protein interaction networks are in agreement with our expectation from a model based on three accepted molecular mechanisms of evolution: duplication, divergence, and the existence of self-interactions. The observed degrees of heterogeneity, small-world property, and modularity, rather than unusual, are the natural outcomes of these evolutionary mechanisms.

Variations in the number of hydrophobic residues on protein surfaces have been proposed as an alternative mechanism to explain the degree heterogeneity of protein

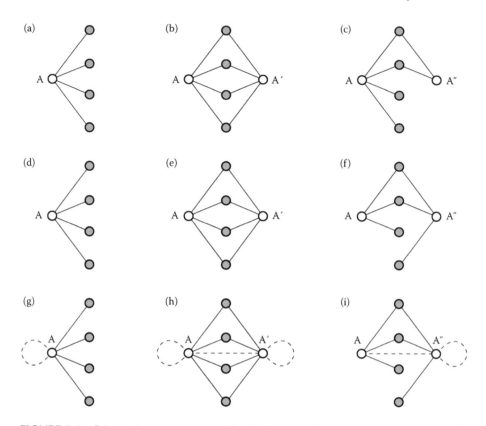

FIGURE 8.2 Schematic representation of the duplication–divergence mechanism. a: Protein A (*empty circle*) and its interacting partners (*filled circles*). b: The protein A is duplicated, generating a new copy A′ and new interactions between A′ and the interacting partners of A. c: Following mutation and selection some of the interactions are lost. a–c: Asymmetric mode of evolution, only one duplicate loses interactions. d–f: Symmetric model of evolution, both duplicates may lose interactions. g–i: Contribution of self-interactions. When the original protein A forms a homodimer, its duplication may result in an interaction between the duplicates.

interaction networks (47). Indeed, there is a positive correlation between the degree of a protein and the magnitude of its structural disorder (48). There is no indication, however, that the structural disorder and the duplication-divergence mechanisms are antagonists. Their interrelation should be the subject of further study. Finally, it is worth pointing out that connectivity is not the only constraint in protein evolution. Other factors such as protein structure, genomic position of the encoding genes, and their expression patterns are important as well (49).

8.4   FUNCTION PREDICTION

Protein interactions have been traditionally used to uncover novel proteins and their functions. The snowball approach has been extensively used to find interacting partners of a protein of known function, followed by an investigation of the functions of the interacting partners and further continuing the process by looking at the next layer of interacting partners. These methods are inspired in the *guilty-by-association* principle: if protein A interacts with protein B performing function {*f*}, it is likely that protein B performs function {*f*} as well. With the development of high-throughput techniques to map protein interaction networks, we are able to apply the guilty-by-association principle at the network scale (50).

A common problem is to make putative predictions for unclassified proteins given the functions of classified proteins and the protein interaction network of the organism of interest (Figure 8.3). A simple approach is to use a *majority rule* (51), assigning to each unclassified protein the most abundant function among its interacting partners (Figure 8.3). This method does not allow, however, making predictions for unclassified proteins whose interacting partners are unclassified, and it does not take advantage of the interaction among unclassified proteins. The simplest extension is a recursive application of the majority rule. A first round of the majority rule results in putative classifications for all unclassified proteins with at least one classified interacting partner. After assuming the latter proteins as classified we can proceed to

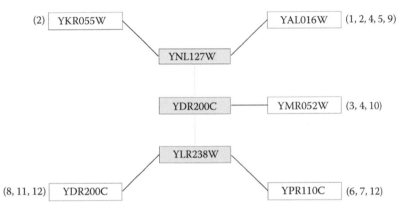

FIGURE 8.3 Protein function prediction. Subnet of the protein interaction network of the yeast *Saccharomyces cerevisiae*. Proteins in gray boxes are unclassified (unknown function) while the others are classified proteins with their functions indicated within the brackets, labeled according to the following: 1—cell growth; 2—budding, cell polarity, and filament formation; 3—pheromone response, mating-type determination, sex-specific proteins; 4—cell cycle check point proteins; 5—cytokinesis; 6—rRNA synthesis; 7—tRNA synthesis; 8—transcriptional control; 9—other transcription activities; 10—other pheromone response activities; 11—stress response; and 12—nuclear organization. Applying the majority rule method we obtain the putative classifications YNL127W (2), YDR200C (3,4,10) and YLR238W (12). Furthermore, iterating once more the majority rule by taking into account the interactions of the three unclassified proteins, we obtain the following classification: YNL127W (2,4), YDR200C (3,4,10) and YLR238W (12). In this way we find another putative function for YNL127W. This is the spirit of global optimization methods, taking advantage of the prediction we are making for proteins of unknown function. *Source*: Adapted from Vazquez, A., Flammini, A., Maritan, A. and Vespignani, A. (2002) Modeling of protein interaction networks, *ComPlexUs* 1:38–44.

a second round of the majority rule and assign putative classifications to unclassified proteins at a distance two from the starting classified proteins. This recursive approach is then continued until a prediction has been made for all unclassified proteins with a path to at least one starting classified protein. Another strategy consists of extending the neighborhood of unclassified proteins beyond proteins at distance one (its interacting partners), to include proteins at distance two and higher. The majority rule is then applied within this extended neighborhood, without (52) or with (53) different weights according to the distance from the target unclassified protein.

The majority rule can also be interpreted as a local maximization of the number of interacting proteins sharing at least a common function, where by *local* we mean an unclassified protein and its interacting partners. This new interpretation provides a hint for an intuitive extension of the majority rule to the whole protein interaction network. *Global optimization* assigns a function to all unclassified proteins such that the total number of proteins sharing at least one common function is maximized (16) (Figure 8.3). This method significantly increases the success rate over the majority rule approach (16) and can be efficiently solved using a belief propagation algorithm (54).

A completely different approach attempts to identify modules within the PIN and then assign a putative function to the module, consequently to all proteins within the module. This direction is complemented by several available algorithms to find modules within a network [see reference (55) for a review].

8.5   SUMMARY AND OUTLOOK

The data currently available indicate that protein interaction networks are characterized by degree heterogeneity, the small-world property, and modularity. The first two properties imply the resilience of the network to random disruptions of proteins or interactions due to mutations, and the fragility to the disruption of hub proteins. The modularity is thought to be both cause and effect of tinkering evolution. The modules represent protein complexes, signaling cascades, and other cell components that evolve partially independent.

Protein interaction network evolution models based on the concept of gene duplication and divergence provide a convincing explanation to the observed network properties. Thus, rather than unexpected, the three properties mentioned above are the natural outcome of evolution. This does not exclude, however, the potential existence of false positive and false negative interactions in current network maps. It only means that from the statistical point of view we do not expect any significant deviation.

Beyond the descriptive analysis, protein interaction networks are increasingly used as scaffolds of statistical inference. Algorithms for protein function prediction based on the correlation between protein–protein interaction and protein function have proven to be successful. The integration of protein interaction network data with other high-throughput biological data such as gene expression have been shown to increase significantly the positive predictive power and reduce false predictions (56–58). Future developments in this area will certainly increase our predictive power.

The contribution of genes to malfunction or disease can be inferred using protein interaction networks. In the context of neurodegenerative diseases, protein interaction networks have been used to infer candidate genes for inherited ataxias (59), Huntington's disease (60,61), and a compendium of neurodegenerative disorders (62) (see Chapter 9 for a detailed discussion). With the increasing coverage of the protein interaction network in humans (5,6) these approaches will play an important role in the future to uncover new genes associated with neurodegenerative disorders and other human diseases.

REFERENCES

1. Fields, S. and Song, O. (1989) A novel genetic system to detect protein-protein interactions, *Nature* 340:245–6.
2. Giot, L., Bader, J. S., Brouwer, C. et al. (2003) A protein interaction map of *Drosophila melanogaster*, *Science* 302:1727–36.
3. Ito, T., Tashiro, K., Muta, S. et al. (2000) Toward a protein-protein interaction map of the budding yeast: A comprehensive system to examine two-hybrid interactions in all possible combinations between the yeast proteins, *Proc Natl Acad Sci U S A* 97:1143–7.
4. Li, S., Armstrong, C. M., Bertin, N. et al. (2004) A map of the interactome network of the metazoan *C. elegans*, *Science* 303:540–3.
5. Rual, J. F., Venkatesan, K., Hao, T. et al. (2005) Towards a proteome-scale map of the human protein-protein interaction network, *Nature* 437:1173–8.
6. Stelzl, U., Worm, U., Lalowski, M. et al. (2005) A human protein-protein interaction network: A resource for annotating the proteome, *Cell* 122:957–68.
7. Uetz, P., Giot, L., Cagney, G. et al. (2000) A comprehensive analysis of protein-protein interactions in *Saccharomyces cerevisiae*, *Nature* 403:623–7.
8. Gavin, A. C., Bosche, M., Krause, R. et al. (2002) Functional organization of the yeast proteome by systematic analysis of protein complexes, *Nature* 415:141–7.
9. Ho, Y., Gruhler, A., Heilbut, A. et al. (2002) Systematic identification of protein complexes in *Saccharomyces cerevisiae* by mass spectrometry, *Nature* 415:180–3.
10. Jeong, H., Mason, S. P., Barabasi, A. L. and Oltvai, Z. N. (2001) Lethality and centrality in protein networks, *Nature* 411:41–2.
11. Wagner, A. (2001) The yeast protein interaction network evolves rapidly and contains few redundant duplicate genes, *Mol Biol Evol* 18:1283–92.
12. Peri, S., Navarro, J. D., Amanchy, R. et al. (2003) Development of human protein reference database as an initial platform for approaching systems biology in humans, *Genome Res* 13:2363–71.
13. Barabasi, A. L. and Albert, R. (1999) Emergence of scaling in random networks, *Science* 286:509–12.
14. Tanaka, R., Yi, T. M. and Doyle, J. (2005) Some protein interaction data do not exhibit power law statistics, *FEBS Lett* 579:5140–4.
15. Wuchty, S. (2004) Evolution and topology in the yeast protein interaction network, *Genome Res* 14:1310–4.
16. Vazquez, A., Flammini, A., Maritan, A. and Vespignani, A. (2003) Global protein function prediction from protein-protein interaction networks, *Nat Biotechnol* 21:697–700.
17. Calderwood, M. A., Venkatesan, K., Xing, L. et al. (2007) Epstein-Barr virus and virus human protein interaction maps, *Proc Natl Acad Sci U S A* 104:7606–11.
18. Watts, D. J. and Strogatz, S. H. (1998) Collective dynamics of 'small-world' networks, *Nature* 393:440–2.

19. Callaway, D. S., Newman, M. E., Strogatz, S. H. and Watts, D. J. (2000) Network robustness and fragility: Percolation on random graphs, *Phys Rev Lett* 85:5468–71.
20. Cohen, R., Erez, K., ben-Avraham, D. and Havlin, S. (2000) Resilience of the internet to random breakdowns, *Phys Rev Lett* 85:4626–8.
21. Molloy, M. and Reed, B. (1995) A critical point for random graphs with a given degree sequence, *Random Structures and Algorithms* 6:161–79.
22. Pastor-Satorras, R. and Vespignani, A. (2001) Epidemic spreading in scale-free networks, *Phys Rev Lett* 86:3200–3.
23. Vazquez, A. (2006) Polynomial growth in branching processes with diverging reproductive number, *Phys Rev Lett* 96:038702.
24. Milo, R., Itzkovitz, S., Kashtan, N. et al. (2004) Superfamilies of evolved and designed networks, *Science* 303:1538–42.
25. Palla, G., Derenyi, I., Farkas, I. and Vicsek, T. (2005) Uncovering the overlapping community structure of complex networks in nature and society, *Nature* 435:814–8.
26. Yook, S. H., Oltvai, Z. N. and Barabasi, A. L. (2004) Functional and topological characterization of protein interaction networks, *Proteomics* 4:928–42.
27. Vazquez, A., Flammini, A., Maritan, A. and Vespignani, A. (2002) Modeling of protein interaction networks, *ComPlexUs* 1:38–44.
28. Sole, R. V. and Valverde, S. (2008) Spontaneous emergence of modularity in cellular networks, *J R Soc Interface* 5:129–33.
29. Maslov, S. and Sneppen, K. (2002) Specificity and stability in topology of protein networks, *Science* 296:910–3.
30. Gentleman, R. and Huber, W. (2007) Making the most of high-throughput protein-interaction data, *Genome Biol* 8:112.
31. Han, J. D., Dupuy, D., Bertin, N., Cusick, M. E. and Vidal, M. (2005) Effect of sampling on topology predictions of protein-protein interaction networks, *Nat Biotechnol* 23:839–44.
32. Albert, R. and Barabasi, A. L. (2002) Statistical mechanics of complex networks, *Rev Mod Phys* 74:47–97.
33. Dorogovtsev, S. N. and Mendes, J. F. F. (2002) Evolution of networks, *Adv Phys* 51:1079–1187.
34. Krapivsky, P. L., Redner, S. and Leyvraz, F. (2000) Connectivity of growing random networks, *Phys Rev Lett* 85:4629–33.
35. Eisenberg, E. and Levanon, E. (2003) Preferential attachment in the protein network evolution, *Phys Rev Lett* 91:138701.
36. Kellis, M., Birren, B. W. and Lander, E. S. (2004) Proof and evolutionary analysis of ancient genome duplication in the yeast *Saccharomyces cerevisiae*, *Nature* 428:617–24.
37. Taylor, J. S. and Raes, J. (2004) Duplication and divergence: The evolution of new genes and old ideas, *Annu Rev Genet* 38:615–43.
38. Ohono, S. (1970) *Evolution by gene duplication* (Berlin: Springer-Verlag).
39. Force, A., Lynch, M., Pickett, F. B. et al. (1999) Preservation of duplicate genes by complementary, degenerative mutations, *Genetics* 151:1531–45.
40. Sole, R. V., Pastor-Satorras, R., Smith, E. D. and Kepler, T. (2002) A model of large-scale proteome evolution, *Adv Comp Syst* 5:43–54.
41. Wagner, A. (2003) How the global structure of protein interaction networks evolves, *Proc Biol Sci* 270:457–66.
42. Pastor-Satorras, R., Smith, E. and Sole, R. V. (2003) Evolving protein interaction networks through gene duplication, *J Theor Biol* 222:199–210.
43. Pereira-Leal, J. B., Levy, E. D., Kamp, C. and Teichmann, S. A. (2007) Evolution of protein complexes by duplication of homomeric interactions, *Genome Biol* 8:R51.

44. Middendorf, M., Ziv, E. and Wiggins, C. H. (2005) Inferring network mechanisms: The *Drosophila melanogaster* protein interaction network, *Proc Natl Acad Sci U S A* 102:3192–7.

45. Vazquez, A. (2003) Growing networks with local rules: preferential attachment, clustering hierarchy and degree correlations, *Phys Rev E* 67:056104.

46. Chung, F., Lu, L., Dewey, T. G. and Galas, D. J. (2003) Duplication models for biological networks, *J Comput Biol* 10:677–87.

47. Deeds, E. J., Ashenberg, O. and Shakhnovich, E. I. (2006) A simple physical model for scaling in protein-protein interaction networks, *Proc Natl Acad Sci U S A* 103:311–6.

48. Haynes, C., Oldfield, C. J., Ji, F. et al. (2006) Intrinsic disorder is a common feature of hub proteins from four eukaryotic interactomes, *PLoS Comput Biol* 2:e100.

49. Pal, C., Papp, B. and Lercher, M. J. (2006) An integrated view of protein evolution, *Nat Rev Genet* 7:337–48.

50. Mayer, M. L. and Hieter, P. (2000) Protein networks-built by association, *Nat Biotechnol* 18:1242–3.

51. Schwikowski, B., Uetz, P. and Fields, S. (2000) A network of protein-protein interactions in yeast, *Nat Biotechnol* 18:1257–61.

52. Hishigaki, H., Nakai, K., Ono, T., Tanigami, A. and Takagi, T. (2001) Assessment of prediction accuracy of protein function from protein-protein interaction data, *Yeast* 18:523–31.

53. Chua, H. N., Sung, W. K. and Wong, L. (2006) Exploiting indirect neighbours and topological weight to predict protein function from protein-protein interactions, *Bioinformatics* 22:1623–30.

54. Leone, M. and Pagnani, A. (2005) Predicting protein functions with message passing algorithms, *Bioinformatics* 21:239–47.

55. Sharan, R., Ulitsky, I. and Shamir, R. (2007) Network-based prediction of protein function, *Mol Syst Biol* 3:88.

56. Deng, M., Tu, Z., Sun, F. and Chen, T. (2004) Mapping gene ontology to proteins based on protein-protein interaction data, *Bioinformatics* 20:895–902.

57. Joshi, T., Chen, Y., Becker, J. M., Alexandrov, N. and Xu, D. (2004) Genome-scale gene function prediction using multiple sources of high-throughput data in yeast *Saccharomyces cerevisiae*, *Omics* 8:322–33.

58. Marcotte, E. M., Pellegrini, M., Thompson, M. J., Yeates, T. O. and Eisenberg, D. (1999) A combined algorithm for genome-wide prediction of protein function, *Nature* 402:83–6.

59. Lim, J., Hao, T., Shaw, C. et al. (2006) A protein-protein interaction network for human inherited ataxias and disorders of Purkinje cell degeneration, *Cell* 125:801–14.

60. Giorgini, F. and Muchowski, P. J. (2005) Connecting the dots in Huntington's disease with protein interaction networks, *Genome Biol* 6:210.

61. Goehler, H., Lalowski, M., Stelzl, U. et al. (2004) A protein interaction network links GIT1, an enhancer of huntingtin aggregation, to Huntington's disease, *Mol Cell* 15:853–65.

62. Limviphuvadh, V., Tanaka, S., Goto, S., Ueda, K. and Kanehisa, M. (2007) The commonality of protein interaction networks determined in neurodegenerative disorders (NDDs), *Bioinformatics* 23:2129–38.

9 Knowledge-Based Analysis of Protein Interaction Networks in Neurodegenerative Diseases

Minoru Kanehisa, Vachiranee Limviphuvadh, and Mao Tanabe

CONTENTS

9.1 INTRODUCTION

The large-scale datasets generated by gene sequencing, proteomics, and other high-throughput experimental technologies are the bases for understanding life as a molecular system and for developing medical, industrial, and other practical applications. In order to facilitate bioinformatics analysis of such large-scale datasets, it is essential to organize our knowledge on higher levels of systemic functions in a computable form, so that it can be used as a reference for inferring molecular systems from the information contained in the building blocks. Thus, we have been developing the KEGG (Kyoto Encyclopedia of Genes and Genomes) database (http://www.genome.jp/kegg/), an integrated resource of about 20 databases (1). The main component is the KEGG PATHWAY database, consisting of manually drawn graphical diagrams of molecular networks, called *pathway maps*, and representing various cellular processes and organism behaviors. KEGG PATHWAY is a reference database for pathway mapping, which is the process to match, for example, a genomic or transcriptomic content of genes against KEGG reference pathway maps to infer systemic functions of the cell or the organism.

As part of the KEGG PATHWAY database, we organize disease pathway maps representing our knowledge of causative genes and molecular networks related to them for human diseases, including cancers, immune disorders, neurodegenerative diseases, metabolic disorders, and infectious diseases. Here we focus on neurodegenerative diseases, which were among the first to be made available on the KEGG PATHWAY database. A diverse range of neurodegenerative diseases is commonly characterized by the accumulation of abnormal protein aggregates. Causative genes, including those that produce abnormal proteins, have been identified in various neurodegenerative diseases. The current information is not sufficient to find common molecular mechanisms of the diseases. In this chapter we first present an overview of KEGG, including the KEGG DISEASE and KEGG DRUG databases, and describe the KEGG PATHWAY maps for six neurodegenerative diseases: Alzheimer's disease (AD), Parkinson's disease (PD), amyotrophic lateral sclerosis (ALS), Huntington's disease (HD), dentatorubropallidoluysian atrophy (DRPLA), and prion diseases (PRION). We then present bioinformatics analysis to combine and expand these pathway maps toward identification of common proteins and common interactions, which may lead to a better understanding of common molecular pathogenic mechanisms (2).

9.2   OVERVIEW OF KEGG

9.2.1   KNOWLEDGE REPRESENTATION

KEGG (the Kyoto Encyclopedia of Genes and Genomes) is a biological systems database that integrates genomic, chemical, and systemic functional information for cells and organisms (1). Table 9.1 shows 12 core databases of KEGG, which are all manually curated. Our knowledge of higher level systemic functions is represented in two ways, *pathway maps* and *molecular lists*, and then stored in KEGG PATHWAY, and in the other three databases in the systems information category of Table 9.1. The KEGG pathway maps are manually drawn graphical diagrams representing networks (graphs) of molecular interactions and reactions for metabolism, signal transduction, other cellular processes and organism behaviors—including human diseases. Certain categories of pathway maps are hierarchically organized and/or presented as large global maps that can be manipulated by the "KEGG Atlas viewer" with zooming and navigation capabilities.

TABLE 9.1
Core Databases of KEGG

Category	Database	Content	Identifier
Systems information	KEGG PATHWAY	Pathway maps	map number
	KEGG BRITE	Functional hierarchies	br number
	KEGG MODULE	Pathway modules	M number
	KEGG DISEASE	Diseases	H number
Genomic information	KEGG ORTHOLOGY	KEGG orthology (KO) groups	K number
	KEGG GENES	Genes in complete genomes	locus_tag
	KEGG GENOME	Organisms with complete genomes	three-letter code
Chemical information	KEGG COMPOUND	Metabolites and other chemical compounds	C number
	KEGG GLYCAN	Glycans	G number
	KEGG DRUG	Drugs	D number
	KEGG REACTION	Enzymatic reactions	R number
	KEGG ENZYME	Enzyme nomenclature	EC number

Pathway maps present detailed pictures of molecular networks. Often our knowledge is too fragmentary to be represented as pathway maps. Molecular lists are less detailed but more general representations. KEGG BRITE is an ontology database where molecular lists are hierarchically categorized, representing our knowledge on protein families, chemical compound families, and drug classifications, among others. KEGG MODULE is a supplement to KEGG PATHWAY indicating tighter functional units of pathways and complexes, represented simply as lists of molecules without specifying connection patterns. For human diseases our knowledge on genetic and molecular factors is organized in KEGG DISEASE. More detailed molecular mechanisms, whenever known, are represented as disease pathway maps in KEGG PATHWAY. Each KEGG DISEASE entry is characterized by a list of known causative genes and other lists of molecules such as environmental factors, diagnostic markers, therapeutic drugs, etc. These lists are prepared such that they contain aspects of molecular systems behind the diseases, which may be used as a starting point of bioinformatics analysis for integration with other data and knowledge.

9.2.2 Mapping Procedures

KEGG is widely used as a reference database for biological interpretation of large-scale datasets. This can be achieved by three types of mapping procedures for genes, proteins, and/or chemical substances against the four databases mentioned above (Figure 9.1). First, KEGG pathway mapping is used to map a genomic or transcriptomic content of genes or a metabolomic content of chemical compounds, for instance, to KEGG reference pathway maps. This mapping is used to infer systemic functional information. In general, the pathway mapping is a process to infer

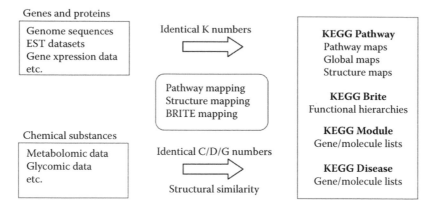

FIGURE 9.1   KEGG pathway mapping and other mapping procedures.

molecular networks with or without graphical diagrams. The mapping against the molecular lists of KEGG MODULE and KEGG DISEASE may also be called pathway mapping. Second, BRITE mapping enables similar inference by functional enrichment of BRITE hierarchies (ontologies). In addition, BRITE mapping enables understanding of relationships between genomic and chemical data, such as possible ligand–protein interactions or drug–target relationships, based on the interrelated BRITE hierarchies. Third, chemical structure mapping is a variant of KEGG pathway mapping against a special class of pathway maps, where biosynthetic pathways are represented as the chemical structures of synthesized molecules. This enables linking the repertoire of genes in the genome or the transcriptome to the diversity of chemical structures for specific classes of endogenous chemical substances, such as glycans, lipids, and plant secondary metabolites (3–5).

The three types of mapping procedures are actually performed as a set of operations between molecular lists, one list given as a query dataset and another corresponding to each of the database entries: pathway maps, BRITE hierarchies, and module/disease entries. The matching is based on the identifiers of KEGG pathway nodes and BRITE hierarchy nodes, namely, K numbers for genes and proteins and C/D/G numbers for chemical substances (Table 9.1; Figure 9.1). The matching C/D/G numbers imply identical chemical structures for chemical compounds, drugs, or glycans. Thus, computational tools have been developed to compare molecular lists based on chemical structural similarities, in order to identify groups of related chemical substances (6,7). For genes and proteins the K numbers represent ortholog groups in the KEGG Orthology system described below. In other words, sequence variabilities of orthologous genes and proteins among different organisms are already incorporated in the K numbers.

9.2.3   KEGG ORTHOLOGY SYSTEM

The KEGG Orthology (KO) system is the basis for representing systems information (Table 9.1) by grouping orthologous genes in different organisms. Orthologous genes or proteins are supposed to have the same biological function, which is defined here in the context of KEGG pathways or BRITE hierarchies. The KEGG pathway maps and the BRITE functional hierarchies are created in terms of the orthologs, rather than individual genes or proteins, to represent biological systems across organisms or organism groups. The K number, which is the entry identifier of the KO database, represents an ortholog group corresponding to a node of the KEGG pathway or a bottom leaf of the BRITE hierarchy. Thus, once genes in a genome are assigned to ortholog groups and given the K numbers, organism-specific pathways and organism-specific hierarchies can be computationally generated through the mapping procedures described above. For the organisms stored in KEGG (completely sequenced genomes), genes are annotated with the K numbers by using both manual and computational procedures. For other organisms, computational tools can be used to automatically assign ortholog groups (K numbers) based on genome sequence data or EST datasets (8).

9.2.4   DISEASE AND DRUG INFORMATION RESOURCES

Cells and organisms are biological systems consisting of multiple molecular building blocks with complex wiring diagrams. A disease should be considered as a perturbed state of the biological system (or a perturbation in the graph map), and a drug is a perturbant bringing the biological system back to the stable state. Thus, in order to understand systemic behaviors of cells and organisms, including human diseases, the molecular wiring diagrams must be extended to molecules in the environment. We are developing knowledge bases and computational methods for integrated analysis of genomic and chemical information, such as genetic and environmental factors of common diseases, drug targets and drug leads, enzyme-associated genes, and xenobiotic compounds in microbial biodegradation pathways. Here we briefly summarize the features of the KEGG DISEASE and KEGG DRUG databases (Table 9.2).

KEGG DRUG is a chemical-structure-based information resource for all approved drugs in Japan and the U.S.A. Each chemical structure is identified by the D number, and is associated with generic names, trade names, efficacy, target information, and

---

TABLE 9.2
Disease and Drug Information Resources in KEGG

	KEGG Disease	KEGG Drug
URL	http://www.genome.jp/kegg/disease/	http://www.genome.jp/kegg/drug/
Content	Lists of disease genes and molecular factors	Chemical-structure-based collection of all approved drugs in Japan and the U.S.A.
Pathway	KEGG pathway maps for human diseases	KEGG DRUG structure maps for drug development
BRITE hierarchy	Disease classifications including: Pathogens and infectious diseases Human diseases ICD-10 disease classification	Drug classifications including: Therapeutic category of drugs (Japan) USP drug classification (USA) ATC classification (WHO) TCM drugs (Japan)

drug classifications. Sometimes these data are linked to the KEGG DRUG structure map. The generic names are standardized names supposed to correspond to unique chemical structures, but the correspondences can be different in different countries or organizations. KEGG DRUG incorporates variations among Japan, the U.S.A., and selected EU countries. Trade names are linked, whenever available, to package insert (labels) information in outside resources. The drug classifications are part of the KEGG BRITE functional hierarchies, including the therapeutic category of drugs in Japan and the Anatomical Therapeutic Chemical (ATC) classification by the World Health Organization. The target information is presented in the context of KEGG pathway maps, enabling integrated analysis of genomic and chemical information. The KEGG DRUG structure maps are manually drawn graphical maps representing experts' knowledge on drug developments, including chronology of chemical structure transformation patterns, target-based structure classifications, and skeleton-based structure classifications.

KEGG DISEASE is designed for in silico analysis of molecular networks that are associated with human diseases. Each entry consists of lists of genes and molecules that are highly integrated with other KEGG resources including the PATHWAY, DRUG, GENES, and BRITE databases. At the time of this writing KEGG DISEASE is still a preliminary database, except for cancers that are relatively well organized. There are 55 entries for cancers and 14 cancer pathway maps available. In addition, a global map for the pathways in cancer has been created by manually combining 14 pathway maps, for common signaling pathways involved in different types of cancers. Figure 9.2 illustrates an example of cancer in the nervous system: glioma with known disease genes, markers, and drugs. The figure also illustrates the relationships among DISEASE, DRUG, and PATHWAY entries.

Figure 9.3 is a representation of this particular type of cancer in the global cancer map, which also indicates a result of KEGG pathway mapping. The KEGG reference pathway map is manually drawn to indicate molecular interaction/reaction networks, where gene products (proteins) and chemical substances (compounds, glycans, and drugs) are denoted by boxes and circles, respectively, and identified by the K numbers and the C/G/D numbers, respectively. The organism-specific pathways are computationally generated by converting the K numbers to corresponding gene identifiers in specific organisms. Here we prepared a list of genes involved in glioma using the human gene identifiers, and mapped against the reference pathway of the global cancer map. This type of analysis reveals overall features and relationships among different cancers.

9.2.5 NEURODEGENERATIVE DISEASES IN KEGG

The pathway maps for neurodegenerative diseases were created before the cancer maps and they are currently under revision. Here we use the original version of the pathway maps for six neurodegenerative diseases: Alzheimer's disease (AD), Parkinson's disease (PD), amyotrophic lateral sclerosis (ALS), Huntington's disease (HD), dentatorubropallidoluysian atrophy (DRPLA), and prion diseases (PRION). The pathway maps were manually created by examining over 70 review articles and extracting protein interactions associated with pathological mechanisms, especially

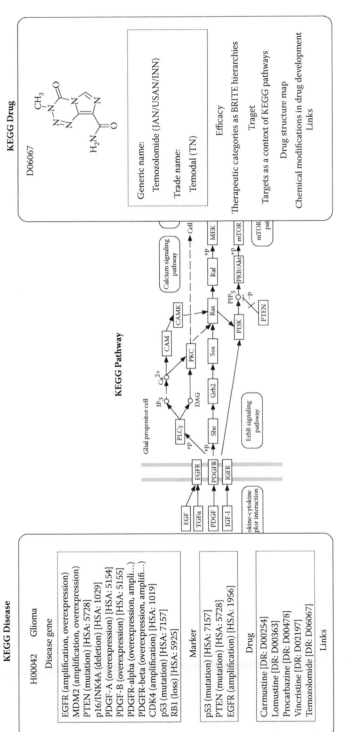

FIGURE 9.2   KEGG DISEASE and KEGG DRUG databases. This example shows interrelationships for glioma among the DISEASE, PATHWAY, and DRUG entries.

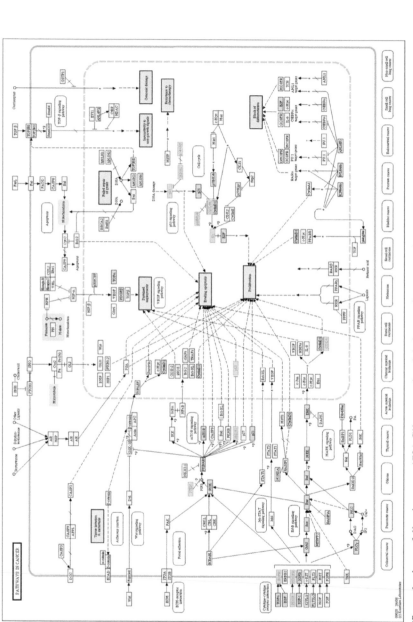

FIGURE 9.3 (See color insert following page 172.) Glioma on the global map of cancer pathways. The global map is a manually combined map of 14 existing cancer pathway maps in KEGG. Proteins are represented by boxes with the default coloring of green and those appearing in glioma are colored blue. Boxes marked with red are causative gene products for this cancer.

TABLE 9.3

Six Neurodegenerative Diseases Analyzed

Disease	Causative Genes	Protein Aggregates	Cellular Location of Aggregates
Alzheimer's disease (AD)	APP (mutation) APOE PSEN1 (mutation) PSEN2 (mutation)	Amyloid beta peptide, Tau	Extracellular, cytoplasmic
Parkinson's disease (PD)	SNCA (mutation/triplication) Parkin (mutation, genomic rearrangement) UCHL1 (mutation) PINK1 DJ1 (mutation) LRRK2 (mutation) NR4A2 (mutation)	Alpha-synuclein	Cytoplasmic
Amyotrophic lateral sclerosis (ALS)	SOD1 (mutation) ALS2 (mutation) SETX (mutation) VAPG (mutation) NEFH (deletion/insertion in KSP repeat motif)	Superoxide dismutase-1	Cytoplasmic
Huntington's disease (HD)	Huntingtin (CAG repeat expansion)	Huntingtin	Nuclear, cytoplasmic
Dentatorubropallidoluysian atrophy (DRPLA)	ATN1 (CAG repeat expansion)	Atrophin-1	Nuclear
Prion diseases	PRNP (mutation)	Prion	Extracellular

protein aggregation and/or neuronal cell loss (2). As summarized in Table 9.3 some gene products and their proteolytic fragments are known to be involved in aggregation and deposition causing neuronal damage (9–15). In a clear contrast to the 14 pathway maps for cancer that were manually combined into a single global map, there was almost no overlap among the six pathway maps for neurodegenerative diseases. Thus, we undertook computational extension of molecular networks to see whether there would be any commonality among the six neurodegenerative diseases.

9.3 NETWORK ANALYSIS OF NEURODEGENERATIVE DISEASES

9.3.1 Protein–Protein Interaction Dataset from Literature

The computational extension of molecular networks is based on a list of binary relations (protein–protein interactions or protein–compound interactions), rather than a list of nodes (proteins or compounds). They can be mapped or compared against the KEGG reference pathways to examine the possibilities of additional connections. Here we focus only on the protein–protein interaction datasets, which were prepared

TABLE 9.4

Number of Proteins and Interactions in Extended Network

Disease	Causative Gene Products	Number of Proteins	Number of Interactions
AD	APP, APOE, PSEN1, PSEN2	96	144
PD	SNCA, Parkin, UCHL1, DJ1	34	36
ALS	SOD1, ALS2	16	15
HD	Huntingtin	42	43
DRPLA	ATN1	16	15
Prion	PRNP	21	29
Total	13	201[a]	282

[a] Number of unique proteins.

in two ways. One dataset is constructed from the survey of literature in PubMed (http://www.ncbi.nlm.nih.gov/sites/entrez), and the other is obtained from outside databases. For the dataset from the literature we performed the PubMed search with keywords such as "bind" and "interact" and manually selected 233 articles describing experimental results on protein–protein interactions involving causative gene products. As shown in Table 9.4, we identified a total of 282 interactions involving 201 unique proteins, which had been confirmed in neurons or with proteins derived from mammalian brains.

9.3.2 COMMON PROTEINS LINKING DISEASE AND NORMAL PATHWAYS

The networks extended from the six neurodegenerative disease maps are superimposed to determine common proteins and common interactions, which are defined as intersections of two or more extended molecular networks as illustrated in Figure 9.4. We identified 19 common proteins, most of which appear in normal pathways for signaling and cellular processes in the KEGG PATHWAY database. These processes include apoptosis (eight proteins), MAPK signaling pathway (four proteins), Jak-STAT signaling pathway (three proteins), and focal adhesion (three proteins). The eight proteins in the KEGG apoptosis map were the members of the CASP family (CASP8, CASP3, CASP6, and CASP7) and the Bcl-2 family (BCL2, BCL2L1, BAD, and BAX). The four proteins in the MAPK signaling pathway map were GRB2, HSPA5, MAPT, and CASP3. The three proteins in the Jak-STAT signaling pathway map were CREBBP, GRB2, and BCL2L1; and the three proteins in the focal adhesion map were GRB2, BAD, and BCL2L1. In terms of the number of diseases, CASP8 (an initiator caspase) was found in most of them, and found to be related to four diseases: AD, HD, DRPLA, and PRION. Four proteins, BCL2, CASP3, GAPD, and GRB2, were found to be common in three diseases. We could not find any common protein that was related to all the diseases.

The 19 common proteins were relatively well-known proteins and it was reasonable to search for links from the neurodegenerative diseases to apoptosis, cell cycle

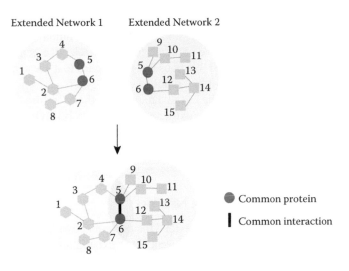

FIGURE 9.4 Common proteins and common interactions in the extended protein interaction networks.

regulation, and other cellular processes. A variety of neurodegenerative diseases are known to involve excessive apoptosis in distinct brain areas. The Bcl-2 protein family and caspase 7 are involved in ER-stress-induced cell death, which is known to protect cells against the toxic buildup of misfolded proteins. Misfolded proteins, and the associated endoplasmic reticulum (ER) stress, are emerging as a clue to understanding common mechanisms of neurodegenerative diseases (16). GRB2 is known to be involved in cell cycle, oncogenic proliferation, neuronal development, cell differentiation, and apoptosis. In our analysis GRB2 was a common protein related to AD, HD, and PRION and it was linked to the three pathway maps: MAPK signaling, Jak-STAT signaling, and focal adhesion. The cell cycle regulators including GRB2 and MAPT (tau) are expressed in post-mitotic neurons of affected or susceptible brain regions in neurodegenerative diseases, and it has been suggested that aberrant activation of cell cycle leads to neuronal cell death (17).

9.3.3 LARGE-SCALE PROTEIN–PROTEIN INTERACTION DATASET

The analysis shown above identified potentially interesting links between disease pathways and normal pathways for cellular processes. Understanding these normal pathways will be a key issue to elucidate the biological functions of common mechanisms of neurodegenerative diseases. However, our findings were not really surprising because the common proteins were members of the pathways already implicated in neurodegenerative diseases. We thus included an additional protein–protein interaction dataset for extension of the KEGG neurodegenerative disease pathway maps.

We used HPRD (Human Protein Reference Database, http://www.hprd.org/), a database of protein interactions and other annotations for human proteins, and obtained 14,023 protein–protein interactions consisting of 4734 proteins that are linked to Entrez Gene IDs (and thus KEGG GENES IDs). In addition, we used

HugeIndex (Human Gene Expression Index, http://www.biotechnologycenter.org/hio/), which is a repository for gene expression data on normal human tissues using high-density oligonucleotide arrays. We obtained 567 brain-selective genes that are highly expressed only in the brain and 391 housekeeping genes that are expressed in all tissue types, with a total number of 894 unique genes.

We combined 282 interactions in the initial dataset from the literature and 14,023 interactions taken from HPRD. We then extracted interactions involving either 201 proteins in the initial dataset or 894 proteins (gene products) taken from HugeIndex. As a result, we obtained the second dataset of 1041 protein–protein interactions involving 528 proteins.

9.3.4 Extended Protein Interaction Network

As before, the second dataset was used to extend the protein networks in the six neurodegenerative disease maps, where no apparent links existed among the 13 causative gene products. The extended network was a large network, which contained 196 proteins (247 interactions) that were directly linked from the 13 causative gene products and 385 proteins (703 interactions) that were found within two links. Some proteins were highly connected. Twenty-eight proteins, including nine causative gene products, were hubs with 10 or more interactions. The top five hubs were causative gene products: APP, PSEN1, Huntingtin, SNCA, and PSEN2, with 47, 46, 32, 28, and 27 interaction partners, respectively. The following hubs, such as CTNNB1, DLG4 and BCL2, were not causative gene products but were related to signal transduction and apoptosis.

By distinguishing each of the neurodegenerative diseases for the network extension, we found 174 common proteins and 202 common interactions (Figure 9.4) shared in at least two neurodegenerative diseases. Figure 9.5 illustrates a combined picture of the extended network involving these proteins and interactions. Of the 174 common proteins, 19 proteins were already found in the analysis with the initial dataset from literature. The other 155 proteins were identified from this extended network, 74 of which were already included in the initial dataset but not identified as common proteins. Thus, 81 proteins were newly found common proteins from the additional dataset of HPRD and HugeIndex. One of the causative gene products of Alzheimer's disease, PSEN1, became a common protein that was linked to all six neurodegenerative diseases. Five proteins, BCAP31 (B-cell receptor associated protein), DNCL1 (dynein light chain), HSPCA (heat shock protein), NR3C1 (nuclear receptor), and PIN1 (peptidylprolyl cis/trans isomerase), which were among the 81 proteins, were linked to five diseases. All these proteins seem to have relevance in understanding neurodegenerative diseases as discussed before (2). According to the KEGG PATHWAY database the 81 newly identified common proteins were involved in focal adhesion, regulation of actin cytoskeleton, adherens junction, axon guidance, tight junction, and leukocyte transendothelial migration, thus expanding potential links from neurodegenerative diseases to these pathways.

In contrast to the common proteins, the common interactions found in the extended network enhance the importance of pathways already known, especially apoptosis. Among the 202 common interactions, we found that one interaction, BCL2–CASP3,

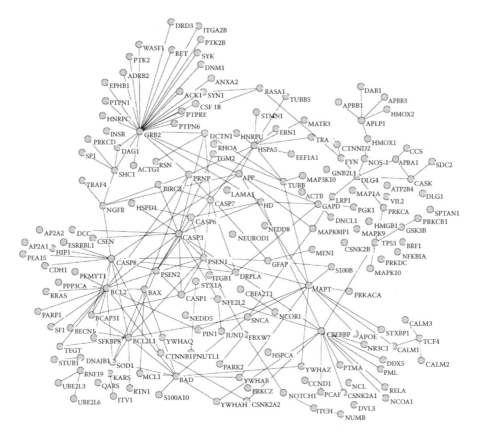

FIGURE 9.5   A portion of the extended network of neurodegenerative diseases involving 174 common proteins and 202 common interactions.

was common to five diseases except Parkinson's disease, and that five interactions, BAD–BCL2, BCL2–CASP8, BCL2L1–CASP8, CASP3–HSPD1, and CASP3–NFE2L2, were common to four diseases. The KEGG pathways containing more than five common interactions were apoptosis, Wnt signaling pathway, and focal adhesion. There is emerging evidence that focal adhesion and Wnt signaling are associated with some neurodegenerative diseases (18,19). Alteration or impairment of such pathways and their downstream interactions may contribute to the common pathogenic mechanisms in neurodegenerative diseases.

9.3.5   PROTEIN DOMAIN ANALYSIS

In order to obtain further functional clues of the extended protein network of neurodegenerative diseases, we performed domain analysis of 146 proteins constituting the 202 common interactions. We used the Pfam database (http://pfam.sanger.ac.uk/) and calculated the relative frequency of Pfam domains in the 146 proteins relative to the entire HPRD dataset. The characteristic domains with the frequency of more than five times than HPRD were: 14-3-3 protein, phosphotyrosine interaction domain (PTB/PID), caspase domain (peptidase C14), apoptosis regulator proteins Bcl-2 family, and WW domain, which were all implicated in neurodegenerative diseases (2).

9.3.6 DOMAIN-BASED SIMILARITY OF NEURODEGENERATIVE DISEASES

Increasing evidence suggests considerable overlap of clinicopathological features among neurodegenerative diseases (20). We investigated the overlap among the six neurodegenerative diseases based on the domain analysis as follows. First, the proteins linked from the causative gene products of each disease were extracted up to two steps away in the extended network. The number of proteins was 237, 137, 46, 145, 55, and 104 for AD, PD, ALS, HD, DRPLA, and PRION, respectively. Then, Pfam domains were assigned to each extracted protein with the HMMER program using the E-value cutoff of 0.001. The number of different domains was 266, 157, 54, 194, 89, and 137 for AD, PD, ALS, HD, DRPLA, and PRION, respectively. For each disease, we constructed a profile vector representing the domain distribution, where each element was defined as the ratio of the number of proteins containing the domain against the number of all proteins related to the disease. The domain-based similarity of diseases was then defined by the correlation coefficient of the profile vectors.

Table 9.5 shows the correlation coefficients of the profile vectors among the six neurodegenerative diseases. AD, PD, HD, and PRION showed relatively high correlations with each other, but ALS and DRPLA seemed to be different from the other diseases. This may be because the protein–protein interactions of ALS and DRPLA are less studied, and consequently the number of identified domains is smaller than the others.

We found that PD and HD showed the highest correlation. Common domains that contributed to this high correlation were cation transporter/ATPase C-terminus domain, cation transporter/ATPase N-terminus domain, E1-E2 ATPase domain, helix–loop–helix DNA-binding domain, and haloacid dehalogenase-like hydrolase. The KEGG pathways related to these domains were tight junction, long-term potentiation, calcium signaling pathway, phosphatidyl inositol signaling system, olfactory transduction, and insulin signaling pathway. To our knowledge tight junction has not previously been associated with the

TABLE 9.5
Domain-Based Similarity of Neurodegenerative Diseases

	PD	HD	PRION	DRPLA	ALS
AD	0.84	0.84	0.83	0.57	0.40
PD		0.87	0.78	0.48	0.42
HD			0.82	0.59	0.35
PRION				0.53	0.46
DRPLA					0.33

mechanism of either disease. The motor symptoms such as tremor, rigidity, and bradykinesia are common between PD and HD, both of which are classified as diseases of basal ganglia. The basal ganglia are involved in muscle-driven movements of the body and there may be a link between the neuronal cell loss in the basal ganglia and dysfunction of tight junctions, for example, in the blood–brain barrier (21).

9.4   CONCLUDING REMARKS

Bioinformatics approaches have been used for integrated analysis of different types of large-scale datasets generated by high-throughput experiments in genomics, transcriptomics, proteomics, metabolomics, glycomics, chemical genomics, etc. Here we emphasize the importance of properly computerizing more traditional data and knowledge in the literature, which often represent human integration and interpretation of low-throughput, but high-quality experimental data. In KEGG such data and knowledge are computerized in two ways, as molecular networks and molecular lists, enabling integration with large-scale datasets, which can also be treated as a molecular list or a molecular network, such as a list of genes in the genome or a set of protein–protein interactions that is essentially a molecular network. We showed an example of extending KEGG pathways (molecular networks) by superimposing a dataset of protein–protein interactions. A similar approach may also be taken to incorporate protein–compound interactions including drug–target relationships for extension of KEGG pathways.

There are already many databases available containing information about genetic and molecular factors of human diseases. The OMIM database (22) is the most comprehensive and contains detailed descriptions about genes and phenotypes of genetic disorders. In contrast, we attempt to develop a computable disease information resource that can be integrated with other data and knowledge. Thus, the disease information is represented in terms of the gene/molecule lists in the KEGG DISEASE database and, whenever more detail is known, in terms of the molecular networks in the disease maps of the KEGG PATHWAY database. Understanding molecular networks associated with normal brain function and with human diseases is a step forward to understand causative genetic interactions. However, there still remains a huge gap from molecular networks to phenotypes. In the KEGG database we have initiated an effort to capture still higher-level knowledge, including brain maps and computable representation of cellular networks, toward computational prediction of human diseases.

REFERENCES

1. Kanehisa, M., Araki, M., Goto, S. et al. (2008) KEGG for linking genomes to life and the environment, *Nucleic Acids Res* 36:D480–4.
2. Limviphuvadh, V., Tanaka, S., Goto, S., Ueda, K. and Kanehisa, M. (2007) The commonality of protein interaction networks determined in neurodegenerative disorders (NDDs), *Bioinformatics* 23:2129–38.

3. Hashimoto, K., Goto, S., Kawano, S. et al. (2006) KEGG as a glycome informatics resource, *Glycobiology* 16:63R–70R.

4. Hashimoto, K., Yoshizawa, A. C., Okuda, S. et al. (2008) The repertoire of desaturases and elongases reveals fatty acid variations in 56 eukaryotic genomes, *J Lipid Res* 49:183–91.

5. Kawano, S., Hashimoto, K., Miyama, T., Goto, S. and Kanehisa, M. (2005) Prediction of glycan structures from gene expression data based on glycosyltransferase reactions, *Bioinformatics* 21:3976–82.

6. Aoki, K. F., Yamaguchi, A., Ueda, N. et al. (2004) KCaM (KEGG Carbohydrate Matcher): A software tool for analyzing the structures of carbohydrate sugar chains, *Nucleic Acids Res* 32:W267–72.

7. Hattori, M., Okuno, Y., Goto, S. and Kanehisa, M. (2003) Development of a chemical structure comparison method for integrated analysis of chemical and genomic information in the metabolic pathways, *J Am Chem Soc* 125:11853–65.

8. Moriya, Y., Itoh, M., Okuda, S., Yoshizawa, A. C. and Kanehisa, M. (2007) KAAS: An automatic genome annotation and pathway reconstruction server, *Nucleic Acids Res* 35:W182–5.

9. Aguzzi, A. and Polymenidou, M. (2004) Mammalian prion biology: One century of evolving concepts, *Cell* 116:313–27.

10. Gardian, G. and Vecsei, L. (2004) Huntington's disease: Pathomechanism and therapeutic perspectives, *J Neural Transm* 111:1485–94.

11. Hague, S. M., Klaffke, S. and Bandmann, O. (2005) Neurodegenerative disorders: Parkinson's disease and Huntington's disease, *J Neurol Neurosurg Psychiatry* 76:1058–63.

12. Julien, J. P. (2001) Amyotrophic lateral sclerosis. Unfolding the toxicity of the misfolded, *Cell* 104:581–91.

13. Pardo, L. M. and van Duijn, C. M. (2005) In search of genes involved in neurodegenerative disorders, *Mutat Res* 592:89–101.

14. Rocchi, A., Pellegrini, S., Siciliano, G. and Murri, L. (2003) Causative and susceptibility genes for Alzheimer's disease: A review, *Brain Res Bull* 61:1–24.

15. Rudnicki, D. D. and Margolis, R. L. (2003) Repeat expansion and autosomal dominant neurodegenerative disorders: Consensus and controversy, *Expert Rev Mol Med* 5:1–24.

16. Rao, R. V. and Bredesen, D. E. (2004) Misfolded proteins, endoplasmic reticulum stress and neurodegeneration, *Curr Opin Cell Biol* 16:653–62.

17. Vincent, I., Pae, C. I. and Hallows, J. L. (2003) The cell cycle and human neurodegenerative disease, *Prog Cell Cycle Res* 5:31–41.

18. Caltagarone, J., Jing, Z. and Bowser, R. (2007) Focal adhesions regulate Abeta signaling and cell death in Alzheimer's disease, *Biochim Biophys Acta* 1772:438–45.

19. Caricasole, A., Bakker, A., Copani, A. et al. (2005) Two sides of the same coin: Wnt signaling in neurodegeneration and neuro-oncology, *Biosci Rep* 25:309–27.

20. Armstrong, R. A., Lantos, P. L. and Cairns, N. J. (2005) Overlap between neurodegenerative disorders, *Neuropathology* 25:111–24.

21. Whitton, P. S. (2007) Inflammation as a causative factor in the aetiology of Parkinson's disease, *Br J Pharmacol* 150:963–76.

22. Hamosh, A., Scott, A. F., Amberger, J. S., Bocchini, C. A. and McKusick, V. A. (2005) Online Mendelian Inheritance in Man (OMIM), a knowledgebase of human genes and genetic disorders, *Nucleic Acids Res* 33:D514–7.

10 Redox Proteomics of Oxidatively Modified Brain Proteins in Mild Cognitive Impairment

Tanea T. Reed, Rukhsana Sultana, and D. Allan Butterfield

CONTENTS

10.1   MILD COGNITIVE IMPAIRMENT

Mild cognitive impairment (MCI) is generally referred to as the transitional zone between normal cognitive aging and early dementia or clinically probable Alzheimer's disease (AD) (1), although not all AD patients pass through an MCI stage. The term was first coined by Petersen (2). Most individuals with MCI eventually develop AD, which suggests MCI may be the earliest phase of the AD (3–6). MCI can be divided into two broad subtypes: amnestic (memory-affecting) MCI or non-amnestic MCI (2,7). Other functions, such as language, attention, and visuospatial skills, may be impaired in either type. Amnestic mild MCI patients characteristically have subtle but measurable memory disorder not associated with dementia. Individuals with MCI are at an increased risk of developing AD, or another form of dementia with a rate of progression between 10% and 15% per year (8,9), although there have been cases where patients have reverted to normal (1,10,11).

Criteria for MCI include (a) a memory complaint corroborated by an informant; (b) objective memory test impairment (age and education adjusted); (c) general normal global intellectual function; (d) activities of daily living not disturbed; (e) clinical dementia rating (CDR) score of 0.0 to 0.5; (f) no dementia; and (g) a clinical evaluation that revealed no other cause for memory decline (12). Moreover, neuroimaging studies by magnetic resonance imaging (MRI) demonstrate the atrophy of the hippocampus or entorhinal cortex in MCI patients, indicating the relationships with transition of normal aging to MCI, then later to clinical AD (13). Pathologically, MCI brain shows mild degradation of the hippocampus, entorhinal cortex, sulci, and gyri using MRI (14,15). These aforementioned areas undergo considerable degradation in AD (16–20). Since the hippocampus is the region of the brain primarily responsible for processing memory, it is clearly understandable why those persons with AD and MCI have memory loss.

10.2   OXIDATIVE STRESS

Under normal physiological conditions, there is equilibrium between the level of antioxidants and pro-oxidants in a cell. However, when environmental factors, stressors, or disease occur, this homeostasis can become imbalanced in favor of pro-oxidants, resulting in a phenomenon known as oxidative stress (21). Oxidative stress can also transpire through an antioxidant deficiency (22) or excess reactive oxygen species/ reactive nitrogen species (ROS/RNS) production. Moreover, oxidatively damaged proteins are often removed by the 20S proteosome. Defects in the proteosome system would lead to elevated levels of oxidatively modified proteins and neurotoxicity (21). Oxidative stress plays a significant role in neurodegenerative disease (23–27). If oxidative stress is involved in the progression of AD and is not a consequence of this disease, there should be evidence of elevated oxidative stress in the beginning stages of the disease (28). Manifested by elevated levels of nucleic acid oxidation, protein oxidation, and lipid peroxidation, oxidative damage is most severe in the hippocampus, a brain region that is responsible for memory processing and cognitive function (23,24). Moreover, oxidative stress-mediating entities per se induce neuronal death in vitro, and protein oxidation and lipid peroxidation in the superior and middle temporal gyri (SMTG) of MCI patients are increased (29). All of these studies strongly suggest that oxidative stress is a primary event in the development of AD.

10.3 POST-TRANSLATIONAL MODIFICATIONS

Post-translational modifications (PMTs), such as protein oxidation, are seen in a plethora of neurodegenerative diseases (23,30). A few examples of PMTs are phosphorylation, nitration, carbonylation, 4-hydroxy-2-nonenal (HNE) modification, glycosylation, and S-glutathionylation. Proteins are sensitive to oxidation by ROS and RNS, and oxidative damage is a result from such interaction. Protein oxidation leads to loss of protein function and often cell death via necrotic or apoptotic processes (30). Protein oxidation is increased in MCI (31,32) and will be discussed thoroughly in the next section. A complete list of oxidatively modified proteins in MCI brain determined by redox proteomics can be seen in Table 10.1.

10.4 PROTEIN CARBONYLATION

Protein carbonyl levels have been recognized as the most widely used indicator of oxidative stress (33–35). Protein carbonyl levels increase in aging and age-associated neurodegenerative diseases (36,37). Proteins can be oxidized directly by reactive oxygen species to generate protein carbonyls. This happens by several different mechanisms including amino acid side chain (Lys, Thr, Pro, and Arg) and metal-assisted oxidation (38) (Figure 10.1). Beta scission from the peptide backbone is the second method by which protein carbonyls are created (Figure 10.2). Third, the adduction of carbonyl-containing reactive aldehydes (i.e., acrolein, 4-hydroxynonenal) to amino acids leads to carbonyls (Figure 10.3). Advanced glycation end products resulting from Amadori chemistry can also lead to protein oxidation (39,40). Indexing protein carbonyl content is a fundamental key to understanding protein oxidation. Protein carbonyl levels most often are analyzed by immunochemical detection of the hydrazone formed by reaction of 2,4-dinitrophenylhydrazine with protein carbonyls (30,38).

10.5 HNE MODIFICATION

Lipid peroxidation is a complex process involving the interaction of oxygen-derived free radicals with polyunsaturated fatty acids, resulting in a variety of highly reactive electrophilic aldehydes that are capable of easily attaching covalently to proteins by forming adducts with cysteine, lysine, or histidine residues (41) through Michael addition (30). Among the aldehydes formed, malondialdehyde (MDA) and HNE represent the major products of lipid peroxidation (41). The brain is particularly vulnerable to lipid peroxidation (42) due to its richness in polyunsaturated fatty acids, high oxygen consumption, and abundant quantities of redox transition metals (43,44).

Lipid peroxidation is highly evident in neurodegenerative disease (24,45,46). Lipid peroxidation occurs through ongoing free radical chain reactions until termination occurs (Figure 10.4). Free radicals attack an allylic hydrogen atom to form

TABLE 10.1
Proteins Oxidatively Modified in MCI

Function	Protein	Region	Oxidative Modification
Energy metabolism	Alpha enolase	Hippocampus, IPL	C, H, N
	Glucose-regulated precursor protein	Hippocampus	N
	Aldolase	Hippocampus	N
	Pyruvate kinase	Hippocampus, IPL	C, H
	Phosphoglycerate kinase	Hippocampus	H
	Pyruvate kinase	IPL	H
	Lactate dehydrogenase	Hippocampus	H
Neuroplasticity	Glutamine synthetase	Hippocampus	C
Mitochondrial dysfunction	Malate dehydrogenase	Hippocampus	N
	ATP synthase	Hippocampus, IPL	H
Antioxidant defense	Glutathione S-transferase Mu	IPL	N
	Multidrug-resistant protein 3	IPL	N
	Peroxiredoxin VI	Hippocampus	N
	Heat shock protein 70	Hippocampus, IPL	H, N
	Peptidyl-prolyl *cis/trans* isomerase 1 (Pin-1)	Hippocampus	C
	Carbonyl reductase 1	Hippocampus	H
Structural dysfunction	Dihydropyriminidase-like protein-2	Hippocampus	N
	Fascin 1	Hippocampus	N
	Beta-actin	IPL	H
Signal transduction	14-3-3 gamma	IPL	N
	Neuropolypeptide h3	Hippocampus	H
Protein synthesis	Initiation factor alpha (eIF-α)	IPL	H
	Elongation factor Tu (EF-Tu)	IPL	H

Note: C, carbonylation; N, nitration; H, HNE modification.

a carbon centered radical (step 1). This radical reacts with O_2 to produce peroxyl radicals (step 2). These peroxyl radicals can react with adjacent lipids forming a lipid hydroperoxide repeating the cycle (step 3). The lipid hydroperoxide can decompose to produce multiple reactive products such as acrolein, malondialdehyde, and HNE. In MCI, acrolein and HNE have been found to be significantly elevated (31,47,48). Lipid peroxidation can be terminated by two radicals reacting forming a nonradical and oxygen (step 4). HNE is a major product of lipid peroxidation and causes cell toxicity. Lipids are particularly vulnerable to oxidation due to the fact that polyunsaturated fatty acids are abundant in brain and the presence of oxygen in the lipid bilayer is at millimolar levels. Glutathione has been shown to detoxify HNE in cells (49). Increased levels of HNE cause disruption of Ca^{2+} homeostasis, membrane

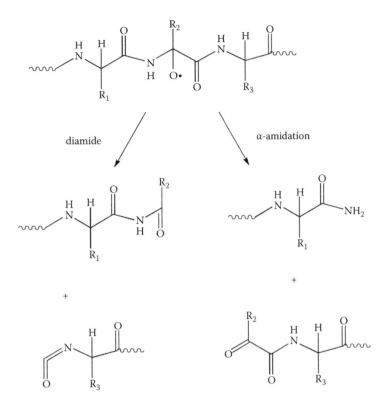

FIGURE 10.1 Beta scission from peptide backbone.

damage, and cell death (41). Vitamin E (alpha-tocopherol) is a "chain breaking" antioxidant and can terminate propagation steps of lipid peroxidation. When the hydrogen is abstracted in step 1, an alpha-tocopherol radical forms that can be reverted back to vitamin E by ascorbic acid (vitamin C) or glutathione (GSH), both potent antioxidants.

HNE is an alpha, beta-unsaturated alkenal product of omega-6 polyunsaturated fatty acids and is a major cytotoxic end product of lipid peroxidation that mediates oxidative stress-induced death in many cell types (41,50,51). HNE accumulates in membranes at concentrations of 10 μM to 5 mM in response to oxidative insults (41) and invokes a wide range of biological activities, including inhibition of protein and DNA synthesis (52–56), disruption of Ca^{2+} homeostasis, membrane damage, cell death (41), and activation of stress signaling pathways (50,57).

Several publications report that the brains of MCI patients are under oxidative stress. Increased levels of thiobarbituric acid reactive substance (TBARS), malondialdehyde (MDA), F_2 isoprostanes and F_4 neuroprostanes, as well as soluble and protein-bound HNE, specific markers of in vivo lipid peroxidation (30), were significantly elevated in cerebrospinal fluid (CSF), plasma, urine, and brains of MCI patients compared with controls (29,31,46,58), suggesting that lipid peroxidation may be an early event in the pathogenesis of the disease.

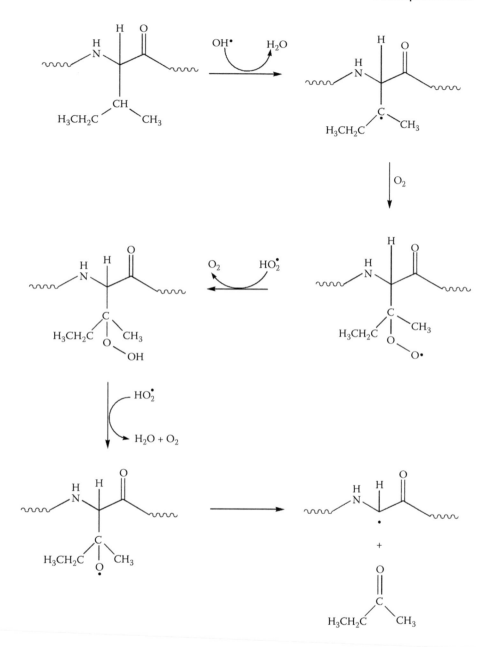

FIGURE 10.2   Amino acid side chain oxidation.

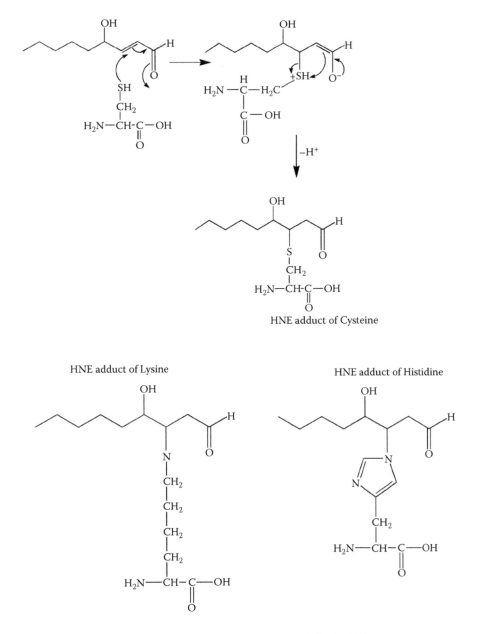

HNE adduct of Cysteine

HNE adduct of Lysine

HNE adduct of Histidine

FIGURE 10.3 Covalent modification of amino acids by Michael addition.

(1) LH + X• → L• + XH

(2) L• + O$_2$ → LOO•

(3) LOO• + LH → LOOH + L•

(4) LOO • + LOO •→ nonradical + O$_2$

FIGURE 10.4   Lipid peroxidation reaction summary.

10.6   PROTEIN NITRATION

In conjunction with the enzyme nitric oxide synthase, arginine is converted into nitric oxide and L-citrulline (Figure 10.5). Nitric oxide can react with superoxide radical anion (O$_2^-$) forming the strong oxidant peroxynitrite (Figure 10.6). Peroxynitrite can oxidize tyrosine, methionine, tryptophan, and cysteine residues to promote protein nitration (59,60). In the presence of CO$_2$, peroxynitrite can exist as an anion (ONOO$^-$) or other reactive intermediates. A nitrosoperoxyl intermediate is formed from the combination of peroxynitrite and carbon dioxide, which rearranges to form nitrocarbonate. This species can be cleaved to form carbonate and NO$_2$ radicals (Figure 10.6); the latter reacts with a tyrosyl free radical in the 3-position to form 3-nitrotyrosine (Figure 10.7).

Previous research has shown that peroxynitrite can interact with proteins (61,62), lipids (63), DNA (64,65), and RNA (66) to promote damage in these biological molecules. Tyrosine nitration is associated with Alzheimer's disease (AD) (67) as well as Parkinson's disease (68). Gamma-glutamylcysteinylethyl ester (GCEE), a derivative of gamma-glutamylcysteine, can cross the blood–brain barrier (BBB) and has been proven to prevent peroxynitrite-induced damage by upregulating glutathione production (69). Lipoic acid has also been used to reduce protein nitration (70) as has gamma-tocopherol (71). Increased protein nitration also can lead to an elevated release of RNS. Nitration of proteins results in the inactivation of several important

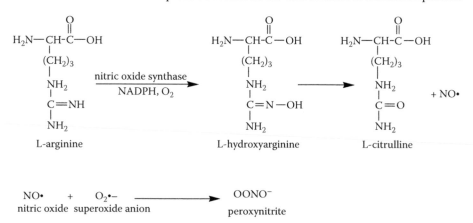

FIGURE 10.5   Production of nitric oxide from L-citrulline.

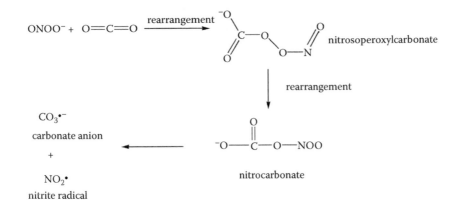

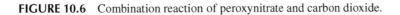

FIGURE 10.6 Combination reaction of peroxynitrate and carbon dioxide.

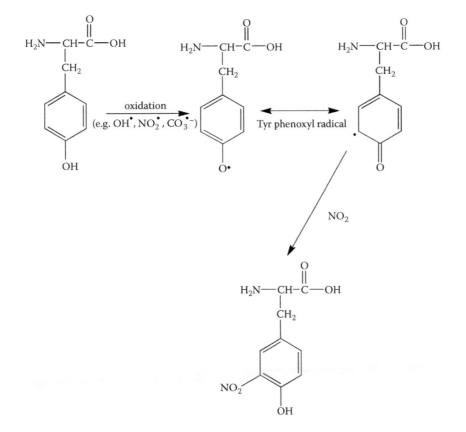

FIGURE 10.7 Formation of 3-nitrotyrosine.

mammalian proteins such as MnSOD (72–75), glyceraldehyde 3-phosphate dehydrogenase (76,77), actin (78,79), synaptic proteins (80), and tyrosine hydroxylase (81,82), among others.

10.7   OXIDATIVE DAMAGE IN MCI BRAIN

Protein oxidation has been observed in brains of patients with MCI (32). Increased protein-bound HNE immunoreactivity is observed in the MCI hippocampus and MCI inferior parietal lobule (IPL) (31) as well as elevated levels of protein carbonyls (83) and 3-nitrotyrosine (32). Since MCI is arguably the earliest form of AD and there is supporting evidence of oxidative damage in the AD brain, these data suggest that elevated levels of oxidative stress in MCI can contribute to the progression of this neurodegenerative disease.

10.8   REDOX PROTEOMICS

The proteome is the complete set of proteins expressed in a cell or organism. This term was first coined by Wilkins et al. in 1996 (84). Proteomics is the analysis of the proteome. Proteomics describes a protein in terms of its amino acid sequence, mass, isoelectric point (pI), and structure. This technique is helpful in determining protein identification, modification, and function. Since the advent of proteomics, this methodology is one of the fastest growing areas of biomedical science. Proteomics is widely used to determine potential biomarkers in neurodegenerative diseases, specifically AD (85–87). Technology in the field of proteomics is rapidly expanding and evolving. Proteomics is now being developed for clinical diagnosis (88,89) and in establishing biomarkers for disease including MCI (87,90,91) and AD (86,92).

10.8.1   2D GEL ELECTROPHORESIS APPROACHES FOR REDOX PROTEOMICS

Proteomics traditionally has employed a two-dimensional protein separation (see Chapter 3 for a discussion of protein separation techniques, and Chapter 4 for a discussion of 2D electrophoresis approaches). There are three basic steps to any proteomics experiment: separation, quantification, and identification. In 2D electrophoresis, proteins are separated on the basis of their isoelectric point (pI) in the first dimension. Protein samples are placed on immobilized pH gradient (IPG) strips composed of a polyacrylamide matrix. Isoelectric focusing (IEF) takes advantage of this principle by applying a potential to the proteins embedded on the IPG strip. The proteins migrate from the cathode to the anode, and at the pH where the protein reaches its specific pI they cease to move.

In the second dimension, proteins are separated according to their migration rate, which is a function of protein mass and protein shape. IPG protein strips are equilibrated in a dithiothreitol (DTT)-containing buffer. DTT reduces the disulfide bonds to thiol groups. After reduction and removal of DTT solution, an iodoacetamide (IA) solution is added to the IPG strip to alkylate the thiol groups and prevent them from recombining. This reaction is performed in the dark as IA is photosensitive. The IPG strip is loaded onto the gel and voltage is applied. Low molecular weight proteins travel faster through the polyacrylamide gel and are located closer to the bottom of the gel. High molecular weight proteins travel slower through the gel and are found toward the upper portion of the gel.

After electrophoresis, 2D gels are fixed and then stained with an appropriate protein stain and subsequently prepared for visualization. SYPRO Ruby is a common fluorescent stain used to detect proteins on a gel. Photons must be excited ($\lambda_{ex} = 470$ nm) so they can emit ($\lambda_{em} = 618$ nm) and be detected. SYPRO Ruby is a sensitive stain as it can detect 1 ng of protein/spot. Other methods of staining for protein visualization include staining with Coomassie blue and silver staining. Although there are various other staining methods, SYPRO Ruby staining is a mass spectrometry-compatible, quick method with shorter preparation time compared to others. After protein visualization, all gels are scanned and images are imported into a software analysis package for comparison studies.

An additional, identical gel is run, and left unstained. This unstained gel is soaked in a transfer buffer, and then sandwiched between filter paper, a nitrocellulose membrane, and filter paper in order for the proteins to transfer onto the nitrocellulose membrane. These components are properly arranged and a thin glass rod is rolled over the top filter paper to remove any possible bubbles that may have formed between the layers. A weighted plate is added to the top of the filter paper and current is applied to allow proper protein transfer.

Typically, Western blots are blocked with BSA to prevent any nonspecific binding. A primary antibody that is specific for the proteins of interest is added to the blot. The antibody will bind to the protein of interest (antigen); the membrane is washed to remove any unbound molecules. A second antibody (attached to an enzyme) is added that will bind to the primary antibody–antigen complex. The protein of interest will be identified through a chemiluminescent or chromophoric reaction, depending on the nature of the secondary antibody. The blots are then dried, scanned, saved as images, and exported into the software analysis program. The Western blots in a particular experiment are compared to each other through computer algorithms based on immunoreactivity levels.

For protein carbonylated proteins, post derivatization after immunoprecipitation can be used as well. In this method, the nitrocellulose membranes are equilibrated in 20% methanol for five minutes. Membranes are then incubated in 2N HCl for five minutes and then in 0.5 mM DNPH solution for exactly five minutes. The membranes are washed three times in 2N HCl and five times in 50% methanol (five minutes each wash), dried, scanned, and saved as images.

10.8.2 Immunochemical Detection

To confirm the correct identification of the proteins detected by mass spectrometry, a representative protein identified by redox proteomics is immunoprecipitated. Brain samples are first pre-cleared using protein A/G–agarose beads for one hour at 4°C and then incubated overnight with an appropriate antibody for the protein of interest. The protein A/G mixture is ideal because protein A-conjugated agarose beads respond to rabbit-raised antibodies, while protein G-conjugated agarose beads are used if the antibodies were raised in goats or mice. The following day, the samples are incubated for one hour with protein A/G–agarose then washed three times with RIPA buffer (93). Proteins are then solubilized in IEF rehydration buffer followed by 2D electrophoresis and 2D Western blot.

10.8.3   IMAGE ANALYSIS

The gels and nitrocellulose blots are scanned and saved as images using a transiluminator and scanner, respectively. Several commercial software packages including PDQuest, Phoretix, and Progenesis can be used for matching and analysis of visualized protein spots among different gels and blots. The principles of measuring intensity values by 2D analysis software are similar to those of densitometric measurement. Although for 2D gel comparison, two-dimensional difference gel electrophoresis (2D-DIGE) could be used since gel-to-gel matching sometimes is not reproducible using other approaches. Spot matching is performed to ascertain the average normalized intensity of gels in an experimental sample to that of spots in a control sample. Only those spots that are considered statistically significant by Student's t-test ($p < 0.05$) are selected for identification. Sophisticated statistical analyses for microarray data are not applicable for proteomics studies (94,95).

10.8.4   TRYPSIN DIGESTION

The selected spots are excised from the 2D gel, transferred into microcentrifuge tubes, and washed with ammonium bicarbonate (NH_4HCO_3) to swell the polyacrylamide gel and push the protein into the gel matrix (for a detailed discussion of protein identification by mass spectrometry, refer to Chapter 5). Acetonitrile is then added to shift equilibrium away from the protein and shrinks the gel matrix. The solvent is removed, and the gel pieces are dried in a flow hood. The protein's disulfide bonds are reduced with DTT at 56°C then alkylated with IA. Acetonitrile and ammonium bicarbonate steps are repeated and protein spots are allowed to dry. The gel pieces are rehydrated with 20 ng/μL modified trypsin in NH_4HCO_3 with the minimal volume to cover the gel pieces. The gel pieces then are chopped into smaller pieces and incubated with shaking overnight at 37°C with trypsin. This technique is termed "ingel" trypsin digestion and has several advantages including higher recovery because the protein is cleaved into many smaller peptides that result from sequence-specific proteolysis, which is an important means of identifying the protein of interest (96).

10.8.5   MASS SPECTROMETRY

These tryptic digests (smaller peptides) constitute mass fingerprints, which are unique to each protein, and the molecular weight of each peptide is determined by the use of mass spectrometry (MS; see Chapters 5 and 7). The two ion sources are matrix-assisted laser desorption ionization (MALDI) and electrospray ionization (ESI). In MALDI the analyte is mixed with a matrix, and this mixture is allowed to dry. A pulsed laser hits the target containing this analyte/matrix mixture and disperses the analyte. The matrix absorbs the energy from the laser pulse, and the matrix molecules containing the analyte are transferred into the gas phase. In the gas phase, H^+ transfer from the acidic matrix to the peptides occurs, thereby putting charges on the peptides in the gas phase. A time of flight (TOF) mass analyzer measures how long it takes for the ions to reach the detector. Electrospray ionization is a technique that provides the transport of ions from solution to the gas phase. In ESI, the peptide in solution flows through a narrow capillary tube into a vacuum and the MS at atmospheric pressure and 4000 V to create ions. The charges on the droplets extend the solution to form a (Taylor) cone, causing the solution to disperse as a mist of fine droplets. As the solvent evaporates, the droplet size decreases the total charges of the proteins in the droplet remain the same but the surface area of each droplet decreases to one ion per droplet. Individual ions then flow into the mass analyzer. ESI can be coupled to other separation techniques (i.e., high performance liquid chromatography [HPLC]) to make it more versatile (as discussed in Chapter 5).

10.8.6 Database Searching

Better known as peptide mass fingerprinting (PMF), this process requires the use of MS analysis to determine the experimental masses. Experimental masses are used by on-line protein databases to compare and match protein-specific mass fingerprints generated by the in silico digestion of proteins. This comparison is used to identify the protein of interest based on the quality of peptide matches (97–99). PMF is used to identify proteins from tryptic peptide fragments by utilizing an appropriate search engine (Table 10.2, and Chapter 5). Database searches are conducted allowing for up to one missed trypsin cleavage and using the assumption that the peptides are monoisotopic, oxidized at methionione residues, and carbamiodomethylated at cysteine residues. Mass tolerance of 150 ppm/g is the window of error allowed for matching the peptide mass values. Probability-based MOWSE scores are estimated by comparison of search results against estimated random match population and are reported as $-10 * \text{Log}_{10} (p)$, where p is the probability that the identification of the protein is incorrect. As a correlate, the higher the MOWSE score, the lower the p value, yielding a higher probability of accurate protein identification. All protein identifications must be reviewed (and in most cases validated; for this see Chapters 11 and 14) to ensure that they are in the expected size and pI range based on position in the gel.

TABLE 10.2
Listing of Online Search Engines

Name	Address
Aldente	http://www.expasy.ch/tools/aldente
Mascot	http://www.matrixscience.com
MS-fit	http://prospector.ucsf.edu/prospector/4.0.8/html/msfit.htm
PeptideSearch	http://www.narrador.embl-heidelberg.de/
Profound	http://prowl.rockefeller.edu/prowl-cgi/profound.exe

10.9   ENZYME ASSAYS

Enzymatic activity is generally lower in oxidatively modified enzymes compared to control (31,100,101). Oxidative stress leads to oxidative modification of key proteins and enzymes (102). Oxidative modification of amino acids, particularly at the active site, can cause loss of enzymatic activity and subsequent protein dysfunction (100,103–105). Therefore, the determination of enzyme activity is frequently used to help validate the identities of the proteomics-identified proteins.

10.10   ENERGY-RELATED PROTEINS

Glycolysis is a metabolic pathway that converts glucose to pyruvate and generates two ATP molecules in the process. ATP, the energy source of the cell, is extremely important at nerve terminals for normal neural communication. Decreased levels of cellular ATP at nerve terminals may lead to loss of synapses and synaptic function, and may ultimately contribute to memory loss in amnestic MCI patients. Synapse loss is an early event in AD, since it is observed in MCI and early AD (106). Energy metabolism alteration is a current major hypothesis of AD and has been reported in brain in an advanced stage of MCI (107–110). Several proteins identified by proteomics as energy-related proteins in MCI brain support this notion. There is an assortment of energy-related proteins that have been identified as being oxidatively modified in MCI hippocampus and IPL in addition to previous proteomics studies of AD and cell culture models of AD, identifying an increased oxidation of enolase (111–116). *Alpha-enolase*, a critical glycolytic protein, interconverts 2-phosphoglycerate to phosphoenolpyruvate (Figure 10.8). The enzymatic activity of alpha-enolase is reduced significantly compared to control samples (83). Reduction in enzymatic activity can lead to protein dysfunction and in this case, lowered ATP production.

 Pyruvate kinase is also a glycolytic enzyme that catalyzes the final step in glycolysis, the conversion of phosphoenolpyruvate to pyruvate with the concomitant transfer of the high-energy phosphate group from phosphoenolpyruvate to ADP, thereby generating ATP (Figure 10.9). Under aerobic conditions, pyruvate can be transported to the mitochondria, where it enters the TCA cycle and is further broken down to produce considerably more ATP through oxidative phosphorylation. Therefore, the

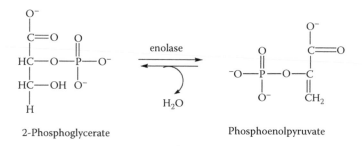

2-Phosphoglycerate   Phosphoenolpyruvate

FIGURE 10.8   Interconversion of 2-phosphoglycerate to phosphoenolpyruvate via enolase.

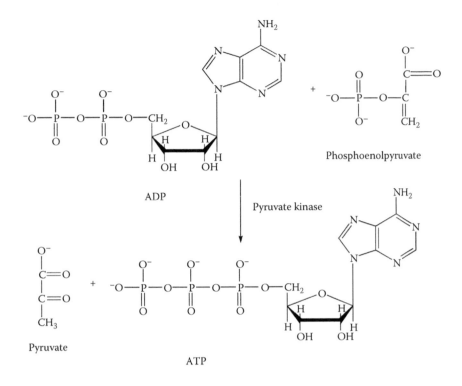

FIGURE 10.9 Pyruvate kinase enzymatic reaction.

oxidative inactivation of this important enzyme in the hippocampi of MCI subjects could conceivably result in the reduced ATP production and alter ATP-dependent processes, such as signal transduction and cell potential maintenance, thereby leading to altered Ca^{2+} homeostasis and neuronal dysfunction.

Lactate dehydrogenase B (LDH) reduces pyruvate to lactate by NADH (Figure 10.10). Lactate is a substrate for gluconeogenesis and since glucose is the major supplier of energy to the brain, proper lactate production is crucial (117). LDH dysfunction and subsequent reduced glucose metabolism are commonly observed in the positron emission tomography (PET) scans of MCI and AD brain (118,119). Enzyme activity of lactate dehydrogenase is significantly reduced in MCI hippocampus, which further correlates protein dysfunction and enzyme activity impairment.

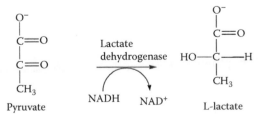

FIGURE 10.10 Conversion of pyruvate to L-lactate via lactate dehydrogenase.

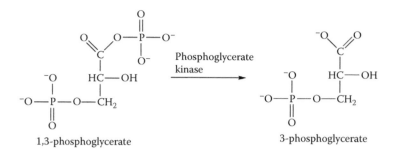

FIGURE 10.11   Phosphoglycerate kinase enzymatic reaction.

Likewise, specific enzyme activity was also lower in MCI hippocampus. Impairment of this enzyme could initiate reduction of glucose production and creation of excess pyruvate.

Phosphoglycerate kinase catalyzes the reaction to convert 1,3-bisphosphoglycerate to 3-phosphoglycerate. This reaction undergoes substrate phosphorylation by phosphoryl transfer from 1,3-bisphosphoglycerate to ADP to produce ATP (Figure 10.11). Additionally, enzyme activity is reduced, thus suggesting that oxidative modification leads to impairment of protein function. Impairment of phosphoglycerate kinase results in decreased energy production and irreversible downstream effects, such as multidrug resistance (120).

Aldolase is an essential protein that cleaves fructose 1,6-bisphosphate to dihydroxyacetonephosphate (DHAP) and glyceraldehyde-3-phosphate (G3P) (Figure 10.12). The production of G3P is essential in continuing glycolysis in order to produce adequate amounts of ATP to maintain ion-motive ATPases, signal transduction, glutamate, and glucose transporters (121,122). Oxidative modification of aldolase can compromise membrane symmetry and render neurons vulnerable to excitotoxicity and apoptosis.

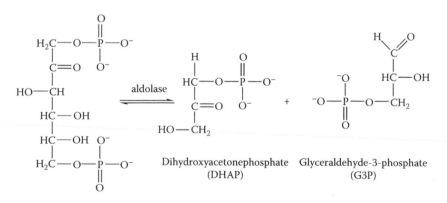

FIGURE 10.12   Aldolase enzymatic reaction.

The impairment of these energy-related proteins would lead to decreased ATP production, with consequent dysfunction in electrochemical gradients, ion pumps, and voltage-gated ion channels, glucose and glutamate transporters, loss of membrane asymmetry, as well as lower efficiency on such processes as Ca^{2+} homeostasis, cell potential, and signal transduction. All these glycolytic protein oxidative modifications observed in the earliest stage of AD support the hypothesis of energy metabolism alteration in Alzheimer's disease and are consistent with the known PET alterations in MCI and AD.

10.11   NEUROPLASTICITY

It is well documented that *glutamine synthetase* (GLUL) activity declines in AD (103,123–125). GLUL catalyzes the rapid amination of glutamate to form the nonneurotoxic amino acid glutamine. This reaction maintains the optimal level of glutamate and ammonia in neurons and modulates excitotoxicity. Together with the action of glutamate receptors and glutamate transporters, this process is important to maintaining neuroplasticity (126). A decline in neuroplasticity is suggested to correlate with the progression of AD from MCI (127). Therefore, oxidative inactivation of GLUL suggests the glutamate–glutamine cycles in hippocampi of MCI subjects are impaired, which may contribute to excitoxicity and impairment of neuroplasticity during the development of AD (128).

10.12   MITOCHONDRIAL DYSFUNCTION

Mitochondrial dysfunction is associated with AD; therefore, the proteomic identification of ATP synthase in MCI IPL and hippocampus contributes to evidence suggesting a role of mitochondrial dysfunction in the progression of AD. *ATP synthase alpha-chain* is a mitochondrial regulating subunit of complex V, which plays a key role in energy production. Alteration of complex V of the electron transport chain, ATP synthase, results in impaired ATP production in the mitochondria, the ATP "powerhouse." ATP synthase goes through a sequence of coordinated conformational changes of its major subunits (alpha and beta) to produce ATP. ATP synthase, alpha-subunit, is the only HNE-modified protein observed by redox proteomics in both MCI hippocampus and IPL. At an early stage, ATP synthase is tightly associated with tau aggregated proteins in neurofibrillary tangles in AD (129), making ATP synthase a potential target for AD therapeutics. The oxidation of ATP synthase leads to the inactivation of this mitochondrial complex. Failure of ATP synthase could contribute to a decrease in the activity of the entire electron transport chain and impaired ATP production, resulting in possible electron leakage from their carrier molecules to generate ROS, suggesting an alternative rationalization for the well-documented existence of oxidative stress in AD and MCI (24,26,31,32,83,130,131). The specific activity of ATP synthase was reduced in MCI hippocampus and IPL compared to age-matched controls, providing insight into enzyme efficiency in MCI (28). Additionally, *malate dehydrogenase* is oxidatively modified in MCI; this enzyme is responsible for oxidation of L-malate to oxaloacetate in the citric acid cycle, another ATP-producing metabolic pathway (Figure 10.13).

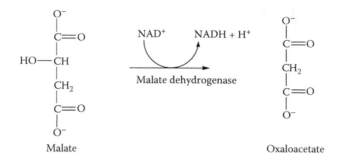

FIGURE 10.13 Malate dehydrogenase enzymatic reaction.

As chaperones, *heat shock proteins* assist in establishing proper protein confor-mation in an effort to prevent protein aggregation by repairing misfolded proteins or guiding misfolding proteins to the proteosome for degradation (132). Numerous heat shock proteins (Hsp70, Hsc71, and Hsp90 and 60, respectively) have been found to be oxidatively modified in neurodegenerative diseases including MCI (133), AD (112), and HD (134). Oxidative modification of heat shock proteins may exacerbate protein misfolding and protein aggregation, leading to reduced effec-tive proteosomal activity and eventual proteosomal overload and dysfunction, known to occur in AD (135).

10.13 ANTIOXIDANT DEFENSE

Peptidyl-prolyl cis/trans isomerase (Pin-1) is a regulatory protein that reversibly alters the conformation of proline residues in proteins from *cis* to *trans* (136). The WW domain of Pin-1 recognizes phosphorylated Ser-Pro and phosphorylated Thr-Pro motifs in proteins, and thereby binds to many cell cycle-regulating proteins, APP, and tau protein. Pin-1 is co-localized with phosphorylated tau and also shows an inverse relationship to the expression of tau in AD brains (137–139). Moreover, Pin-1 is oxidatively inhibited in AD brain (140) and is able to restore the function of tau protein in AD (141). Therefore, the oxidative inactivation of Pin-1 in the hip-pocampi of MCI subjects could be one of the initial events that trigger tangle forma-tion, oxidative damage, cell cycle alterations, and eventual neuronal death in AD brains. Pin-1 specifically recognizes pSer-Pro and pThr-Pro motifs. Pin-1 is exces-sively carbonylated in MCI (83,142) and AD hippocampus (104). Moreover, Pin-1 binds to APP and affects production of amyloid beta peptide. Pin-1 enzymatic activ-ity is significantly decreased in MCI and AD brain, contributing to previous research showing loss of enzymatic activity in oxidatively modified proteins (103,143). This decrease in Pin-1 activity may be responsible for the increased accumulation of both phosphorylated tau protein in AD hippocampus, which is rich in neurofibrillary tan-gles (NFT) and senile plaques (SP), rich in amyloid beta.

Glutathione S-transferase mu and *multidrug resistance protein 3* are nitrated in MCI IPL. These two proteins play an important role in regulating cellular pro-cesses by decreasing the levels of oxidants or by removing toxic compounds that are generated in the cell. Glutathione S-transferases (GSTs) catalyze the binding of HNE to glutathione, resulting in the formation of a GSH-HNE conjugate that is exported out from the cell by the multidrug resistant protein-1 (MRP1). This process thereby plays an important role in cellular protection against oxidative stress. In AD brain, GST protein levels and activity were reported to be decreased; in addition, GST was found to be oxidatively modified by HNE (144). GSTs have a high catalytic activity against HNE and are oxidatively modified and downregulated in AD brain (144). As a corollary, overexpression of GST can combat the effects of HNE toxicity in culture (145).

Peroxiredoxin 6 (Prx VI) is found to be nitrated in MCI hippocampus. Peroxiredoxins remove toxic hydrogen peroxide from the cell, resulting in reduced ROS production. Peroxiredoxins can reduce peroxynitrite at a high catalytic rate, which may modulate protein nitration and cell damage (146). Prx VI is the only peroxiredoxin that uses glutathione as an electron donor, while all other peroxiredoxins (PrxI–PrxV) use thioredoxin. In addition, peroxiredoxins play a role in cell differentiation and apoptosis. The decrease in the activity of this enzyme may also lead to decreased phospholipase A2 activity, one of the target proteins regulated by peptidyl prolyl *cis/trans* isomerase (Pin-1), a protein that has been reported to be downregulated and have decreased activity in MCI and AD brain (104,141,142).

The enzyme GST forms a complex with Prx VI in order to modulate both enzyme activities. These proteins work in coordination with one another either directly or indirectly, thereby protecting the cell from toxicants. These results provide insight into how the changes of these proteins may contribute to tau hyperphosphorylation and neurofibrillary tangle formation, in addition to development of oxidative stress. This result could conceivably be related to the identification of MRP1 as a protein with elevated HNE binding in AD (144). This protein was also mapped to undergo DNA damage by oxidative stress mechanisms (147).

Carbonyl reductase is a vital enzyme that can reduce carbonyl-containing compounds to their resultant alcohols, thereby reducing protein carbonyl levels. Subsequent malfunction or downregulation of this enzyme could cause an increase in protein carbonyls, which because of the polarity of the carbonyl moiety could expose ordinarily buried hydrophobic amino acids to the solvent (i.e., disrupt conformation). Carbonyl reductase has been shown to reduce the lipid peroxidation product, HNE (148). Carbonyl reductase expression is altered in Down's syndrome and AD patients (149). The gene for carbonyl reductase is located in close proximity to the gene for Cu/Zn superoxide dismutase (SOD1) (150). Interestingly, the genes for SOD1, carbonyl reductase, and APP are located on chromosome 21, which is a trisomy in Down's syndrome patients (151). The link between increased amyloid deposition and decreased carbonyl reductase enzymatic activity is unclear. The current research posits a possible intriguing relationship among amyloid beta, Down's syndrome, and carbonyl reductase in neurodegeneration.

Oxidative stress is the imbalance of pro-oxidants and antioxidants, where equilibrium is favoring the side of pro-oxidants. This can also mean antioxidant defense is low due to protein modification. Increased protein expression of peroxiredoxins (I and II), Cu/ZnSOD, carbonyl reductase, and alcohol dehydrogenase, glutathione S-transferase (GST), and multidrug resistant protein-1 (MRP1) in AD brain has been reported (149,152–155). Similarly, antioxidant defense is altered in MCI (156). Several antioxidant proteins undergo oxidative modification in MCI brain (133).

10.14   STRUCTURAL DYSFUNCTION

Dihydropyriminidase related protein 2 (DRP-2) and *fascin 1* are identified as nitrated proteins in MCI hippocampus. DRP-2 is a member of the dihydropy-rimidinase-related protein family that is involved in axonal outgrowth and path-finding through transmission and modulation of extracellular signals (157,158). Additionally, DRP-2 interacts with and modulates collapsin, which aids in dendrite elongation, guidance, and growth cone collapse. DRP-2 has been reported to be associated with NFT, which may lead to decreased levels of cytosolic DRP-2 in AD (159). This, in turn, would eventually lead to abnormal neuritic and axonal growth, thus accelerating neuronal degeneration in AD (160), which is one of the characteristic hallmarks of AD pathology. Increased oxidation (41) and decreased expression of DRP-2 protein was observed in AD. In adult Down's syndrome (DS) (159), fetal DS (161), schizophrenia, and affective disorders, DRP-2 has lower levels in brain. Since memory and learning are associated with synaptic remodeling, nitration and subsequent loss of function of this protein could conceivably be involved in the observed memory decline in MCI. Moreover, the decreased function of DRP-2 could be involved in the shortened dendritic length and synapse loss observed in AD (162). Maintaining neuronal communication is essential in learning and memory. Since memory is greatly depreciated in MCI (163,164), certain proteins involved in neuronal communication are oxidatively modified in MCI and AD (112,133,155).

Fascin 1 (FSCN1) is an actin-bundling, structural protein also known as p55 (165) and is involved in cell adhesion (166) and cell motility (167). It is a marker for dendritic functionality (168). Addition of p55 has been shown to protect cells from oxidative stress produced by an insult (169). The identification of this protein as nitrated in MCI brain is consistent with the notion that loss of function of this protein lessens protection against oxidative damage and could be an important event in the transition of MCI to AD. Fascin 1 has also been found to interact with protein kinase C alpha (PKCα), which regulates focal adhesions (170). Impairment of this protein can be related to faulty neurotransmission from the affected dendritic projections. In a beagle dog model of AD, in which beagle amyloid beta has the same sequence as human amyloid beta, fascin was a protein protected by a diet rich in antioxidants and in a program of environmental enrichment, i.e., making new synapses (171).

Actin is a principal protein playing a central role in maintaining cellular integrity, morphology, and the structure of the plasma membrane. Actin microfilaments play a role in the neuronal membrane cytoskeleton by maintaining the distribution of membrane proteins, and segregating axonal and dendritic proteins (172). In the CNS, actin is distributed widely in neurons, astrocytes, and blood vessels (173) and is particularly concentrated in presynaptic terminals, dendritic spines, and growth cones. Oxidation of actin, by HNE modification, can lead to loss of membrane cytoskeletal structure, decreased membrane fluidity, and trafficking of synaptic proteins and mitochondria.

Moreover, actin is involved in the elongation of the growth cone, and loss of function of actin could play a role in synapse loss and neuronal communication, which may be associated with progressive memory loss documented in AD (174).

10.15   SIGNAL TRANSDUCTION

14-3-3-protein gamma is found to be nitrated in MCI IPL. 14-3-3 gamma is a member of the 14-3-3 protein families, which are highly expressed in the brain (175,176). These proteins are involved in a number of cellular functions including signal transduction, protein trafficking, and metabolism (175). 14-3-3 gamma is associated with neurofibrillary tangles in AD (177) and levels of 14-3-3 proteins are increased in MCI brain (155), AD brain (178,179), AD CSF (180), calorically restricted rats (181), and ICV and neuronal models of AD (116,182). The nitration of 14-3-3 gamma could change its conformation, which conceivably could lead to altered binding to two of its normal binding partners, glycogen synthase kinase 3 beta (GSK3β) and tau. One of the isoforms of 14-3-3 can act as a scaffolding protein and simultaneously bind to tau and GSK3β in a multiprotein tau phosphorylation complex (183). This complex may promote tau phosphorylation and polymerization (184,185), leading to the formation of tangles and further leading to neurodegeneration in AD.

Neuropolypeptide h3 is critical for modulation of the enzyme choline acetyltransferase, which is vital in signal transduction and cell communication. The loss of choline acetyltransferase leads to reduced levels of the neurotransmitter acetylcholine, causing poor neurotransmission (186). NMDA receptors activate the production of this enzyme, and alteration of the NMDA receptor mediates cholinergic deficits (187). AD has cholinergic deficits, consistent with dysregulation in acetylcholine levels and loss of cholinergic neurons (188–191). Consequently, the oxidative modification of this protein in MCI and AD (113) highlights the role of cholinergism in the development of this dementing disease.

10.16   PROTEIN SYNTHESIS

EF-Tu and *eIF*-alpha are intimately involved in protein synthesis machinery. Oxidation, and possibly impairment, of initiation factors of protein synthesis in MCI patients were shown by redox proteomics (28). Human mitochondrial EF-Tu (EF-Tu) is a nuclear-encoded protein and functions in the translational apparatus of mitochondria (192). Like its prokaryotic homolog, the mammalian EF-Tu GTPase hydrolyzes a molecule of GTP each time an amino-acylated tRNA is accommodated on the A site of the ribosome, and its recycling depends on the exchange factor EF-Ts. Nuclear genes encode most respiratory chain subunits and all protein components necessary for maintenance and expression of mtDNA. Mitochondria play pivotal roles in eukaryotic cells in producing cellular energy and essential metabolites as well as in controlling apoptosis by integrating numerous death signals (193). Mitochondrial protein synthesis inhibition is associated with the impairment of differentiation in different cell types, including neurons (194). The coordination of mitochondrial and nuclear genetic systems in the cell is necessary for proper mitochondrial biogenesis and cellular functioning. eIF-α is an abundant protein required to bind aminoacyl-tRNA to acceptor sites of ribosomes in a GTP-dependent manner during protein synthesis (195). Recently, eIF-α has been shown to be involved in cytoskeletal organization by binding and bundling actin filaments and microtubules. eIF-α is an important determinant of cell proliferation and senescence (196), since it is regulated in aging, transformation, and growth arrest. Inhibition of eIF-α induces apoptosis (197), indicating that eIF-α activity is critical to normal cell function. Numerous studies have provided indirect evidence that suggests alterations in protein synthesis may occur in AD (53,198–201) and decreased protein synthesis in AD and MCI (53,202). The dysfunction of the protein synthesis apparatus, mediated in part by oxidative stress, could compromise the ability of cells to generate the various factors needed to regulate cell homeostasis, thus contributing to impaired neuronal function and to the development of neuropathology in MCI patients.

10.17 CONCLUSIONS

In conclusion, the redox proteomics-identified brain proteins in MCI brain play important roles in different neuronal functions and are directly or indirectly linked to AD pathology. Comparative analysis of these proteins between MCI IPL and hippocampus brain regions showed alpha enolase as a common target of protein oxidation by all three modifications mentioned. This suggests that energy metabolism may be among the first cellular properties that become severely affected in MCI. A similar sensitivity to energy metabolism was observed in AD brain (112,113,116,203). Future studies using animal models of the different stages of this dementing disorder should help in further delineating the mechanisms of MCI pathogenesis and to develop effective therapies to combat conversion of MCI to AD. Several HNE-bound and nitrated proteins identified in MCI were identical to those found in not only AD, but other neurodegenerative diseases. These enzymes are involved in cholinergic processes, energy metabolism, transport, detoxification, protein synthesis, and stress response properties. Loss of function or modification to any protein can be catastrophic for neuronal communication, ATP production, and other essential cell functions. Because oxidative stress is involved in arguably the earliest phase of AD (31,32,83), and because MCI may be considered a prodromal phase of AD, the identification of specific target proteins that are oxidatively modified by products of lipid peroxidation and protein nitration may provide vital insights into the role of oxidative damage in mechanisms of neuronal death in AD. The current work forms a framework for subsequent experiments and provides potential new targets for neuroprotective therapeutic intervention in MCI.

ACKNOWLEDGMENTS

The authors thank the University of Kentucky A.D.C. Clinical and Neuropathology Cores for providing the brain specimens used for this study. This research was supported in part by grants from NIH (AG-10836; AG-05119; AG-029839).

REFERENCES

1. Winblad, B., Palmer, K., Kivipelto, M. et al. (2004) Mild cognitive impairment—Beyond controversies, towards a consensus: Report of the International Working Group on Mild Cognitive Impairment, *J Intern Med* 256:240–6.
2. Petersen, R. C. (2004) Mild cognitive impairment as a diagnostic entity, *J Intern Med* 256:183–94.
3. Almkvist, O., Basun, H., Backman, L. et al. (1998) Mild cognitive impairment—An early stage of Alzheimer's disease? *J Neural Transm Suppl* 54:21–9.
4. Flicker, C., Ferris, S. H. and Reisberg, B. (1991) Mild cognitive impairment in the elderly: Predictors of dementia, *Neurology* 41:1006–9.
5. Luis, C. A., Loewenstein, D. A., Acevedo, A., Barker, W. W. and Duara, R. (2003) Mild cognitive impairment: Directions for future research, *Neurology* 61:438–44.
6. Morris, J. C., Storandt, M., Miller, J. P. et al. (2001) Mild cognitive impairment represents early-stage Alzheimer disease, *Arch Neurol* 58:397–405.
7. Portet, F., Ousset, P. J. and Touchon, J. (2005) What is a mild cognitive impairment? *Rev Prat* 55:1891–4.
8. Maioli, F., Coveri, M., Pagni, P. et al. (2007) Conversion of mild cognitive impairment to dementia in elderly subjects: A preliminary study in a memory and cognitive disorder unit, *Arch Gerontol Geriatr* 44 Suppl 1:233–41.
9. Rozzini, L., Chilovi, B. V., Conti, M. et al. (2007) Conversion of amnestic mild cognitive impairment to dementia of Alzheimer type is independent to memory deterioration, *Int J Geriatr Psychiatry* 22:1217–22.
10. Apostolova, L. G., Dutton, R. A., Dinov, I. D. et al. (2006) Conversion of mild cognitive impairment to Alzheimer disease predicted by hippocampal atrophy maps, *Arch Neurol* 63:693–9.
11. Petersen, R. C. (2000) Mild cognitive impairment: Transition between aging and Alzheimer's disease, *Neurologia* 15:93–101.
12. Petersen, R. C. (2003) Mild cognitive impairment clinical trials, *Nat Rev Drug Discov* 2:646–53.
13. de Leon, M. J., DeSanti, S., Zinkowski, R. et al. (2004) MRI and CSF studies in the early diagnosis of Alzheimer's disease, *J Intern Med* 256:205–23.
14. Devanand, D. P., Pradhaban, G., Liu, X. et al. (2007) Hippocampal and entorhinal atrophy in mild cognitive impairment: Prediction of Alzheimer disease, *Neurology* 68:828–36.
15. Jack, C. R., Jr., Petersen, R. C., Xu, Y. C. et al. (1999) Prediction of AD with MRI-based hippocampal volume in mild cognitive impairment, *Neurology* 52:1397–1403.
16. Barnes, J., Godbolt, A. K., Frost, C. et al. (2007) Atrophy rates of the cingulate gyrus and hippocampus in AD and FTLD, *Neurobiol Aging* 28:20–8.
17. Du, A. T., Schuff, N., Amend, D. et al. (2001) Magnetic resonance imaging of the entorhinal cortex and hippocampus in mild cognitive impairment and Alzheimer's disease, *J Neurol Neurosurg Psychiatry* 71:441–7.
18. Du, A. T., Schuff, N., Kramer, J. H. et al. (2004) Higher atrophy rate of entorhinal cortex than hippocampus in AD, *Neurology* 62:422–7.
19. Mevel, K., Chetelat, G., Desgranges, B. and Eustache, F. (2006) Alzheimer's disease, hippocampus and neuroimaging, *Encephale* 32 Pt 4:S1149–54.
20. Mori, E. (2005) Hippocampal atrophy and memory disturbance, *No To Shinkei* 57:1067–78.
21. Halliwell, B. (2006) Oxidative stress and neurodegeneration: Where are we now? *J Neurochem* 97:1634–58.
22. Brown, L. A., Harris, F. L. and Jones, D. P. (1997) Ascorbate deficiency and oxidative stress in the alveolar type II cell, *Am J Physiol* 273:L782–8.

23. Butterfield, D. A. and Kanski, J. (2001) Brain protein oxidation in age-related neuro-degenerative disorders that are associated with aggregated proteins, *Mech Aging Dev* 122:945–62.

24. Butterfield, D. A. and Lauderback, C. M. (2002) Lipid peroxidation and protein oxidation in Alzheimer's disease brain: potential causes and consequences involving amyloid beta-peptide-associated free radical oxidative stress, *Free Radic Biol Med* 32:1050–60.

25. Giasson, B. I., Ischiropoulos, H., Lee, V. M. and Trojanowski, J. Q. (2002) The relationship between oxidative/nitrative stress and pathological inclusions in Alzheimer's and Parkinson's diseases, *Free Radic Biol Med* 32:1264–75.

26. Markesbery, W. R. (1997) Oxidative stress hypothesis in Alzheimer's disease, *Free Radic Biol Med* 23:134–47.

27. Zhu, X., Smith, M. A., Perry, G. and Aliev, G. (2004) Mitochondrial failures in Alzheimer's disease, *Am J Alzheimers Dis Other Demen* 19:345–52.

28. Reed, T., Perluigi, M., Sultana, R. et al. (2008) Redox proteomic identification of 4-hydroxy-2-nonenal-modified brain proteins in amnestic mild cognitive impairment: Insight into the role of lipid peroxidation in the progression and pathogenesis of Alzheimer's disease, *Neurobiol Dis* 30:107–20.

29. Keller, J. N., Schmitt, F. A., Scheff, S. W. et al. (2005) Evidence of increased oxidative damage in subjects with mild cognitive impairment, *Neurology* 64:1152–6.

30. Butterfield, D. A. and Stadtman, E. R. (1997) Protein oxidation processes in aging brain, *Adv Cell Aging Gerontol* 2:161–91.

31. Butterfield, D. A., Reed, T., Perluigi, M. et al. (2006) Elevated protein-bound levels of the lipid peroxidation product, 4-hydroxy-2-nonenal, in brain from persons with mild cognitive impairment, *Neurosci Lett* 397:170–3.

32. Butterfield, D. A., Reed, T. T., Perluigi, M. et al. (2007) Elevated levels of 3-nitrotyrosine in brain from subjects with amnestic mild cognitive impairment: Implications for the role of nitration in the progression of Alzheimer's disease, *Brain Res* 1148:243–8.

33. Dalle-Donne, I., Rossi, R., Colombo, R., Giustarini, D. and Milzani, A. (2006) Biomarkers of oxidative damage in human disease, *Clin Chem* 52:601–23.

34. Dalle-Donne, I., Scaloni, A., Giustarini, D. et al. (2005) Proteins as biomarkers of oxidative/nitrosative stress in diseases: the contribution of redox proteomics, *Mass Spectrom Rev* 24:55–99.

35. Levine, R. L., Garland, D., Oliver, C. N. et al. (1990) Determination of carbonyl content in oxidatively modified proteins, *Methods Enzymol* 186:464–78.

36. Aksenov, M. Y., Aksenova, M. V., Butterfield, D. A., Geddes, J. W. and Markesbery, W. R. (2001) Protein oxidation in the brain in Alzheimer's disease, *Neuroscience* 103:373–83.

37. Stadtman, E. R. (2001) Protein oxidation in aging and age-related diseases, *Ann NY Acad Sci* 928:22–38.

38. Stadtman, E. R. and Berlett, B. S. (1997) Reactive oxygen-mediated protein oxidation in aging and disease, *Chem Res Toxicol* 10:485–94.

39. Loske, C., Neumann, A., Cunningham, A. M. et al. (1998) Cytotoxicity of advanced glycation endproducts is mediated by oxidative stress, *J Neural Transm* 105:1005–15.

40. Munch, G., Schinzel, R., Loske, C. et al. (1998) Alzheimer's disease—Synergistic effects of glucose deficit, oxidative stress and advanced glycation endproducts, *J Neural Transm* 105:439–61.

41. Esterbauer, H., Schaur, R. J. and Zollner, H. (1991) Chemistry and biochemistry of 4-hydroxynonenal, malonaldehyde and related aldehydes, *Free Radic Biol Med* 11:81–128.

42. Pamplona, R., Dalfo, E., Ayala, V. et al. (2005) Proteins in human brain cortex are modified by oxidation, glycoxidation, and lipoxidation. Effects of Alzheimer disease and identification of lipoxidation targets, *J Biol Chem* 280:21522–30.

43. O'Brien, J. S. and Sampson, E. L. (1965) Lipid composition of the normal human brain: Gray matter, white matter, and myelin, *J Lipid Res* 6:537–44.
44. Skinner, E. R., Watt, C., Besson, J. A. and Best, P. V. (1993) Differences in the fatty acid composition of the grey and white matter of different regions of the brains of patients with Alzheimer's disease and control subjects, *Brain* 116 (Pt 3):717–25.
45. Montine, T. J., Neely, M. D., Quinn, J. F. et al. (2002) Lipid peroxidation in aging brain and Alzheimer's disease, *Free Radic Biol Med* 33:620–6.
46. Perluigi, M., Sultana, R., Cenini, G., Di Domenico, F., Memo, M., Pierce, W. M., Coccia, R., and Butterfield, D. A. (2009) Redox Proteomics Identification of HNE-Modified Brain Proteins in Alzheimer's Disease: Role of Lipid Peroxidaton in Alzheimer's Disease Pathogenesis, *Proteomics—Clin Appl* 118:131–150.
47. Bader Lange, M. L., Cenini, G., Piroddi, M. et al. (2008) Loss of phospholipid asymmetry and elevated brain apoptotic protein levels in subjects with amnestic mild cognitive impairment and Alzheimer disease, *Neurobiol Dis* 29:456–64.
48. Williams, T. I., Lynn, B. C., Markesbery, W. R. and Lovell, M. A. (2006) Increased levels of 4-hydroxynonenal and acrolein, neurotoxic markers of lipid peroxidation, in the brain in mild cognitive impairment and early Alzheimer's disease, *Neurobiol Aging* 27:1094–9.
49. Subramaniam, R., Roediger, F., Jordan, B. et al. (1997) The lipid peroxidation product, 4-hydroxy-2-trans-nonenal, alters the conformation of cortical synaptosomal membrane proteins, *J Neurochem* 69:1161–9.
50. Tamagno, E., Robino, G., Obbili, A. et al. (2003) H2O2 and 4-hydroxynonenal mediate amyloid beta-induced neuronal apoptosis by activating JNKs and p38MAPK, *Exp Neurol* 180:144–55.
51. Uchida, K. (2003) 4-Hydroxy-2-nonenal: A product and mediator of oxidative stress, *Prog Lipid Res* 42:318–43.
52. Camandola, S., Poli, G. and Mattson, M. P. (2000) The lipid peroxidation product 4-hydroxy-2,3-nonenal inhibits constitutive and inducible activity of nuclear factor kappa B in neurons, *Brain Res Mol Brain Res* 85:53–60.
53. Ding, Q., Markesbery, W. R., Chen, Q., Li, F. and Keller, J. N. (2005) Ribosome dysfunction is an early event in Alzheimer's disease, *J Neurosci* 25:9171–5.
54. Drake, J., Petroze, R., Castegna, A. et al. (2004) 4-Hydroxynonenal oxidatively modifies histones: Implications for Alzheimer's disease, *Neurosci Lett* 356:155–8.
55. Poot, M., Verkerk, A., Koster, J. F., Esterbauer, H. and Jongkind, J. F. (1988) Reversible inhibition of DNA and protein synthesis by cumene hydroperoxide and 4-hydroxy-nonenal, *Mech Ageing Dev* 43:1–9.
56. Uchida, K. and Stadtman, E. R. (1992) Modification of histidine residues in proteins by reaction with 4-hydroxynonenal, *Proc Natl Acad Sci U S A* 89:4544–8.
57. Okada, K., Wangpoengtrakul, C., Osawa, T. et al. (1999) 4-Hydroxy-2-nonenal-mediated impairment of intracellular proteolysis during oxidative stress. Identification of proteasomes as target molecules, *J Biol Chem* 274:23787–93.
58. Pratico, D., Clark, C. M., Liun, F. et al. (2002) Increase of brain oxidative stress in mild cognitive impairment: A possible predictor of Alzheimer disease, *Arch Neurol* 59:972–6.
59. Alvarez, B. and Radi, R. (2003) Peroxynitrite reactivity with amino acids and proteins, *Amino Acids* 25:295–311.
60. Tien, M., Berlett, B. S., Levine, R. L., Chock, P. B. and Stadtman, E. R. (1999) Peroxynitrite-mediated modification of proteins at physiological carbon dioxide concentration: pH dependence of carbonyl formation, tyrosine nitration, and methionine oxidation, *Proc Natl Acad Sci U S A* 96:7809–14.

61. Hazen, S. L., Gaut, J. P., Hsu, F. F. et al. (1997) p-Hydroxyphenylacetaldehyde, the major product of L-tyrosine oxidation by the myeloperoxidase-H2O2-chloride system of phagocytes, covalently modifies epsilon-amino groups of protein lysine residues, *J Biol Chem* 272:16990–8.

62. Smith, M. A., Richey Harris, P. L., Sayre, L. M., Beckman, J. S. and Perry, G. (1997) Widespread peroxynitrite-mediated damage in Alzheimer's disease, *J Neurosci* 17:2653–7.

63. Botti, H., Trostchansky, A., Batthyany, C. and Rubbo, H. (2005) Reactivity of peroxynitrite and nitric oxide with LDL, *IUBMB Life* 57:407–12.

64. Niles, J. C., Wishnok, J. S. and Tannenbaum, S. R. (2006) Peroxynitrite-induced oxidation and nitration products of guanine and 8-oxoguanine: Structures and mechanisms of product formation, *Nitric Oxide* 14:109–21.

65. Szabo, C. (1996) DNA strand breakage and activation of poly-ADP ribosyltransferase: A cytotoxic pathway triggered by peroxynitrite, *Free Radic Biol Med* 21:855–69.

66. Masuda, M., Nishino, H. and Ohshima, H. (2002) Formation of 8-nitroguanosine in cellular RNA as a biomarker of exposure to reactive nitrogen species, *Chem Biol Interact* 139:187–97.

67. Good, P. F., Werner, P., Hsu, A., Olanow, C. W. and Perl, D. P. (1996) Evidence of neuronal oxidative damage in Alzheimer's disease, *Am J Pathol* 149:21–8.

68. Good, P. F., Hsu, A., Werner, P., Perl, D. P. and Olanow, C. W. (1998) Protein nitration in Parkinson's disease, *J Neuropathol Exp Neurol* 57:338–42.

69. Drake, J., Kanski, J., Varadarajan, S., Tsoras, M. and Butterfield, D. A. (2002) Elevation of brain glutathione by gamma-glutamylcysteine ethyl ester protects against peroxynitrite-induced oxidative stress, *J Neurosci Res* 68:776–84.

70. Whiteman, M., Tritschler, H. and Halliwell, B. (1996) Protection against peroxynitrite-dependent tyrosine nitration and alpha 1-antiproteinase inactivation by oxidized and reduced lipoic acid, *FEBS Lett* 379:74–6.

71. Christen, S., Woodall, A. A., Shigenaga, M. K. et al. (1997) gamma-tocopherol traps mutagenic electrophiles such as NO(X) and complements alpha-tocopherol: Physiological implications, *Proc Natl Acad Sci U S A* 94:3217–22.

72. Anantharaman, M., Tangpong, J., Keller, J. N. et al. (2006) Beta-amyloid mediated nitration of manganese superoxide dismutase: Implication for oxidative stress in a APPNLH/NLH X PS-1P264L/P264L double knock-in mouse model of Alzheimer's disease, *Am J Pathol* 168:1608–18.

73. Aoyama, K., Matsubara, K., Fujikawa, Y. et al. (2000) Nitration of manganese superoxide dismutase in cerebrospinal fluids is a marker for peroxynitrite-mediated oxidative stress in neurodegenerative diseases, *Ann Neurol* 47:524–7.

74. Ischiropoulos, H., Zhu, L., Chen, J. et al. (1992) Peroxynitrite-mediated tyrosine nitration catalyzed by superoxide dismutase, *Arch Biochem Biophys* 298:431–7.

75. MacMillan-Crow, L. A. and Thompson, J. A. (1999) Tyrosine modifications and inactivation of active site manganese superoxide dismutase mutant (Y34F) by peroxynitrite, *Arch Biochem Biophys* 366:82–8.

76. Hara, M. R., Cascio, M. B. and Sawa, A. (2006) GAPDH as a sensor of NO stress, *Biochim Biophys Acta* 1762:502–9.

77. Souza, J. M. and Radi, R. (1998) Glyceraldehyde-3-phosphate dehydrogenase inactivation by peroxynitrite, *Arch Biochem Biophys* 360:187–94.

78. Clements, M. K., Siemsen, D. W., Swain, S. D. et al. (2003) Inhibition of actin polymerization by peroxynitrite modulates neutrophil functional responses, *J Leukoc Biol* 73:344–55.

79. Neumann, P., Gertzberg, N., Vaughan, E. et al. (2006) Peroxynitrite mediates TNF-{alpha}-induced endothelial barrier dysfunction and nitration of actin, *Am J Physiol Lung Cell Mol Physiol* 290:L674–84.

80. Di Stasi, A. M., Mallozzi, C., Macchia, G. et al. (2002) Peroxynitrite affects exocytosis and SNARE complex formation and induces tyrosine nitration of synaptic proteins, *J Neurochem* 82:420–9.
81. Blanchard-Fillion, B., Souza, J. M., Friel, T. et al. (2001) Nitration and inactivation of tyrosine hydroxylase by peroxynitrite, *J Biol Chem* 276:46017–23.
82. Gow, A. J., Duran, D., Malcolm, S. and Ischiropoulos, H. (1996) Effects of peroxynitrite-induced protein modifications on tyrosine phosphorylation and degradation, *FEBS Lett* 385:63–6.
83. Butterfield, D. A., Poon, H. F., St Clair, D. et al. (2006) Redox proteomics identification of oxidatively modified hippocampal proteins in mild cognitive impairment: Insights into the development of Alzheimer's disease, *Neurobiol Dis* 22:223–32.
84. Wilkins, M. R., Sanchez, J. C., Gooley, A. A. et al. (1996) Progress with proteome projects: Why all proteins expressed by a genome should be identified and how to do it, *Biotechnol Genet Eng Rev* 13:19–50.
85. Davidsson, P. and Sjogren, M. (2005) The use of proteomics in biomarker discovery in neurodegenerative diseases, *Dis Markers* 21:81–92.
86. Galasko, D. (2005) Biomarkers for Alzheimer's disease—Clinical needs and application, *J Alzheimers Dis* 8:339–46.
87. Lee, J. W., Namkoong, H., Kim, H. K. et al. (2007) Fibrinogen gamma—A chain precursor in CSF: A candidate biomarker for Alzheimer's disease, *BMC Neurol* 7:14.
88. Hortin, G. L., Jortani, S. A., Ritchie, J. C., Jr., Valdes, R., Jr. and Chan, D. W. (2006) Proteomics: A new diagnostic frontier, *Clin Chem* 52:1218–22.
89. Solassol, J., Boulle, N., Maudelonde, T. and Mange, A. (2005) Clinical proteomics: Towards early detection of cancers, *Med Sci (Paris)* 21:722–9.
90. Ho, L., Sharma, N., Blackman, L. et al. (2005) From proteomics to biomarker discovery in Alzheimer's disease, *Brain Res Rev* 48:360–9.
91. Simonsen, A. H., McGuire, J., Hansson, O. et al. (2007) Novel panel of cerebrospinal fluid biomarkers for the prediction of progression to Alzheimer dementia in patients with mild cognitive impairment, *Arch Neurol* 64:366–70.
92. Zhang, J., Goodlett, D. R., Quinn, J. F. et al. (2005) Quantitative proteomics of cerebrospinal fluid from patients with Alzheimer disease, *J Alzheimers Dis* 7:125–33; discussion 173–80.
93. Poon, H. F., Vaishnav, R. A., Getchell, T. V., Getchell, M. L. and Butterfield, D. A. (2006) Quantitative proteomics analysis of differential protein expression and oxidative modification of specific proteins in the brains of old mice, *Neurobiol Aging* 27:1010–9.
94. Boguski, M. S. and McIntosh, M. W. (2003) Biomedical informatics for proteomics, *Nature* 422:233–7.
95. Maurer, H. H. and Peters, F. T. (2005) Toward high-throughput drug screening using mass spectrometry, *Ther Drug Monit* 27:686–8.
96. Thongboonkerd, V., McLeish, K. R., Arthur, J. M. and Klein, J. B. (2002) Proteomic analysis of normal human urinary proteins isolated by acetone precipitation or ultracentrifugation, *Kidney Int* 62:1461–9.
97. Aebersold, R. and Goodlett, D. R. (2001) Mass spectrometry in proteomics, *Chem Rev* 101:269–95.
98. Domon, B. and Aebersold, R. (2006) Mass spectrometry and protein analysis, *Science* 312:212–7.
99. Gygi, S. P. and Aebersold, R. (2000) Mass spectrometry and proteomics, *Curr Opin Chem Biol* 4:489–94.
100. Aksenov, M., Aksenova, M., Butterfield, D. A. and Markesbery, W. R. (2000) Oxidative modification of creatine kinase BB in Alzheimer's disease brain, *J Neurochem* 74:2520–7.

101. Poon, H. F., Frasier, M., Shreve, N. et al. (2005) Mitochondrial associated metabolic proteins are selectively oxidized in A30P alpha-synuclein transgenic mice—A model of familial Parkinson's disease, *Neurobiol Dis* 18:492–8.

102. Butterfield, D. A. and Castegna, A. (2003) Proteomic analysis of oxidatively modified proteins in Alzheimer's disease brain: Insights into neurodegeneration, *Cell Mol Biol (Noisy-le-grand)* 49:747–51.

103. Hensley, K., Hall, N., Subramaniam, R. et al. (1995) Brain regional correspondence between Alzheimer's disease histopathology and biomarkers of protein oxidation, *J Neurochem* 65:2146–56.

104. Sultana, R., Boyd-Kimball, D., Poon, H. F. et al. (2006) Oxidative modification and down-regulation of Pin1 in Alzheimer's disease hippocampus: A redox proteomics analysis, *Neurobiol Aging* 27:918–25.

105. Tangpong, J., Cole, M. P., Sultana, R. et al. (2007) Adriamycin-mediated nitration of manganese superoxide dismutase in the central nervous system: insight into the mechanism of chemobrain, *J Neurochem* 100:191–201.

106. Scheff, S. W., Price, D. A., Schmitt, F. A. and Mufson, E. J. (2006) Hippocampal synaptic loss in early Alzheimer's disease and mild cognitive impairment, *Neurobiol Aging* 27:1372–84.

107. Cao, Q., Jiang, K., Zhang, M. et al. (2003) Brain glucose metabolism and neuropsychological test in patients with mild cognitive impairment, *Chin Med J (Engl)* 116:1235–8.

108. Geddes, J. W., Pang, Z. and Wiley, D. H. (1996) Hippocampal damage and cytoskeletal disruption resulting from impaired energy metabolism. Implications for Alzheimer disease, *Mol Chem Neuropathol* 28:65–74.

109. Messier, C. and Gagnon, M. (1996) Glucose regulation and cognitive functions: Relation to Alzheimer's disease and diabetes, *Behav Brain Res* 75:1–11.

110. Vanhanen, M. and Soininen, H. (1998) Glucose intolerance, cognitive impairment and Alzheimer's disease, *Curr Opin Neurol* 11:673–7.

111. Boyd-Kimball, D., Castegna, A., Sultana, R. et al. (2005) Proteomic identification of proteins oxidized by Abeta(1-42) in synaptosomes: Implications for Alzheimer's disease, *Brain Res* 1044:206–15.

112. Castegna, A., Aksenov, M., Thongboonkerd, V. et al. (2002) Proteomic identification of oxidatively modified proteins in Alzheimer's disease brain. Part II: Dihydropyrimidinase-related protein 2, alpha-enolase and heat shock cognate 71, *J Neurochem* 82:1524–32.

113. Castegna, A., Thongboonkerd, V., Klein, J. B. et al. (2003) Proteomic identification of nitrated proteins in Alzheimer's disease brain, *J Neurochem* 85:1394–401.

114. Poon, H. F., Castegna, A., Farr, S. A. et al. (2004) Quantitative proteomics analysis of specific protein expression and oxidative modification in aged senescence-accelerated-prone 8 mice brain, *Neuroscience* 126:915–26.

115. Sultana, R., Boyd-Kimball, D., Poon, H. F. et al. (2006) Redox proteomics identification of oxidized proteins in Alzheimer's disease hippocampus and cerebellum: An approach to understand pathological and biochemical alterations in AD, *Neurobiol Aging* 27:1564–76.

116. Sultana, R., Perluigi, M. and Butterfield, D. A. (2006) Redox proteomics identification of oxidatively modified proteins in Alzheimer's disease brain and in vivo and in vitro models of AD centered around Abeta(1-42), *J Chromatogr B Analyt Technol Biomed Life Sci* 833:3–11.

117. Kida, K., Nishio, T., Nagai, K., Matsuda, H. and Nakagawa, H. (1982) Gluconeogenesis in the kidney in vivo in fed rats. Circadian change and substrate specificity, *J Biochem (Tokyo)* 91:755–60.

118. Hoyer, S. (2004) Glucose metabolism and insulin receptor signal transduction in Alzheimer disease, *Eur J Pharmacol* 490:115–25.

119. Rapoport, S. I. (1999) Functional brain imaging in the resting state and during activation in Alzheimer's disease. Implications for disease mechanisms involving oxidative phosphorylation, *Ann N Y Acad Sci* 893:138–53.
120. Duan, Z., Lamendola, D. E., Yusuf, R. Z. et al. (2002) Overexpression of human phosphoglycerate kinase 1 (PGK1) induces a multidrug resistance phenotype, *Anticancer Res* 22:1933–41.
121. Keller, J. N., Mark, R. J., Bruce, A. J. et al. (1997) 4-Hydroxynonenal, an aldehydic product of membrane lipid peroxidation, impairs glutamate transport and mitochondrial function in synaptosomes, *Neuroscience* 80:685–96.
122. Mattson, M. P. (2004) Metal-catalyzed disruption of membrane protein and lipid signaling in the pathogenesis of neurodegenerative disorders, *Ann N Y Acad Sci* 1012:37–50.
123. Aksenov, M. Y., Aksenova, M. V., Butterfield, D. A. et al. (1996) Glutamine synthetase-induced enhancement of beta-amyloid peptide A beta (1-40) neurotoxicity accompanied by abrogation of fibril formation and A beta fragmentation, *J Neurochem* 66:2050–6.
124. Butterfield, D. A., Hensley, K., Cole, P. et al. (1997) Oxidatively induced structural alteration of glutamine synthetase assessed by analysis of spin label incorporation kinetics: Relevance to Alzheimer's disease, *J Neurochem* 68:2451–7.
125. Howard, B. J., Yatin, S., Hensley, K. et al. (1996) Prevention of hyperoxia-induced alterations in synaptosomal membrane-associated proteins by N-tert-butyl-alpha-phenylnitrone and 4-hydroxy-2,2,6,6-tetramethylpiperidin-1-oxyl (Tempol), *J Neurochem* 67:2045–50.
126. Lamprecht, R. and LeDoux, J. (2004) Structural plasticity and memory, *Nat Rev Neurosci* 5:45–54.
127. Ikonomovic, M. D., Mufson, E. J., Wuu, J. et al. (2003) Cholinergic plasticity in hippocampus of individuals with mild cognitive impairment: Correlation with Alzheimer's neuropathology, *J Alzheimers Dis* 5:39–48.
128. Lee, H. G., Zhu, X., Ghanbari, H. A. et al. (2002) Differential regulation of glutamate receptors in Alzheimer's disease, *Neurosignals* 11:282–92.
129. Sergeant, N., Wattez, A., Galvan-valencia, M. et al. (2003) Association of ATP synthase alpha-chain with neurofibrillary degeneration in Alzheimer's disease, *Neuroscience* 117:293–303.
130. Butterfield, D. A., Castegna, A., Lauderback, C. M. and Drake, J. (2002) Evidence that amyloid beta-peptide-induced lipid peroxidation and its sequelae in Alzheimer's disease brain contribute to neuronal death, *Neurobiol Aging* 23:655–64.
131. Butterfield, D. A., Drake, J., Pocernich, C. and Castegna, A. (2001) Evidence of oxidative damage in Alzheimer's disease brain: Central role for amyloid beta-peptide, *Trends Mol Med* 7:548–54.
132. Hinault, M. P., Ben-Zvi, A. and Goloubinoff, P. (2006) Chaperones and proteases: Cellular fold-controlling factors of proteins in neurodegenerative diseases and aging, *J Mol Neurosci* 30:249–65.
133. Sultana, R., Reed, T., Perluigi, M. et al. (2007) Proteomic identification of nitrated brain proteins in amnestic mild cognitive impairment: A regional study, *J Cell Mol Med* 11:839–51.
134. Perluigi, M., Fai Poon, H., Hensley, K. et al. (2005) Proteomic analysis of 4-hydroxy-2-nonenal-modified proteins in G93A-SOD1 transgenic mice—A model of familial amyotrophic lateral sclerosis, *Free Radic Biol Med* 38:960–8.
135. Magrane, J., Smith, R. C., Walsh, K. and Querfurth, H. W. (2004) Heat shock protein 70 participates in the neuroprotective response to intracellularly expressed beta-amyloid in neurons, *J Neurosci* 24:1700–6.

136. Schutkowski, M., Bernhardt, A., Zhou, X. Z. et al. (1998) Role of phosphorylation in determining the backbone dynamics of the serine/threonine-proline motif and Pin1 substrate recognition, *Biochemistry* 37:5566–75.

137. Holzer, M., Gartner, U., Stobe, A. et al. (2002) Inverse association of Pin1 and tau accumulation in Alzheimer's disease hippocampus, *Acta Neuropathol* 104:471–81.

138. Kurt, M. A., Davies, D. C., Kidd, M., Duff, K. and Howlett, D. R. (2003) Hyperphosphorylated tau and paired helical filament-like structures in the brains of mice carrying mutant amyloid precursor protein and mutant presenilin-1 transgenes, *Neurobiol Dis* 14:89–97.

139. Ramakrishnan, P., Dickson, D. W. and Davies, P. (2003) Pin1 colocalization with phosphorylated tau in Alzheimer's disease and other tauopathies, *Neurobiol Dis* 14:251–64.

140. Sultana, R., Boyd-Kimball, D., Poon, H. F. et al. (2005) Redox proteomics identification of oxidized proteins in Alzheimer's disease hippocampus and cerebellum: An approach to understand pathological and biochemical alterations in AD, *Neurobiol Aging* 27:1564–76.

141. Lu, P. J., Wulf, G., Zhou, X. Z., Davies, P. and Lu, K. P. (1999) The prolyl isomerase Pin1 restores the function of Alzheimer-associated phosphorylated tau protein, *Nature* 399:784–8.

142. Butterfield, D. A., Abdul, H. M., Opii, W. et al. (2006) Pin1 in Alzheimer's disease, *J Neurochem* 98:1697–706.

143. Lauderback, C. M., Hackett, J. M., Huang, F. F. et al. (2001) The glial glutamate transporter, GLT-1, is oxidatively modified by 4-hydroxy-2-nonenal in the Alzheimer's disease brain: The role of Abeta1-42, *J Neurochem* 78:413–6.

144. Sultana, R. and Butterfield, D. A. (2004) Oxidatively modified GST and MRP1 in Alzheimer's disease brain: Implications for accumulation of reactive lipid peroxidation products, *Neurochem Res* 29:2215–20.

145. Xie, C., Lovell, M. A., Xiong, S. et al. (2001) Expression of glutathione-S-transferase isozyme in the SY5Y neuroblastoma cell line increases resistance to oxidative stress, *Free Radic Biol Med* 31:73–81.

146. Peshenko, I. V. and Shichi, H. (2001) Oxidation of active center cysteine of bovine 1-Cys peroxiredoxin to the cysteine sulfenic acid form by peroxide and peroxynitrite, *Free Radic Biol Med* 31:292–303.

147. Akman, S. A., O'Connor, T. R. and Rodriguez, H. (2000) Mapping oxidative DNA damage and mechanisms of repair, *Ann N Y Acad Sci* 899:88–102.

148. Doorn, J. A., Maser, E., Blum, A., Claffey, D. J. and Petersen, D. R. (2004) Human carbonyl reductase catalyzes reduction of 4-oxonon-2-enal, *Biochemistry* 43:13106–14.

149. Balcz, B., Kirchner, L., Cairns, N., Fountoulakis, M. and Lubec, G. (2001) Increased brain protein levels of carbonyl reductase and alcohol dehydrogenase in Down syndrome and Alzheimer's disease, *J Neural Transm Suppl* 6:193–201.

150. Lemieux, N., Malfoy, B. and Forrest, G. L. (1993) Human carbonyl reductase (CBR) localized to band 21q22.1 by high-resolution fluorescence in situ hybridization displays gene dosage effects in trisomy 21 cells, *Genomics* 15:169–72.

151. Korenberg, J. R., Bradley, C. and Disteche, C. M. (1992) Down syndrome: Molecular mapping of the congenital heart disease and duodenal stenosis, *Am J Hum Genet* 50:294–302.

152. Kim, S. H., Fountoulakis, M., Cairns, N. and Lubec, G. (2001) Protein levels of human peroxiredoxin subtypes in brains of patients with Alzheimer's disease and Down syndrome, *J Neural Transm Suppl* 6:223–35.

153. Krapfenbauer, K., Engidawork, E., Cairns, N., Fountoulakis, M. and Lubec, G. (2003) Aberrant expression of peroxiredoxin subtypes in neurodegenerative disorders, *Brain Res* 967:152–60.

154. Schonberger, S. J., Edgar, P. F., Kydd, R., Faull, R. L. and Cooper, G. J. (2001) Proteomic analysis of the brain in Alzheimer's disease: molecular phenotype of a complex disease process, *Proteomics* 1:1519–28.
155. Sultana, R., Boyd-Kimball, D., Cai, J. et al. (2007) Proteomics analysis of the Alzheimer's disease hippocampal proteome, *J Alzheimers Dis* 11:153–64.
156. Mecocci, P. (2004) Oxidative stress in mild cognitive impairment and Alzheimer disease: A continuum, *J Alzheimers Dis* 6:159–63.
157. Hamajima, N., Matsuda, K., Sakata, S. et al. (1996) A novel gene family defined by human dihydropyrimidinase and three related proteins with differential tissue distribution, *Gene* 180:157–63.
158. Kato, Y., Hamajima, N., Inagaki, H. et al. (1998) Post-meiotic expression of the mouse dihydropyrimidinase-related protein 3 (DRP-3) gene during spermiogenesis, *Mol Reprod Dev* 51:105–11.
159. Lubec, G., Nonaka, M., Krapfenbauer, K. et al. (1999) Expression of the dihydropyrimidinase related protein 2 (DRP-2) in Down syndrome and Alzheimer's disease brain is downregulated at the mRNA and dysregulated at the protein level, *J Neural Transm Suppl* 57:161–77.
160. Yoshida, H., Watanabe, A. and Ihara, Y. (1998) Collapsin response mediator protein-2 is associated with neurofibrillary tangles in Alzheimer's disease, *J Biol Chem* 273:9761–8.
161. Weitzdoerfer, R., Fountoulakis, M. and Lubec, G. (2001) Aberrant expression of dihydropyrimidinase related proteins-2,-3 and -4 in fetal Down syndrome brain, *J Neural Transm Suppl*:95–107.
162. Coleman, P. D. and Flood, D. G. (1987) Neuron numbers and dendritic extent in normal aging and Alzheimer's disease, *Neurobiol Aging* 8:521–45.
163. Arnaiz, E. and Almkvist, O. (2003) Neuropsychological features of mild cognitive impairment and preclinical Alzheimer's disease, *Acta Neurol Scand Suppl* 179:34–41.
164. Rami, L., Molinuevo, J. L., Sanchez-Valle, R., Bosch, B. and Villar, A. (2007) Screening for amnestic mild cognitive impairment and early Alzheimer's disease with M@T (Memory Alteration Test) in the primary care population, *Int J Geriatr Psychiatry* 22:294–304.
165. Yamashiro, S., Yamakita, Y., Ono, S. and Matsumura, F. (1998) Fascin, an actin-bundling protein, induces membrane protrusions and increases cell motility of epithelial cells, *Mol Biol Cell* 9:993–1006.
166. Adams, J. C. (1995) Formation of stable microspikes containing actin and the 55 kDa actin bundling protein, fascin, is a consequence of cell adhesion to thrombospondin-1: Implications for the anti-adhesive activities of thrombospondin-1, *J Cell Sci* 108 (Pt 5):1977–90.
167. Adams, J. C. (2004) Roles of fascin in cell adhesion and motility, *Curr Opin Cell Biol* 16:590–6.
168. Pinkus, G. S., Lones, M. A., Matsumura, F. et al. (2002) Langerhans cell histiocytosis immunohistochemical expression of fascin, a dendritic cell marker, *Am J Clin Pathol* 118:335–43.
169. Graziewicz, M. A., Day, B. J. and Copeland, W. C. (2002) The mitochondrial DNA polymerase as a target of oxidative damage, *Nucleic Acids Res* 30:2817–24.
170. Anilkumar, N., Parsons, M., Monk, R., Ng, T. and Adams, J. C. (2003) Interaction of fascin and protein kinase Calpha: a novel intersection in cell adhesion and motility, *Embo J* 22:5390–402.
171. Opii, W. O., Joshi, G., Head, E. et al. (2008) Proteomic identification of brain proteins in the canine model of human aging following a long-term treatment with antioxidants and a program of behavioral enrichment: Relevance to Alzheimer's disease, *Neurobiol Aging* 29:51–70.

172. Battaini, F., Pascale, A., Lucchi, L., Pasinetti, G. M. and Govoni, S. (1999) Protein kinase C anchoring deficit in postmortem brains of Alzheimer's disease patients, *Exp Neurol* 159:559–64.

173. Goldman, J. E. (1983) Immunocytochemical studies of actin localization in the central nervous system, *J Neurosci* 3:1952–62.

174. Masliah, E., Mallory, M., Hansen, L. et al. (1994) Synaptic and neuritic alterations during the progression of Alzheimer's disease, *Neurosci Lett* 174:67–72.

175. Dougherty, M. K. and Morrison, D. K. (2004) Unlocking the code of 14-3-3, *J Cell Sci* 117:1875–84.

176. Takahashi, Y. (2003) The 14-3-3 proteins: Gene, gene expression, and function, *Neurochem Res* 28:1265–73.

177. Sugimori, K., Kobayashi, K., Kitamura, T., Sudo, S. and Koshino, Y. (2007) 14-3-3 protein beta isoform is associated with 3-repeat tau neurofibrillary tangles in Alzheimer's disease, *Psychiatry Clin Neurosci* 61:159–67.

178. Frautschy, S. A., Baird, A. and Cole, G. M. (1991) Effects of injected Alzheimer beta-amyloid cores in rat brain, *Proc Natl Acad Sci U S A* 88:8362–6.

179. Layfield, R., Fergusson, J., Aitken, A. et al. (1996) Neurofibrillary tangles of Alzheimer's disease brains contain 14-3-3 proteins, *Neurosci Lett* 209:57–60.

180. Burkhard, P. R., Sanchez, J. C., Landis, T. and Hochstrasser, D. F. (2001) CSF detection of the 14-3-3 protein in unselected patients with dementia, *Neurology* 56:1528–33.

181. Poon, H. F., Shepherd, H. M., Reed, T. T. et al. (2006) Proteomics analysis provides insight into caloric restriction mediated oxidation and expression of brain proteins associated with age-related impaired cellular processes: Mitochondrial dysfunction, glutamate dysregulation and impaired protein synthesis, *Neurobiol Aging* 27:1020–34.

182. Boyd-Kimball, D., Sultana, R., Poon, H. F. et al. (2005) Proteomic identification of proteins specifically oxidized by intracerebral injection of amyloid beta-peptide (1-42) into rat brain: Implications for Alzheimer's disease, *Neuroscience* 132:313–24.

183. Agarwal-Mawal, A., Qureshi, H. Y., Cafferty, P. W. et al. (2003) 14-3-3 connects glycogen synthase kinase-3 beta to tau within a brain microtubule-associated tau phosphorylation complex, *J Biol Chem* 278:12722–8.

184. Hashiguchi, M., Sobue, K. and Paudel, H. K. (2000) 14-3-3 zeta is an effector of tau protein phosphorylation, *J Biol Chem* 275:25247–54.

185. Hernandez, F., Cuadros, R. and Avila, J. (2004) Zeta 14-3-3 protein favours the formation of human tau fibrillar polymers, *Neurosci Lett* 357:143–6.

186. Ojika, K., Tsugu, Y., Mitake, S., Otsuka, Y. and Katada, E. (1998) NMDA receptor activation enhances the release of a cholinergic differentiation peptide (HCNP) from hippocampal neurons in vitro, *Brain Res Dev Brain Res* 106:173–80.

187. Jouvenceau, A., Dutar, P. and Billard, J. M. (1998) Alteration of NMDA receptor-mediated synaptic responses in CA1 area of the aged rat hippocampus: contribution of GABAergic and cholinergic deficits, *Hippocampus* 8:627–37.

188. Davies, P. and Terry, R. D. (1981) Cortical somatostatin-like immunoreactivity in cases of Alzheimer's disease and senile dementia of the Alzheimer type, *Neurobiol Aging* 2:9–14.

189. Davis, B. M., Mohs, R. C., Greenwald, B. S. et al. (1985) Clinical studies of the cholinergic deficit in Alzheimer's disease. I. Neurochemical and neuroendocrine studies, *J Am Geriatr Soc* 33:741–8.

190. Perry, E. K., Perry, R. H., Smith, C. J. et al. (1986) Cholinergic receptors in cognitive disorders, *Can J Neurol Sci* 13:521–7.

191. Rossor, M. N., Iversen, L. L., Johnson, A. J., Mountjoy, C. Q. and Roth, M. (1981) Cholinergic deficit in frontal cerebral cortex in Alzheimer's disease is age dependent, *Lancet* 2:1422.

192. Ling, M., Merante, F., Chen, H. S. et al. (1997) The human mitochondrial elongation factor tu (EF-Tu) gene: cDNA sequence, genomic localization, genomic structure, and identification of a pseudogene, *Gene* 197:325–36.
193. Orrenius, S., Burgess, D. H., Hampton, M. B. and Zhivotovsky, B. (1997) Mitochondria as the focus of apoptosis research, *Cell Death Differ* 4:427–8.
194. Vayssiere, J. L., Cordeau-Lossouarn, L., Larcher, J. C. et al. (1992) Participation of the mitochondrial genome in the differentiation of neuroblastoma cells, *In Vitro Cell Dev Biol* 28A:763–72.
195. Pestova, T. V. and Hellen, C. U. (2000) The structure and function of initiation factors in eukaryotic protein synthesis, *Cell Mol Life Sci* 57:651–74.
196. Thompson, J. E., Hopkins, M. T., Taylor, C. and Wang, T. W. (2004) Regulation of senescence by eukaryotic translation initiation factor 5A: Implications for plant growth and development, *Trends Plant Sci* 9:174–9.
197. Tome, M. E., Fiser, S. M., Payne, C. M. and Gerner, E. W. (1997) Excess putrescine accumulation inhibits the formation of modified eukaryotic initiation factor 5A (eIF-5A) and induces apoptosis, *Biochem J* 328 (Pt 3):847–54.
198. Chang, R. C., Wong, A. K., Ng, H. K. and Hugon, J. (2002) Phosphorylation of eukaryotic initiation factor-2 alpha (eIF2α) is associated with neuronal degeneration in Alzheimer's disease, *Neuroreport* 13:2429–32.
199. Ferrer, I. (2002) Differential expression of phosphorylated translation initiation factor 2 alpha in Alzheimer's disease and Creutzfeldt-Jakob's disease, *Neuropathol Appl Neurobiol* 28:441–51.
200. Li, X., An, W. L., Alafuzoff, I. et al. (2004) Phosphorylated eukaryotic translation factor 4E is elevated in Alzheimer brain, *Neuroreport* 15:2237–40.
201. Sajdel-Sulkowska, E. M. and Marotta, C. A. (1984) Alzheimer's disease brain: Alterations in RNA levels and in a ribonuclease-inhibitor complex, *Science* 225:947–9.
202. Ding, Q., Markesbery, W. R., Cecarini, V. and Keller, J. N. (2006) Decreased RNA, and increased RNA oxidation, in ribosomes from early Alzheimer's disease, *Neurochem Res* 31:705–10.
203. Sultana, R., Poon, H. F., Cai, J. et al. (2006) Identification of nitrated proteins in Alzheimer's disease brain using a redox proteomics approach, *Neurobiol Dis* 22:76–87.

11 Neuroproteomics in the Neocortex of Mammals
Molecular Fingerprints of Cortical Plasticity

Lieselotte Cnops, Tjing-Tjing Hu,
Gert Van den Bergh, and Lutgarde Arckens

CONTENTS

11.1 INTRODUCTION

With the accumulation of vast amounts of DNA sequences in databases, researchers are realizing more and more that having complete sequences of genomes is not sufficient to elucidate biological function, not even if complemented with detailed information on the dynamics of the transcriptome. A cell or tissue is dependent on a huge number of metabolic and regulatory pathways for its normal functioning and there is no strict linear relationship between gene expressions and the protein complements or "proteome." Proteomics is therefore complementary to genomics and transcriptomics because it focuses on the gene products, the true active agents in a cell or tissue at any given time during a given physiological state. The advent of proteomics techniques has been enthusiastically accepted in most areas of biology and medicine. In neuroscience a host of applications was proposed ranging from neurotoxicology, neurometabolism, and determination of the proteomes of individual brain regions in health and disease, to name a few. We have implemented functional proteomics to help unravel the molecular pathways of brain plasticity that drive the response of the mammalian sensory neocortex to stimulus deprivation during development and adulthood, with a special emphasis on the visual system of the cat.

11.2   FACTS ABOUT CAT VISUAL SYSTEM

The visual cortex of mammals is immature at birth, both anatomically and physiologically, and develops gradually in the first weeks and months of postnatal life. During this period, spontaneous brain activity and visual stimulation induce specific patterns of neuronal activity in the central visual system that contribute to the establishment of visual perception and visually guided behavior. Vision-associated neurons will eventually be tuned by the quantity and the quality of the visual stimuli through both eyes to adequately interpret the information encoded in environmental stimulation patterns. To achieve this, a substantial anatomical organization takes place. Strengthening, remodeling, and eliminating synapses help create the adult-specific neuronal circuitry.

In cat, primary visual cortex alternating bands, called ocular dominance columns, contain neurons that are preferentially activated by either the left or the right eye (1–3). One of the seminal discoveries in developmental neuroscience is that monocular deprivation during the critical period (around postnatal day 30 [P30], for example, by mask rearing or eyelid suture) can modify both the typical physiology of the visual cortex and the anatomical representation of the two eyes in the cortex. The closure or damage of one eye leads to expansion of the columns serving the open eye at the expense of those responding to the deprived eye, which become reduced in size and afferent complexity (1,3–5). This results in poor visual acuity (amblyopia) and contrast sensitivity and loss of depth perception (5–7). Apparently, such cortical structural rearrangements induced by monocular deprivation are restricted to a certain developmental stage, the critical period that starts around eye opening and ends well before adulthood (7–9).

Visual acuity is poor at birth and cannot improve in the absence of patterned visual input as established in congenital cataract patients. Early binocular deprivation appears to prevent the normal organization of the cortical neural architecture necessary for the later development of sensitivity to fine detail (10). Thus, equal presentation of visual stimulation to both eyes is in itself insufficient to accurately shape the primary visual cortex. Binocular mask rearing from eye opening in cats, the animal models for congenital cataract, has been documented to lead also to behavioral changes in adulthood, such as deficits in global motion and to a lesser extent global form processing, just like in congenital cataract patients (11,12). Binocular deprivation of patterned vision by mask rearing undeniably affects the functionality of the visual system later in life, emphasizing the importance of the quality of the visual stimulation in shaping the cortex during development, and highlighting the extremely fragile nature of developing binocular circuits soon after birth.

Two decades of research have further shown that even beyond the critical period the visual cortex of the animal does not become a static entity but instead remains malleable throughout its life. Bilateral retinal lesions initially lead to functional deficits. Immediately after causing small lesions on homonymous parts of the retina in both eyes, neurons in the primary visual cortex with a receptive field situated in the center of the damaged region in the retina no longer respond to stimulation within their original receptive field (13–21). Yet neurons at the edges of the cortical lesion projection zone (LPZ) remain responsive but display enlarged receptive fields that are slightly shifted toward the intact retina surrounding the lesion (18). In ensuing weeks and months, recovery of the lost functions occurs. Much larger receptive field shifts take place, which can ultimately result in the complete filling in of the LPZ depending on the size of the retinal lesions (14,18,22,23). This extensive remodeling of cortical topography thus involves an enlargement of the representation of peri-lesion retina at the expense of the cortical area previously dedicated to the lesioned retina (18). The cortical long-range horizontal connections are recognized as the structural mediators of topographic map reorganization in visual cortex, first through the "unmasking" of existing sub-threshold connections, in a second phase by experience-dependent strengthening of their synapses and finally through sprouting of new collaterals (13,15,22,24–29).

How these rearrangements in strength and anatomy of cortical connections in response to abnormal visual stimulation, either during the critical period or in adulthood, depend on changes in the genome and the proteome expressed by the relevant brain cells, has been the subject of intensive investigations and the first molecular models of brain plasticity are now emerging.

11.3 NEUROPROTEOMICS OF CAT VISUAL CORTEX PLASTICITY

We performed large-scale proteomics studies by comparing protein expression profiles of the primary visual cortex of cats during early postnatal development with those of fully mature adult animals (30–32). This started from the rationale that proteins involved in visual cortex plasticity will demonstrate an increase or decrease in expression level with age in order to assist the refinement of visual cortical synaptic connectivity (7,8). To identify as many putative molecular determinants as possible a two-dimensional difference gel electrophoresis (2D-DIGE) approach was applied directly or in combination with a third separation dimension (hydrophobicity) allowing for the selective enrichment and identification of low-abundance proteins using a reversed-phase chromatography pre-fractionation step (see Chapters 3 and 4 for discussion of these techniques) (Figure 11.1). Together these large-scale screenings provided a list of proteins with roles in metabolism, neurite growth and guidance, synapse formation, cytoskeleton stabilization, and neurotransmitter release (30–32).

For a detailed understanding of the specific role of each of the proteins in the process of cortical plasticity, a deep analysis is needed about their expression under different physiological conditions using complementary technologies. Because neurite growth and guidance, and formation and maturation of synapses underlies cortical plasticity we specifically investigated the plasticity-regulated expression levels of molecules with a relevant meaning in such structural and cellular processes (33–36). We considered two members of the collapsin response mediator protein family (CRMP2 and CRMP4) and two molecules that are involved in the exocytosis-endocytosis machinery at the neuronal synapse, namely dynamin I (Dyn I) and synaptotagmin I (Syt I). CRMPs are phosphoproteins implicated in neuronal differentiation, axon guidance, and growth cone collapse through the signal cascade

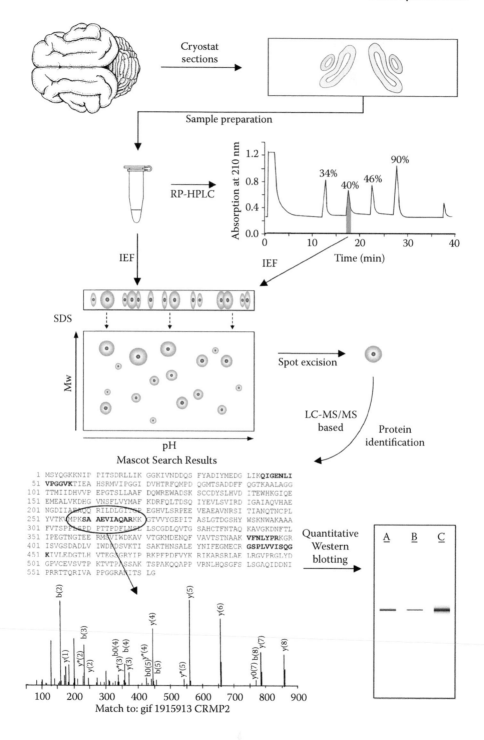

Mascot Search Results

```
  1 MSYQGKKNIP PITSDRLLIK GGKIVNDDQS FYADIYMEDG LIKQIGENLI
 51 VPGGVKTIEA HSRMVIPGGI DVHTRFQMPD QGMTSADDFF QGTKAALAGG
101 TTMIIDHVVP EPGTSLLAAF DQWREWADSK SCCDYSLHVD ITEWHKGIQE
151 EMEALVKDHG VNSFLVYMAF KDRFQLTDSQ IYEVLSVIRD IGAIAQVHAE
201 NGDIIAEAQQ RILDLGITSP EGHVLSRPEE VEAEAVNRSI TIANQTNCPL
251 YVTKVMPKSA AEVIAQARKK GTVVYGEPIT ASLGTDGSHY WSKNWAKAAA
301 FVTSPPLSPD PTTPDFLNSL LSCGDLQVTG SAHCTFNTAQ KAVGKDNFTL
351 IPEGTNGTEE RMSVIWDKAV VTGKMDENQF VAVTSTNAAK VFNLYPRKGR
401 ISVGSDADLV IWDPDSVKTI SAKTHNSALE YNIFEGMECR GSPLVVISQG
451 KIVLEDGTLH VTEGAGRYIP RKPFPDFVYK RIKARSRLAE LRGVPRGLYD
501 GPVCEVSVTP KTVTPASSAK TSPAKQQAPP VRNLHQSGFS LSGAQIDDNI
551 PRRTTQRIVA PPGGRANITS LG
```

Match to: gif 1915913 CRMP2

pathway of Sema3A/collapsin-1 (37–39). Dyn I plays a role in synaptic vesicle recycling by clathrin-mediated endocytosis (40), while Syt I functions as a calcium sensor that triggers an essential step in exocytosis, namely the fusion of vesicle and plasma membrane during neurotransmission (41).

Based on the 2D-DIGE spot patterns, multiple isoforms of CRMP2 and CRMP4 were detected as differentially expressed between kitten and adult cat area 17 (Figures 11.2 and 11.3). All CRMP4 isoforms showed a similar developmental expression, while the CRMP2 isoforms displayed a more diverse age-dependent pattern (Figure 11.3A and 11.3B). Dynamin I and synaptotagmin I were each detected in one differential spot and showed an opposite expression (Figures 11.2, 11.3C and 11.3D). Dynamin I levels increased with age while synaptotagmin I levels decreased with age. We thus selected four proteins with a clearly different expression profile for further analysis.

11.4 CRMPS AND CORTICAL PLASTICITY

Using Western blotting we demonstrated that at least the 64 kDa isoform of CRMP2 and all CRMP4 levels are high early in life (P10, P30) when anatomical and structural rearrangements are necessary for the normal build-up of the visual cortex, in line with a role for these molecules in neuronal differentiation and dendritic and axonal guidance [Figure 11.4; references (42–44)]. Their expression can, however, not be strictly regulated by visual experience since it was already high at eye opening (P10). Also the formation of ocular dominance columns, in contrast to what was believed for a long time, starts experience-independently since clustering of lateral geniculate nucleus (LGN) arbors occurs at least a week before the critical period as recently demonstrated by optical imaging techniques (45,46). Indeed, between postnatal day 10 and 30, the so-called "pre-critical period" (46), spontaneous activity already shapes the visual cortex connectivity and visual experience has then little influence on the cortical organization (47,48). CRMPs may thus, together with other genes, form a molecular cascade that drives the growth of axons and dendrites in the developing cortex during this experience-independent time window.

The strong downregulation of CRMP4 with visual cortex maturation (Figures 11.3 and 4) correlates with the fact that the closure of the critical period is accompanied by the formation of perineuronal nets through extracellular matrix proteins like

FIGURE 11.1 Work-flow of the proteomic screening experiments performed on cat visual cortex samples. The consecutive steps depicted: collection of the brain; production of cryostat sections; sample preparation: manual isolation of specific regions of visual area 17 using a razor blade followed by direct protein extraction or else interspersed by an extra purification step via RP-HPLC; Cye labeling and mixing of the protein samples prior to gel separation (see Chapters 3, 4, and 5 for discussion of these techniques; and (97) for details); first dimension separation: IEF of the protein sample; second dimension separation based on Mw; spot excision and tryptic digest; LC-MS/MS based protein identification: as an example the Mascot search results for one of the CRMP2 spots (see Figures 11.2 and 11.3) are inserted: peptides in bold as identified in the PMF (peptide mass fingerprint) with one peptide sequence checked in MS/MS mode (see Chapter 5); selection of proteins of interest for in-depth analysis via quantitative Western blotting (see Figures 11.4–11.6).

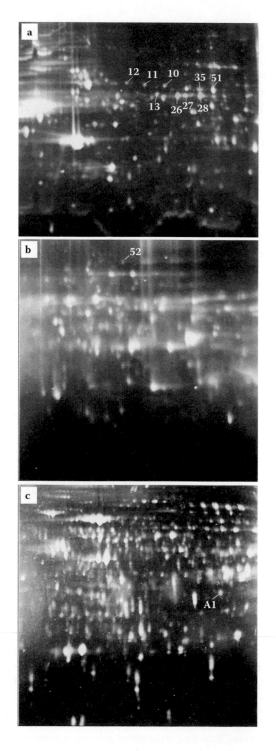

chondroitin sulfate proteoglycans that surround predominantly parvalbumin (PV)-containing interneurons (49–53). CRMP4 is known to react with chondroitin sulfates (54) and is co-localized with PV in adult cat visual cortex (55). The perineuronal nets form a mesh, which holds secreted proteins like Sema3A (47,49), possibly resulting in a selective downregulation of CRMP4, one of the intracellular mediators of the Sema3A signaling cascade. Moreover, exactly these PV-containing interneurons seem to be the most important players in regulating the critical period closure (47,51,56). Thus, reduced CRMP4 expression in normal juvenile and adult subjects correlates well with the stabilization of the network connectivity in the maturing visual cortex by extracellular matrix proteins. Moreover, the fact that CRMP4, in contrast to CRMP2, was not affected by monocular deprivation (Figure 11.5) could probably be related to the cell-type specific expression of CRMP2 in large pyramidal neurons and of CRMP4 in those small PV-positive interneurons (55).

Monocular deprivation influenced the post-translational modification of CRMP2 as only the 62 kDa CRMP2 protein band was significantly affected by this manipulation (Figure 11.5). These observations probably reflect modifications in the activity status of CRMP2, for instance, by dephosphorylation or deglycosylation. CRMP2 glycosylation has indeed been reported to block CRMP2 phosphorylation, thereby regulating its activity (57). Since the expression of CRMP2 is also modulated during regeneration of the olfactory nerve after axotomy and bulbectomy (58), we suggest that during development CRMP2 could induce rearrangements to LGN axonal arbors and intracortical afferents, like retraction of branches and ingrowth of other axonal branches—structural changes that have been implicated in ocular dominance plasticity. On the other hand, binocular deprivation did not affect CRMP2 or CRMP4, indicating that not the loss of quality of vision through visual deprivation, but the disruption of normal binocular visual experience is crucial to induce the observed changes after monocular deprivation. Opposite to our observations in kittens, both CRMP2 and CRMP4 were clearly affected 14 days after binocular retinal lesioning (Figure 11.6). Also, other CRMP studies established the same time-dependent alterations after nervous system injury in adults. By analyzing several post-operative times, a peak of CRMP2 expression was detected in motor neurons of the rat hypoglossal nucleus 14 days after nerve transection (59), while CRMP4 levels were apparent around one or two weeks after ischemia in the striatum of adult rats (60). In

FIGURE 11.2   (See color insert following page 172.) False-colored two-dimensional difference gel electrophoresis (2D-DIGE) images of the comparative analysis of area 17 protein expression patterns between kitten and adult cat. False-colored overlay images of a 2D-DIGE experiment with adult cat, propyl-Cy3-labeled visual area 17 proteins colored in red, and kitten, methyl-Cy5-labeled visual area 17 proteins colored in green. Yellow spots contain proteins that have equal expression levels in the two samples, red spots containing proteins with a higher expression in adult cat visual area 17, and green spots with proteins more abundantly expressed in 30-day-old kitten primary visual cortex. Numbers indicate the spots with statistically significant fluorescence levels, further identified as (A) CRMP2, CRMP4; (B) Dyn I; and (C) Syt I using mass spectrometry. Panels (A) and (B) present protein samples separated according to pI and molecular weight over a pH range of pH 4 to 7 or 5.5–6.7. Panel (C) illustrates proteins separated according to hydrophobicity (HPLC pre-fractionation), pI, and molecular weight.

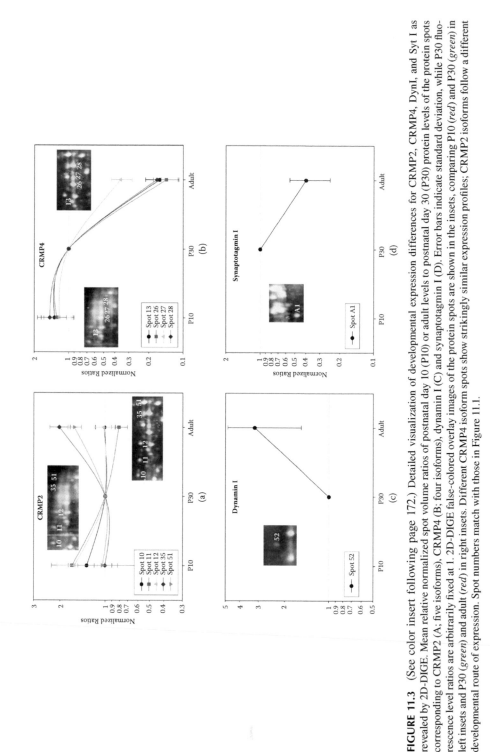

FIGURE 11.3 (See color insert following page 172.) Detailed visualization of developmental expression differences for CRMP2, CRMP4, DynI, and Syt I as revealed by 2D-DIGE. Mean relative normalized spot volume ratios of postnatal day 10 (P10) or adult levels to postnatal day 30 (P30) protein levels of the protein spots corresponding to CRMP2 (A; five isoforms), CRMP4 (B; four isoforms), dynamin I (C) and synaptotagmin I (D). Error bars indicate standard deviation, while P30 fluorescence level ratios are arbitrarily fixed at 1. 2D-DIGE false-colored overlay images of the protein spots are shown in the insets, comparing P10 (red) and P30 (green) in left insets and P30 (green) and adult (red) in right insets. Different CRMP4 isoform spots show strikingly similar expression profiles; CRMP2 isoforms follow a different developmental route of expression. Spot numbers match with those in Figure 11.1.

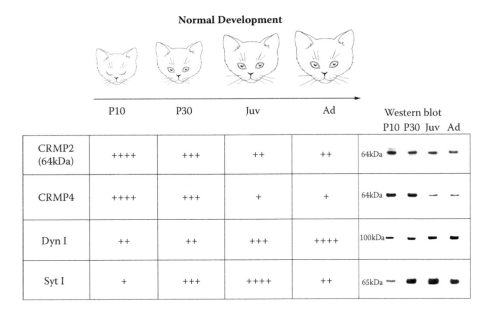

Normal Development

	P10	P30	Juv	Ad	Western blot P10 P30 Juv Ad
CRMP2 (64kDa)	++++	+++	++	++	64kDa
CRMP4	++++	+++	+	+	64kDa
Dyn I	++	++	+++	++++	100kDa
Syt I	+	+++	++++	++	65kDa

FIGURE 11.4 Expression patterns of CRMP2, CRMP4, Dyn I, and Syt I during normal visual cortex development. Western blot analysis for CRMP2, CRMP4, Dyn I, and Syt I was performed on visual cortex samples (area 17) of kittens of postnatal day 10 (P10) and 30 (P30), respectively, corresponding to the time of eye-opening and the peak of the critical period, and of cats at juvenile ages (Juv) and adulthood (Ad) to analyze the molecular expression profiles during normal visual exposure. Expression levels varied between very low (+), low (++), high (+++), and very high (++++).

addition, many other molecules showed major alterations in expression two weeks to one month after retinal lesions (61–66).

CRMP2 and CRMP4 may possibly be involved in a molecular cascade that orchestrates the topographical reorganization of area 17. Indeed, CRMP2 can be phosphorylated by CaMKII (67), a well-known participant in long term potentiation (LTP) processes and already demonstrated to change its phosphorylated state in retinal lesion cats, especially 14 days post-lesion (66). Another factor with a brain-plasticity-related and post-lesion survival time-dependent expression, MEF2C (63), regulates CRMP4 expression (68). Furthermore, also in the adult, both CRMPs may react with structural molecules that coordinate cytoskeleton rearrangements and synaptic plasticity in the adult cortex, like actin or microtubuli bundles and chondroitin-sulfate proteoglycans of the extracellular matrix (54,56,69–71).

CRMPs were implicated in axonal guidance by mediating the chemorepulsive Sema3A signal (72). As suggested for the olfactory system of adult rat, Sema3A could create a molecular barrier to restrict ingrowing olfactory axons (34,58). High CRMP expression in the LPZ of visually deafferented adult cats after two weeks of recovery thus possibly prevents major ingrowth of neuronal collaterals directly into the most inner part of the LPZ. In addition, overexpression of CRMP2 also stimulates the formation of multiple branches on neuronal processes

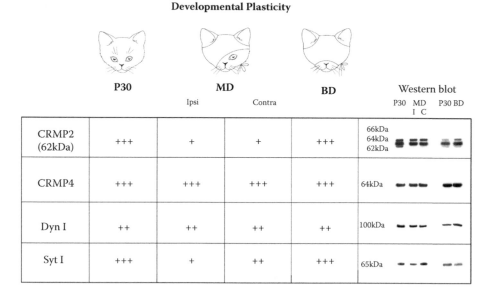

FIGURE 11.5 Manipulation-specific molecular expression changes during monocular and binocular deprivation in kittens. The protein expression levels of CRMP2, CRMP4, Syt I, and Dyn I were determined by Western blotting on normal kittens around the age that the visual cortex is very sensitive to changes in visual experience ("critical period," postnatal day 30 [P30]), after monocular deprivation (MD) of the left eye, and after binocular deprivation (BD). Deprivation was carried out through mask rearing with a linen mask from postnatal day 10 (P10) until P30. In MD kittens, area 17 samples of the hemisphere ispilateral to the deprived eye (ipsi, I) and contralateral to the deprived eye (contra, C) were examined. The 62 kDa isoform of CRMP2 and Syt I were affected by MD, while BD had no effect on any of the molecules.

(69,73) and CRMPs may promote increasing spine dynamics and growth cone formation on existing connections in the LPZ. Indeed, it is known that extending dendritic filopodia may scan the environment for new synaptic partners after sensory deprivation (47,74), while growth cone formation after injury is essential to aid subsequent axonal growth during neuronal regeneration (75). At the same time, CRMP4 possibly plays a neuroprotective role in the LPZ by preventing demyelinization of neurons (34). Later on when CRMP levels decline and thus no longer repel the ingrowth of neuronal connections into the central portion of the LPZ, reorganization by axonal sprouting and synaptogenesis could take place (26,27).

In summary, in the adult, both CRMP2 and CRMP4 may thus promote dendritic spine dynamics or growth cone formation in the LPZ and regulate axonal sprouting in the context of cortical map remodeling in adult area 17. During postnatal development, both CRMPs are involved in structural refinements of the connectivity of the visual cortex, but only changes in CRMP2 activity are necessary for ocular dominance column rearrangements after monocular deprivation.

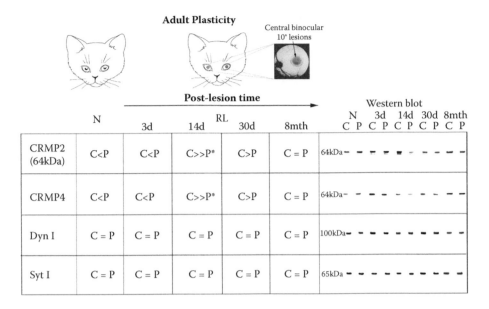

	N	3d	14d	RL 30d	8mth	Western blot
CRMP2 (64kDa)	C<P	C<P	C>>P*	C>P	C = P	64kDa – – – – – – – – – –
CRMP4	C<P	C<P	C>>P*	C>P	C = P	64kDa – – – – – – – – – –
Dyn I	C = P	C = P	C = P	C = P	C = P	100kDa – – – – – – – – – –
Syt I	C = P	C = P	C = P	C = P	C = P	65kDa – – – – – – – – – –

FIGURE 11.6 Western blot analysis to monitor the effect of adult plasticity. The expression of the four molecules of interest was measured in the visual cortex of normal adult cat (N) and adult retinal lesion cats (RL) of different post-lesion survival times: 3 days (3d), 14 days (14d), 30 days (30d), and 8 months (8mth). Central retinal lesions in both eyes specifically deprived central (C) area 17 of visual stimulation and was compared to the remote peripheral (P) region of area 17. CRMP2 and CRMP4, but not Dyn I or Syt I, were affected by the retinal lesions, especially 14 days post-lesion.

11.5 DYNAMIN I AND MATURE CELL SHAPE

For Dynamin I a steady increase in expression with age was observed (Figures 11.3 and 11.4). The expression profile of Dyn I thus showed a clear relationship with its function in the maintenance of the adult cell shape (76–78) in full agreement with previous results obtained in rodent studies. Dyn I is not expressed in prenatal rat cortex, cerebellum, or hippocampus and is preferentially upregulated from P7 with a principal increase between P7–P15 and P23 that is maintained into adulthood (76,77). Its developmental course thus parallels neuronal maturation (40) at a time when axonal growth and synapse formation are already established (76,79). Dyn I seems not involved in developmental or plasticity-related synapse rearrangements in cat visual cortex, as demonstrated here in the monocular and binocular deprived kitten models as well as in the adult retinal lesion model (Figures 11.5 and 11.6). Since Dyn I is suggested to be a microtubule-associated motor protein (80), it is more reasonable to believe that Dyn I stabilizes the microtubule cytoskeleton and modulates the dynamics of actin filaments during endocytosis of mature neurons (76,81). Indeed, increasing levels of Dyn I during visual cortex development correspond to the maturation of the neurotransmission system. Once stable synaptic contacts are established, Dyn I–regulated clathrin-mediated endocytosis and vesicle recycling (82–84) would then be required for continuous cell-to-cell-communication.

11.6 SYNAPTOTAGMIN I AND CORTICAL PLASTICITY

Syt I is the only molecule whose developmental expression profile readily followed the time course of the critical period (Figure 11.4). In the first phase of the postnatal development, Syt I expression was clearly experience-dependently regulated. Around eye opening (P10), very low levels were detected but at the height of the critical period for monocular deprivation (P30), Syt I levels augmented enormously. Syt I protein expression did not immediately decrease afterwards. In juvenile cats, Syt I demonstrated an even more abundant expression while in adults the expression had declined below P30 levels. Since Syt I is an abundant synaptic vesicle protein and essential in mediating calcium-triggered neurotransmitter release (41,85,86), it could play a role in membrane fusion events that contribute to the outgrowth and remodeling of neuronal dendrites (36). Since the postnatal visual cortex still undergoes structural changes, upregulation of Syt I protein expression by visual experience possibly engages refinements in synaptic contacts. Furthermore, Syt I triggers synchronous and suppresses asynchronous neurotransmitter release (87,88). This could be important for the segregation of the LGN afferents in the eye-specific ocular dominance columns within the primary visual cortex. The high Syt I expression levels at juvenile ages have never been reported before. Even if experience-dependent ocular dominance plasticity at that age is already reduced, it is known that around five months of age, extragranular cortical layers are still susceptible to changes in visual input, while layer IV is not plastic any more (48,89–91). Furthermore, at four to six months of age, the kitten visual cortex is still susceptible to effects of monocular deprivation (92). In addition, Winfield (1983) described that the number of synapses increases during development of the cat visual cortex with a maximal density of synapses between P70 and P110 where after the synapse density decreases toward adulthood (93). Also between P8 and P37, an increase in the number of synapses is observed (7) which in turn correlates to the initial increase of Syt I in the visual cortex. Thus, increased Syt I levels correspond well with the boost of synaptic contacts and the correlated increase of neurotransmitter release by Syt I regulated exocytosis.

Upon monocular deprivation, diminished connectivity is detected by reduced size and number of presynaptic terminals, mitochondria, and spines (94). Syt I levels that showed a clear experience-dependent expression pattern during the first stages of normal cat visual cortex development were decreased after monocular deprivation (Figure 11.5), probably in correlation with the reduction of synaptic contacts. As a calcium sensor (41,85,86) it is not unlikely that Syt I detects activity changes in the presynaptic compartment.

However, from our results in adult retinal lesioned cats, we cannot conclude that Syt I was responsible for changes in neurotransmission as observed in retinal lesioned cats [Figure 11.6; references (62,64,65)]. Instead, Syt I may be related to synaptic plasticity only in young animals, in contrast to synapsin, another synaptic vesicle protein that plays a role in both developmental and adult plasticity (6,95). Moreover, some studies doubt the role of Syt I as a Ca^{2+} sensor and indicate only a regulatory role in the membrane fusion of synaptic vesicles (87,96). Furthermore, the observed modifications in GABAergic and glutamatergic neurotransmission as reported after partial visual deafferentation in adult cats (62,64,65) could be explained by changes in amount of neurotransmitter per vesicle, rather than in the number of vesicles that were released under influence of Syt I.

11.7 CONCLUSION

Taken together, specific molecular fingerprints begin to emerge related to the diverse structural and functional response of the visual cortex when coping with alterations in visual input throughout life. Comparative analysis of protein expression patterns in normal, monocular, and binocular deprived animals seems to hold great promise for dissecting the molecular cascades that specifically encode the quality of vision versus binocular competition to guide cortical maturation. Furthermore, comparison of developmental and adult plasticity models will help elucidate to what extent comparable molecular machinery is exploited by visual neurons in reshaping the cortex in response to lesions later in life. Characterization of manipulation-specific molecular sets may hold the key toward understanding the molecular basis of vision. Extracting insights into physiological and pathological mechanisms from complex sets of neuroproteomic experiments is a fascinating challenge. Future integration of genome, transcriptome, proteome, lipidome, and glycome studies with electrophysiological, anatomical, and behavioral investigations should boost our understanding of vision-guided behavior in response to our daily environment.

Identifying key mediators and defining their relationships should be a major goal in the field and will greatly enhance our understanding of the mechanism by which molecular programs regulate cortical plasticity and underlie visual disorders. Finding the keys to cortical plasticity will enable the development of new therapeutic strategies for recovery from sensory loss and brain damage and for goal-directed improvement of post-lesional recovery of brain function throughout life.

11.8 FUTURE CHALLENGES

The many proteomic and genomic data sets that become available are becoming harder to handle and there is a growing need to collate data from an increasing number of studies. Generating lists alone provides little biological insight. Complementing these lists with bioinformatics, protein interaction studies, and functional studies will be crucial in boosting our understanding of the exact relationship between molecular organization and brain function. There is a need for sharing data among laboratories that should be made available freely as was achieved with genome sequences.

ACKNOWLEDGMENTS

We thank past and present members of our laboratory for their contributions and helpful discussions. Work in our laboratory is supported by grants from the FWO Flanders and the Research Council of the K.U. Leuven.

REFERENCES

1. LeVay, S., Stryker, M. P. and Shatz, C. J. (1978) Ocular dominance columns and their development in layer IV of the cat's visual cortex: A quantitative study, *J Comp Neurol* 179:223–44.
2. Payne, B. R., Peters, A. (2002) *The cat primary visual cortex* (San Diego: Academic Press).
3. Shatz, C. J. and Stryker, M. P. (1978) Ocular dominance in layer IV of the cat's visual cortex and the effects of monocular deprivation, *J Physiol* 281:267–83.
4. Hubel, D. H. and Wiesel, T. N. (1963) Receptive fields of cells in striate cortex of very young, visually inexperienced kittens, *J Neurophysiol* 26:994–1002.
5. Wiesel, T. N. and Hubel, D. H. (1965) Comparison of the effects of unilateral and bilateral eye closure on cortical unit responses in kittens, *J Neurophysiol* 28:1029–40.
6. Berardi, N., Pizzorusso, T., Ratto, G. M. and Maffei, L. (2003) Molecular basis of plasticity in the visual cortex, *Trends Neurosci* 26:369–78.
7. Daw, N. W. (1995) *Visual development* (New York: Plenum Press).
8. Hubel, D. H. and Wiesel, T. N. (1970) The period of susceptibility to the physiological effects of unilateral eye closure in kittens, *J Physiol* 206:419–36.
9. Shatz, C. J. and Luskin, M. B. (1986) The relationship between the geniculocortical afferents and their cortical target cells during development of the cat's primary visual cortex, *J Neurosci* 6:3655–68.
10. Lewis, T. L. and Maurer, D. (2005) Multiple sensitive periods in human visual development: Evidence from visually deprived children, *Dev Psychobiol* 46:163–83.
11. Burnat, K., Stiers, P., Arckens, L., Vandenbussche, E. and Zernicki, B. (2005) Global form perception in cats early deprived of pattern vision, *Neuroreport* 16:751–4.
12. Burnat, K., Vandenbussche, E. and Zernicki, B. (2002) Global motion detection is impaired in cats deprived early of pattern vision, *Behav Brain Res* 134:59–65.
13. Chino, Y. M. (1995) Adult plasticity in the visual system, *Can J Physiol Pharmacol* 73:1323–38.
14. Chino, Y. M., Kaas, J. H., Smith, E. L., 3rd, Langston, A. L. and Cheng, H. (1992) Rapid reorganization of cortical maps in adult cats following restricted deafferentation in retina, *Vision Res* 32:789–96.
15. Chino, Y. M., Smith, E. L., 3rd, Kaas, J. H., Sasaki, Y. and Cheng, H. (1995) Receptive-field properties of deafferentated visual cortical neurons after topographic map reorganization in adult cats, *J Neurosci* 15:2417–33.
16. Dreher, B., Burke, W. and Calford, M. B. (2001) Cortical plasticity revealed by circumscribed retinal lesions or artificial scotomas, *Prog Brain Res* 134:217–46.
17. Eysel, U. T. (1992) Cortical plasticity: Remodeling of sensor fields in receptive cortices, *Current Biology* 2:389–91.
18. Gilbert, C. D. and Wiesel, T. N. (1992) Receptive field dynamics in adult primary visual cortex, *Nature* 356:150–2.
19. Kaas, J. H. (1991) Plasticity of sensory and motor maps in adult mammals, *Annu Rev Neurosci* 14:137–67.
20. Kaas, J. H., Krubitzer, L. A., Chino, Y. M. et al. (1990) Reorganization of retinotopic cortical maps in adult mammals after lesions of the retina, *Science* 248:229–31.
21. Schmid, L. M., Rosa, M. G., Calford, M. B. and Ambler, J. S. (1996) Visuotopic reorganization in the primary visual cortex of adult cats following monocular and binocular retinal lesions, *Cereb Cortex* 6:388–405.
22. Chino, Y. M. (1999) The role of visual experience in the cortical topographic map reorganization following retinal lesions, *Restor Neurol Neurosci* 15:165–76.
23. Giannikopoulos, D. V. and Eysel, U. T. (2006) Dynamics and specificity of cortical map reorganization after retinal lesions, *Proc Natl Acad Sci U S A* 103:10805–10.

24. Calford, M. B. (2002) Mechanisms for acute changes in sensory maps, *Adv Exp Med Biol* 508:451–60.
25. Calford, M. B., Wright, L. L., Metha, A. B. and Taglianetti, V. (2003) Topographic plasticity in primary visual cortex is mediated by local corticocortical connections, *J Neurosci* 23:6434–42.
26. Darian-Smith, C. and Gilbert, C. D. (1994) Axonal sprouting accompanies functional reorganization in adult cat striate cortex, *Nature* 368:737–40.
27. Darian-Smith, C. and Gilbert, C. D. (1995) Topographic reorganization in the striate cortex of the adult cat and monkey is cortically mediated, *J Neurosci* 15:1631–47.
28. Das, A. and Gilbert, C. D. (1995) Long-range horizontal connections and their role in cortical reorganization revealed by optical recording of cat primary visual cortex, *Nature* 375:780–4.
29. Das, A. and Gilbert, C. D. (1995) Receptive field expansion in adult visual cortex is linked to dynamic changes in strength of cortical connections, *J Neurophysiol* 74:779–92.
30. Van den Bergh, G., Clerens, S., Cnops, L., Vandesande, F. and Arckens, L. (2003) Fluorescent two-dimensional difference gel electrophoresis and mass spectrometry identify age-related protein expression differences for the primary visual cortex of kitten and adult cat, *J Neurochem* 85:193–205.
31. Van den Bergh, G., Clerens, S., Firestein, B. L., Burnat, K. and Arckens, L. (2006) Development and plasticity-related changes in protein expression patterns in cat visual cortex: A fluorescent two-dimensional difference gel electrophoresis approach, *Proteomics* 6:3821–32.
32. Van den Bergh, G., Clerens, S., Vandesande, F. and Arckens, L. (2003) Reversed-phase high-performance liquid chromatography prefractionation prior to two-dimensional difference gel electrophoresis and mass spectrometry identifies new differentially expressed proteins between striate cortex of kitten and adult cat, *Electrophoresis* 24:1471–81.
33. Byk, T., Ozon, S. and Sobel, A. (1998) The Ulip family phosphoproteins—Common and specific properties, *Eur J Biochem* 254:14–24.
34. Charrier, E., Reibel, S., Rogemond, V. et al. (2003) Collapsin response mediator proteins (CRMPs): Involvement in nervous system development and adult neurodegenerative disorders, *Mol Neurobiol* 28:51–64.
35. Napolitano, M., Marfia, G. A., Vacca, A. et al. (1999) Modulation of gene expression following long-term synaptic depression in the striatum, *Brain Res Mol Brain Res* 72:89–96.
36. Schwab, Y., Mouton, J., Chasserot-Golaz, S. et al. (2001) Calcium-dependent translocation of synaptotagmin to the plasma membrane in the dendrites of developing neurones, *Brain Res Mol Brain Res* 96:1–13.
37. Goshima, Y., Sasaki, Y., Nakayama, T., Ito, T. and Kimura, T. (2000) Functions of semaphorins in axon guidance and neuronal regeneration, *Jpn J Pharmacol* 82:273–9.
38. Minturn, J. E., Fryer, H. J., Geschwind, D. H. and Hockfield, S. (1995) TOAD-64, a gene expressed early in neuronal differentiation in the rat, is related to unc-33, a C. elegans gene involved in axon outgrowth, *J Neurosci* 15:6757–66.
39. Wang, L. H. and Strittmatter, S. M. (1996) A family of rat CRMP genes is differentially expressed in the nervous system, *J Neurosci* 16:6197–6207.
40. Liu, J. P. and Robinson, P. J. (1995) Dynamin and endocytosis, *Endocr Rev* 16:590–607.
41. Südhof, T. C. and Rizo, J. (1996) Synaptotagmins: C2-domain proteins that regulate membrane traffic, *Neuron* 17:379–88.
42. Byk, T., Dobransky, T., Cifuentes-Diaz, C. and Sobel, A. (1996) Identification and molecular characterization of Unc-33-like phosphoprotein (Ulip), a putative mammalian homolog of the axonal guidance-associated unc-33 gene product, *J Neurosci* 16:688–701.

43. Kamata, T., Subleski, M., Hara, Y. et al. (1998) Isolation and characterization of a bovine neural specific protein (CRMP-2) cDNA homologous to unc-33, a *C. elegans* gene implicated in axonal outgrowth and guidance, *Brain Res Mol Brain Res* 54:219–36.

44. Polleux, F., Morrow, T. and Ghosh, A. (2000) Semaphorin 3A is a chemoattractant for cortical apical dendrites, *Nature* 404:567–73.

45. Crair, M. C., Horton, J. C., Antonini, A. and Stryker, M. P. (2001) Emergence of ocular dominance columns in cat visual cortex by 2 weeks of age, *J Comp Neurol* 430:235–49.

46. Feller, M. B. and Scanziani, M. (2005) A precritical period for plasticity in visual cortex, *Curr Opin Neurobiol* 15:94–100.

47. Bence, M. and Levelt, C. N. (2005) Structural plasticity in the developing visual system, *Prog Brain Res* 147:125–39.

48. Katz, L. C. and Shatz, C. J. (1996) Synaptic activity and the construction of cortical circuits, *Science* 274:1133–8.

49. Dityatev, A. and Schachner, M. (2003) Extracellular matrix molecules and synaptic plasticity, *Nat Rev Neurosci* 4:456–68.

50. Guimaraes, A., Zaremba, S. and Hockfield, S. (1990) Molecular and morphological changes in the cat lateral geniculate nucleus and visual cortex induced by visual deprivation are revealed by monoclonal antibodies Cat-304 and Cat-301, *J Neurosci* 10:3014–24.

51. Hensch, T. K. (2005) Critical period plasticity in local cortical circuits, *Nat Rev Neurosci* 6:877–88.

52. Hockfield, S., Kalb, R. G., Zaremba, S. and Fryer, H. (1990) Expression of neural proteoglycans correlates with the acquisition of mature neuronal properties in the mammalian brain, *Cold Spring Harb Symp Quant Biol* 55:505–14.

53. Lander, C., Kind, P., Maleski, M. and Hockfield, S. (1997) A family of activity-dependent neuronal cell-surface chondroitin sulfate proteoglycans in cat visual cortex, *J Neurosci* 17:1928–39.

54. Franken, S., Junghans, U., Rosslenbroich, V. et al. (2003) Collapsin response mediator proteins of neonatal rat brain interact with chondroitin sulfate, *J Biol Chem* 278:3241–50.

55. Cnops, L., Hu, T. T., Burnat, K., Van der Gucht, E. and Arckens, L. (2006) Age-dependent alterations in CRMP2 and CRMP4 protein expression profiles in cat visual cortex, *Brain Res* 1088:109–19.

56. Berardi, N., Pizzorusso, T. and Maffei, L. (2004) Extracellular matrix and visual cortical plasticity: Freeing the synapse, *Neuron* 44:905–8.

57. Cole, R. N. and Hart, G. W. (2001) Cytosolic O-glycosylation is abundant in nerve terminals, *J Neurochem* 79:1080–9.

58. Pasterkamp, R. J., De Winter, F., Holtmaat, A. J. and Verhaagen, J. (1998) Evidence for a role of the chemorepellent semaphorin III and its receptor neuropilin-1 in the regeneration of primary olfactory axons, *J Neurosci* 18:9962–76.

59. Suzuki, Y., Nakagomi, S., Namikawa, K. et al. (2003) Collapsin response mediator protein-2 accelerates axon regeneration of nerve-injured motor neurons of rat, *J Neurochem* 86:1042–50.

60. Liu, P. C., Yang, Z. J., Qiu, M. H., Zhang, L. M. and Sun, F. Y. (2003) Induction of CRMP-4 in striatum of adult rat after transient brain ischemia, *Acta Pharmacol Sin* 24:1205–11.

61. Arckens, L. (2006) The molecular biology of sensory map plasticity in adult mammals. In *Plasticity in the visual system: From genes to circuits*, eds. Pinaud, R., Tremere, L. A. and De Weerd, P., 181–203 (Springer Science + Business Media).

62. Arckens, L., Schweigart, G., Qu, Y. et al. (2000) Cooperative changes in GABA, glutamate and activity levels: The missing link in cortical plasticity, *Eur J Neurosci* 12:4222–32.
63. Leysen, I., Van der Gucht, E., Eysel, U. T. et al. (2004) Time-dependent changes in the expression of the MEF2 transcription factor family during topographic map reorganization in mammalian visual cortex, *Eur J Neurosci* 20:769–80.
64. Massie, A., Cnops, L., Jacobs, S. et al. (2003) Glutamate levels and transport in cat (*Felis catus*) area 17 during cortical reorganization following binocular retinal lesions, *J Neurochem* 84:1387–97.
65. Massie, A., Cnops, L., Smolders, I. et al. (2003) Extracellular GABA concentrations in area 17 of cat visual cortex during topographic map reorganization following binocular central retinal lesioning, *Brain Res* 976:100–8.
66. Van den Bergh, G., Eysel, U. T., Vandenbussche, E., Vandesande, F. and Arckens, L. (2003) Retinotopic map plasticity in adult cat visual cortex is accompanied by changes in Ca2+/calmodulin-dependent protein kinase II alpha autophosphorylation, *Neuroscience* 120:133–42.
67. Yoshimura, Y., Shinkawa, T., Taoka, M. et al. (2002) Identification of protein substrates of Ca(2+)/calmodulin-dependent protein kinase II in the postsynaptic density by protein sequencing and mass spectrometry, *Biochem Biophys Res Commun* 290:948–54.
68. Matsuo, T., Stauffer, J. K., Walker, R. L., Meltzer, P. and Thiele, C. J. (2000) Structure and promoter analysis of the human unc-33-like phosphoprotein gene. E-box required for maximal expression in neuroblastoma and myoblasts, *J Biol Chem* 275:16560–8.
69. Fukata, Y., Kimura, T. and Kaibuchi, K. (2002) Axon specification in hippocampal neurons, *Neurosci Res* 43:305–15.
70. Gu, Y. and Ihara, Y. (2000) Evidence that collapsin response mediator protein-2 is involved in the dynamics of microtubules, *J Biol Chem* 275:17917–20.
71. Quinn, C. C., Gray, G. E. and Hockfield, S. (1999) A family of proteins implicated in axon guidance and outgrowth, *J Neurobiol* 41:158–64.
72. Goshima, Y., Nakamura, F., Strittmatter, P. and Strittmatter, S. M. (1995) Collapsin-induced growth cone collapse mediated by an intracellular protein related to UNC-33, *Nature* 376:509–14.
73. Inagaki, N., Chihara, K., Arimura, N. et al. (2001) CRMP-2 induces axons in cultured hippocampal neurons, *Nat Neurosci* 4:781–2.
74. Lee, W. C., Huang, H., Feng, G. et al. (2006) Dynamic remodeling of dendritic arbors in GABAergic interneurons of adult visual cortex, *PLoS Biol* 4:e29.
75. Geddis, M. S. and Rehder, V. (2003) The phosphorylation state of neuronal processes determines growth cone formation after neuronal injury, *J Neurosci Res* 74:210–20.
76. Faire, K., Trent, F., Tepper, J. M. and Bonder, E. M. (1992) Analysis of dynamin isoforms in mammalian brain: Dynamin-1 expression is spatially and temporally regulated during postnatal development, *Proc Natl Acad Sci U S A* 89:8376–80.
77. Nakata, T., Iwamoto, A., Noda, Y. et al. (1991) Predominant and developmentally regulated expression of dynamin in neurons, *Neuron* 7:461–9.
78. Noda, Y., Nakata, T. and Hirokawa, N. (1993) Localization of dynamin: Widespread distribution in mature neurons and association with membranous organelles, *Neuroscience* 55:113–27.
79. Powell, K. A. and Robinson, P. J. (1995) Dephosphin/dynamin is a neuronal phosphoprotein concentrated in nerve terminals: evidence from rat cerebellum, *Neuroscience* 64:821–33.
80. Shpetner, H. S. and Vallee, R. B. (1989) Identification of dynamin, a novel mechanochemical enzyme that mediates interactions between microtubules, *Cell* 59:421–32.
81. Schafer, D. A. (2002) Coupling actin dynamics and membrane dynamics during endocytosis, *Curr Opin Cell Biol* 14:76–81.

82. Sontag, J. M., Fykse, E. M., Ushkaryov, Y. et al. (1994) Differential expression and regulation of multiple dynamins, *J Biol Chem* 269:4547–54.

83. Urrutia, R., Henley, J. R., Cook, T. and McNiven, M. A. (1997) The dynamins: Redundant or distinct functions for an expanding family of related GTPases?, *Proc Natl Acad Sci U S A* 94:377–84.

84. van der Bliek, A. M., Redelmeier, T. E., Damke, H. et al. (1993) Mutations in human dynamin block an intermediate stage in coated vesicle formation, *J Cell Biol* 122:553–63.

85. Littleton, J. T. and Bellen, H. J. (1995) Synaptotagmin controls and modulates synaptic-vesicle fusion in a Ca(2+)-dependent manner, *Trends Neurosci* 18:177–83.

86. Ullrich, B. and Südhof, T. C. (1995) Differential distributions of novel synaptotagmins: Comparison to synapsins, *Neuropharmacology* 34:1371–7.

87. Yoshihara, M., Adolfsen, B. and Littleton, J. T. (2003) Is synaptotagmin the calcium sensor?, *Curr Opin Neurobiol* 13:315–23.

88. Yoshihara, M. and Littleton, J. T. (2002) Synaptotagmin I functions as a calcium sensor to synchronize neurotransmitter release, *Neuron* 36:897–908.

89. Beaver, C. J., Ji, Q. and Daw, N. W. (2001) Layer differences in the effect of monocular vision in light- and dark-reared kittens, *Vis Neurosci* 18:811–20.

90. Daw, N. W. (1994) Mechanisms of plasticity in the visual cortex. The Friedenwald Lecture, *Invest Ophthalmol Vis Sci* 35:4168–79.

91. Daw, N. W., Fox, K., Sato, H. and Czepita, D. (1992) Critical period for monocular deprivation in the cat visual cortex, *J Neurophysiol* 67:197–202.

92. Cynader, M., Timney, B. N. and Mitchell, D. E. (1980) Period of susceptibility of kitten visual cortex to the effects of monocular deprivation extends beyond six months of age, *Brain Res* 191:545–50.

93. Winfield, D. A. (1983) The postnatal development of synapses in the different laminae of the visual cortex in the normal kitten and in kittens with eyelid suture, *Brain Res* 285:155–69.

94. Tieman, S. B. (1991) Morphological changes in the geniculocortical pathway associated with monocular deprivation, *Ann N Y Acad Sci* 627:212–30.

95. Obata, S., Obata, J., Das, A. and Gilbert, C. D. (1999) Molecular correlates of topographic reorganization in primary visual cortex following retinal lesions, *Cereb Cortex* 9:238–48.

96. Koh, T. W. and Bellen, H. J. (2003) Synaptotagmin I, a Ca2+ sensor for neurotransmitter release, *Trends Neurosci* 26:413–22.

97. Van den Bergh, G. and Arckens, L. (2004) Fluorescent two-dimensional difference gel electrophoresis unveils the potential of gel-based proteomics, *Curr Opin Biotechnol* 15:38–43.

12 A Neuroproteomic Approach to Understanding Visual Cortical Development

Leonard E. White

CONTENTS

12.1 SUMMARY

A comprehensive understanding of how sensorimotor experience shapes brain development in early life necessitates a synthesis of evidence across a broad spectrum of scientific discovery. Phenomenological and mechanistic studies conducted at the level of cellular and systems neuroscience must be integrated with proteomic investigations aimed at understanding the organization and function of the protein networks that instantiate the impact of experience. Recent progress in characterizing the contributions of early visual experience to the development of functional circuits in the visual cortex has laid the foundation for the identification of proteins that are expressed differentially in brain development and modulated by visual experience. Ongoing studies are now uniting the power of proteomics with novel neurophysiological approaches to probe the molecular, cellular, and circuit mechanisms by which experience interacts with endogenous programs of brain development.

12.2   INTRODUCTION

Neural circuits in the cerebral cortex undergo a remarkable phase of growth and maturation in the early postnatal period that is characterized by the emergence of response selectivities and precise patterns of neuroanatomical connections. In the visual cortex, this is a period of brain development when functional properties that are elaborated by cortical circuits, such as a preference for contour orientation and direction of stimulus motion, become organized into columnar patterns of population activity. Such patterns of stimulus-induced population activity may be visualized using optical imaging techniques and represented as functional maps that localize the preferred responses of neural circuits and individual neurons across the visual cortex. In recent years, two functional maps—the map of orientation preference and the map of direction preference—have been especially useful for understanding the development of functional circuits in the visual cortex. Current evidence suggests that intrinsic, activity-dependent mechanisms are sufficient for the self-organization of the map of orientation preference, but this conclusion does not exclude a role for visual experience in early life. Normal patterns of visual experience interact synergistically with these intrinsic mechanisms to promote the full maturation of the map of orientation preference, while abnormal patterns of activity abrogate this synergy, producing broad tuning for contour orientation. The development of response selectivity for the direction of stimulus motion is even more dependent upon normal visual experience in early life; endogenous patterns of neural activity are not sufficient to shape the functional architecture of the neural circuits that give rise to direction selectivity in the visual cortex.

Given the importance of visual experience in early life, studies of the maps of orientation and direction selectivity provide an essential physiological context for understanding the biological processes by which intrinsic mechanisms of neural development are modulated by sensory experience. But to truly build a comprehensive understanding of these mechanisms, it is necessary to integrate knowledge across levels of biological organization and scientific inquiry. Thus, it is essential to unite discovery at the level of systems and cellular neurophysiology with an understanding of the protein networks that mediate the expression of developmental programs at the subcellular level. The challenge before us, therefore, is to apply the power of proteomics and bioinformatics to a functional context that will lead to novel insights into the mechanisms by which experience in early life shapes brain development. We and others have taken up this challenge and begun to approach the development of functional circuits in the visual cortex with just such a multifaceted perspective. Our initial goal has been to identify the proteins that are both regulated by visual experience in early life and implicated in the development of cortical response properties, such as orientation and direction selectivity. Our long-term goal is to understand how these proteins interact in protein–protein networks within neuronal elements in cortical circuits, how their interactions are regulated by vision, and why the orchestration of such networks by experience is essential for the development of computation properties of the cerebral cortex, like direction selectivity, that are especially sensitive to response timing. The importance of doing so is highlighted by the prevalence of deficits in the temporal sequencing of sensory information that are common in a variety of developmental disorders. Fulfilling the promise of improved cognitive and sensorimotor interventions must arise from a deeper, basic understanding of how sensory experience shapes brain development across all levels of organization. The investigative approach outlined in this chapter and in other chapters of this volume (see Chapters 11, 13, and 14) represents an important step in this direction.

12.3 CIRCUIT CONSTRUCTION IN VISUAL CORTICAL DEVELOPMENT

Any approach toward the biological mechanisms that shape brain development in early life must encompass the ongoing construction of cortical circuits that continues well into the postnatal period. When studying visual cortical development, it is worth keeping in mind that at the time of eye-opening (birth in primates), the total number of cortical synapses in the visual cortex is only a fraction of that found in maturity, and that the vast majority of cortical synapses are elaborated when visual experience has the potential to impact the spatial and temporal patterns of neural activity in the visual cortex and thereby modify cortical circuits by activity-dependent mechanisms of plasticity. In the primary division of the visual cortex in rhesus monkeys, for example, a phase of rapid synaptogenesis ensues late in the third trimester before birth and continues exponentially over the first two postnatal months before achieving a stable density of synaptic profiles in neuropil by the third month (1). In carnivores, which open their eyelids some time after birth, a similar phenomenon has been documented: the density of synapses in the visual cortex increases rapidly in the month that follows eye-opening and the onset of patterned visual experience (2,3).

As might be expected, this phase of synaptogenesis in the visual cortex is accompanied by an outgrowth of the axonal projections that establish intracortical networks within the visual cortex. The best studied of these intracortical circuits is the system of long-range horizontal connections in layers 2 and 3 that are known to establish connections among cortical columns with similar response properties (4–6). The axonal projections that define this circuit are elaborated and achieve their mature distributions over this same period of rapid synaptogenesis in postnatal development (7–14). For example, in ferret visual cortex, horizontal connections are limited at about the time of eye-opening, with considerably less spatial coverage across the visual cortex and a lower degree of clustering (10,14). As illustrated in Figure 12.1, adult-like distributions of intrinsic connections are achieved some three to four weeks after eye-opening, at about the same time in postnatal development that the rate of synaptogenesis begins to decline and synaptic densities in layer 2/3 of the visual cortex reach a stable level (15). Although studies of developing horizontal connections in the visual cortex have often emphasized regressive phenomenon, such as collateral pruning and selective synapse elimination as important means of achieving functional maturity (16), the sculpting of intrinsic cortical connectivity occurs in a larger context of net circuit construction [see (17)]. The protein–protein interactions that instantiate this phase of cortical development must, therefore, include the systems of cellular processes that mediate dendritic maturation, axonal elongation, and synaptogenesis.

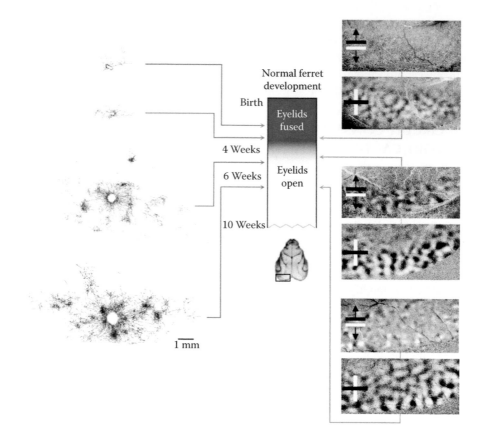

FIGURE 12.1 Summary of visual cortical development in ferrets reared with normal visual experience under standard cycles of light and darkness. Ferrets are born with their eyelids fused and do not open their eyes until about one month of postnatal life (*central graphic*; inset shows dorsal surface of ferret brain and the field of view for optical imaging assessments of population activity in visual cortex). Before the onset of patterned visual experience, intrinsic connections within the visual cortex are short range and sparse (*left*; tracings are reconstructions of axons from tangential sections labeled with the anterograde tracer biocytin injected into the vacant region in the center of the drawing where labeled axons and dendrites were too dense to trace—see White et al., 2001) (46). In the weeks after eye-opening, horizontal connections extend for several millimeters across the visual cortex, giving rise to patchy long-range connections. During this same period, functional maps of orientation and direction preference develop (*right*). Each pair of images was obtained by intrinsic signal optical imaging methods from the same juvenile ferret, with the upper difference image indicating direction selectivity and the lower difference image indicating orientation selectivity (stimulus icons indicate which directions and orientations were used to generate the responses that were then subtracted to produce the difference images; for each image, the dark domains correspond to the dark bar in the icon and the light domains correspond to the light bar). Note the lag in the development of the map of direction preference until after eye-opening and the establishment of the map of orientation preference [see (57)].

12.4 ORIENTATION AND DIRECTION SELECTIVITY IN VISUAL CORTEX

Fundamental to the operations of the visual system is the ability of neuronal networks in the visual cortex to represent the spatial patterns and motion energy engendered by visual scenes. Some five decades ago, D. H. Hubel and T. N. Wiesel discovered the existence of neurons in the visual cortex of cats and monkeys that exhibited receptive field properties that would appear to be integral to the neural representations of form and movement [see (18)]. In particular, they demonstrated that most neurons in the visual cortex respond vigorously to a moving bar of light, but only when the bar is oriented within a narrow range of angles (e.g., near vertical or horizontal). This aspect of response selectivity is termed orientation selectivity, and the angle of best response characterizes the neuron's orientation preference. Additionally, many cortical neurons respond much more strongly when an edge of light or shadow in its optimal orientation is panned in one direction of motion compared to the opposite direction. This property is termed direction selectivity, and the optimal direction of motion for a neuron describes its directional preference.

Hubel and Wiesel also explored the spatial arrangement of neurons with these response preferences across the tangential extent of the visual cortex [see (18)]. Again using cats and monkeys, they discovered that neurons with similar preferences are clustered into radial columns, which are arrayed in a systematic fashion across the extent of the visual cortex. However, the detailed organization of neighboring radial columns and the overall spatial layout of response preference were not well appreciated until the advent of in vivo optical imaging methods that record (indirectly) the summed responses of neurons in upper layers of the visual cortex with columnar (sub-millimeter) resolution. Thus, in the last two decades, it has become possible to use optical means for measuring signals (either intrinsic hemodynamic and light-scattering signals or exogenous fluorescent dye-based signals) that indirectly reflect underlying neuronal selectivities and preferences. These methods have been employed to characterize the spatial layout of such so-called functional maps across the accessible reaches of the visual cortex (19–21). When such methods were applied to the analysis of cortical responses in carnivores and primates, a much more complete picture of the spatial distribution of response preference in the primary division of the visual cortex became known.

Optical imaging methods have been used to characterize the preferred responses to stimulus orientation at every location (pixel or cell, in the case of two-photon functional imaging) in the visual cortex. In each species for which so-called maps of orientation preference have been documented, these maps are characterized by numerous iterations of pinwheel motifs and linear zones that represent orientation preference in a smooth and continuous fashion, save for point discontinuities at the centers of pinwheels (4,22–24). Similarly, the same methods have been employed to characterize the distribution of response preference for the direction of stimulus motion. The results of such studies have demonstrated maps of direction preference in the primary division of the visual cortex in carnivores and in middle temporal visual areas in primates (25–31). The map of direction preference is nested geometrically within the map of orientation preference, such that each iso-orientation domain is subdivided into a pair of smaller domains that represent opposite directions of stimulus motion (28,30,32).

Not surprisingly, the documentation of such maps of response preference in the visual cortex has generated considerable interest and much fruitful debate. For example, although it is conventional to consider the functional maps of the visual cortex to be separable and independent, we and others have argued recently that the maps of computed properties (e.g., orientation selectivity, direction selectivity, and spatial frequency preference) may be understood as a single map of spatiotemporal energy (33–35) [but see also Baker and Issa (36) for a different interpretation]. Furthermore, others have argued for or against the significance of columnar maps of response preference for sensory coding [see, e.g., Chklovskii and Koulakov (37), Swindale (38), and Horton and Adams (39)]. These collateral debates notwithstanding, functional maps in the visual cortex have served as useful models for exploring the neural mechanisms responsible for the development of cortical response properties and the neural circuits from which they arise. Before outlining a neuroproteomic strategy for doing just that, let us first consider the evidence that implicates visual experience in the development and maturation of orientation and direction selectivity in the visual cortex, because any neuroproteomic approach to understanding brain development must consider the impact of sensory experience.

12.5   ROLE FOR SENSORY EXPERIENCE IN DEVELOPMENT OF ORIENTATION AND DIRECTION SELECTIVITY IN VISUAL CORTEX

Until recently, the prevailing view that provided a framework for understanding the mechanisms of functional map formation posited two distinct phases of cortical development: an experience-independent phase in which the basic neural circuits that underlie response selectivities become established and organized into functional maps, and a subsequent phase of plasticity in which initial circuits are elaborated and refined by visual experience (40–46). The central tenet of this framework is that sensory experience has little or no impact on the initial establishment of cortical response properties and the formation of functional maps. This tenet is based on two well-documented observations: (i) cortical maps of ocular dominance (the preferential response of cortical neurons to one eye or the other) and orientation preference are present prior to the onset of visual experience (40–48); and (ii) total visual deprivation has little impact on the establishment and early maturation of either property (46,49). This framework implies that most of the information that is required for map construction—establishing the basic layout of the map and determining which regions will express a particular preference—is innate and that sensory experience plays only a modest role in elaborating a developmental program that has been largely determined by experience-independent mechanisms.

The mechanisms that underlie this developmental program can be roughly divided into two basic classes: molecular recognition mechanisms that rely on gradients of diffusible ligands and cell-surface receptors to specify map topology, and activity-dependent mechanisms that rely on correlated patterns of pre- and post-synaptic activity to guide map formation. The latter can be further refined according to the source of the neural activity patterns: activity that arises endogenously within the developing visual system and, later in development, activity that is driven by visual experience. A finely tuned orchestration of all of these mechanisms is, presumably, essential for the proper establishment and subsequent maturation of functional maps in the visual cortex. This is best exemplified by the development of the map of visual space in the midbrain (tectum), where the mechanisms are best understood [for reviews, see (50–52)]. The initial formation of this map depends on molecular gradients that insure the guidance of axons to the topologically appropriate portions of the tectum; similar molecular mechanisms are likely to operate in the visual cortex (53). However, such mechanisms provide only coarse instruction of map topology; at later stages in development, patterns of retinal activity are required to achieve the finely tuned precision that is characteristic of the map in the mature brain. The precise role of molecular recognition mechanisms in the formation of other functional maps in the visual cortex remains unclear [(54); but see (55)], but a similar sequence—an early stage that specifies the basic structure of the map, followed by a subsequent stage of map refinement—is generally thought to account for the development of all cortical maps.

This account would seem sufficient to explain the development of the map of orientation preference in ferret visual cortex. This map is first recognizable—although often not homogeneous in columnar structure or signal strength—at or just before the time of eye-opening in ferret kits (Figure 12.1) (46,47,56). Over the next several days, columnar structure becomes more uniform across the map and the strength of the mapping signals increases substantially before achieving functional maturity approximately three weeks after eye-opening (46,47,57). This process, as documented with intrinsic signal optical imaging techniques, parallels (with a several day lag) the maturation of neuronal orientation selectivity assessed with unit recording using microelectrodes [(40,47); see also (58)]. Although orientation maps are detectable at the onset of patterned visual experience, the fact that the full maturation of the map occurs after eye-opening raises the possibility that visual experience may play a role in supervising the progression and outcome of circuit construction in the visual cortex (Figure 12.1).

A role for early visual experience in visual cortical development is suggested further by comparable studies of the emergence and maturation of the map of direction preference in ferret visual cortex. The establishment of this map lags behind the development of the orientation preference map by at least a week: the first appearance of directionally selective columnar structure in the population response of ferret visual cortex is not observed until after several days of visual experience following eye-opening (57). Thus, in the youngest ferrets examined, direction selective cortical responses were barely detectable with intrinsic signal imaging and, when differential responses to opposite directions of motion were observed, they were at best only crudely organized into columnar patterns (Figure 12.1). Direction selectivity signals recorded by this means did not differ significantly from non-stimulated background signals until about five days after eye-opening. More uniform, adult-like columnar patterns of direction selective responses were not observed until about one week after eye-opening, well after the establishment of uniform maps of orientation preference (Figure 12.1) (57). These observations are consistent with earlier electro-physiological studies that report the existence of direction selective units in newborn macaques and very young kittens (43,48,59), but the prevalence of sharply tuned neurons in such visually naïve animals is, apparently, quite low (60,61). The delay in the development of direction selectivity until after eye-opening challenges the general notion that functional maps are established early via the operation of experience-independent mechanisms of cortical development. It raises the possibility that the development of the map of direction preference only after the onset of patterned visual experience might not be simply coincidental.

We investigated whether there might be a role for visual experience in the maturation of the map of orientation preference and the establishment of the map of direction preference by testing for the presence of these functional maps in juvenile ferrets that were subject to visual deprivation. Studies of the development of the map of orientation preference in ferrets that were completely deprived of visual experience by dark-rearing provided one key clue that experience has some role to play in the maturation of functional circuits in the visual cortex (46). Thus, the strengthening of the map of orientation preference that occurs after eye-opening in ferrets reared with the benefit of normal day-night cycles did not occur in dark-reared animals: maps of orientation preference were detectable in dark-reared ferrets, but the strength of the maps relative to normally reared, age-matched control ferrets was markedly reduced (Figure 12.2). These results suggest that experience plays an important role in the growth of intracortical circuits, the addition of cortical synapses, and the maturation of the synaptic mechanisms that are required to achieve full map strength. Moreover,

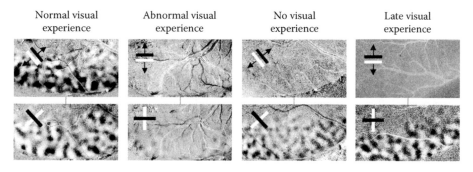

FIGURE 12.2 Impact of visual experience on the development of the maps of orientation and direction preference in ferret visual cortex. Images in each column are difference images indicating direction selectivity (*upper image*) or orientation selectivity (*lower image*), as in Figure 12.1. Rearing animals with abnormal visual experience by keeping their eyelids fused (*second column from left*) prevents the development of direction selectivity and severely weakens orientation selectivity; note the near absence of columnar structure in the images compared to the normal ferret (*far left column*). Rearing animals without visual experience (*second column from right*) is equally effective in preventing the development of direction selectivity, but much less harmful to orientation selectivity. Rearing animals in darkness until two weeks after eye-opening and then allowing two to three weeks of normal vision (*far right column*) allows for the full maturation of orientation selectivity, but does not permit the development of direction selectivity. Early vision in the first two weeks after eye-opening is necessary for the development of direction selectivity in ferret visual cortex [see (57)].

it is not just the quantity of experience that drives map strengthening; the quality of visual experience is critical for the development of orientation selectivity. This conclusion came from parallel studies where we reared ferrets with their eyelids kept closed to engender abnormal patterns of visual experience. In these animals, vision through closed eyelids had a more devastating impact on orientation selectivity and columnar map structure than dark-rearing: the map of orientation preference in the visual cortex was virtually abolished by this treatment (see Figure 12.2; cf. the middle two columns of images). Taken together, these results in normal and visually deprived ferrets suggest that the pattern of visually evoked neural activity, not just its presence, is crucial for guiding the ongoing construction of the neural circuits that instantiate the mature map of orientation preference.

Studies of direction selectivity in dark-reared ferrets make an even more compelling case for the role of experience in this constructive phase of cortical maturation; they indicate that the normal lag in the formation and maturation of the map of direction preference until after the onset of patterned visual experience is not coincidental (57). Animals that were deprived of vision by dark-rearing during the first two weeks after eye-opening failed to develop maps of direction preference, even though (as described above) maps of orientation preference were present in these same ferrets (Figure 12.2). Remarkably, this was also true of animals that were provided normal day-night cycles for some period following deprivation during this early two-week period. These animals with later visual experience following early deprivation showed complete restoration of all known bandwidths of selectivity and that define functional maturity in visual cortical networks, despite the persistence of severe impairments in cortical direction selectivity (Figure 12.2) [see (57)]. These last findings argue that early dark-rearing did not simply delay the normal progression of visual cortical development, as the effect of dark-rearing on critical period plasticity is usually interpreted (62–64). Rather, these results suggest that dark-rearing deprives the developing visual system of an important source of spatial, temporal, and/or luminance cues that are required for the formation of directionally selective circuits in cortical networks.

More generally, these studies of normal and visually deprived ferrets establish model systems for understanding how patterns of neural activity evoked by sensory experience interact with endogenous mechanisms of neural development to promote the construction of functional neural circuits in the cerebral cortex. Thus, these studies have characterized four distinct phenotypic groups of animals that are differentiated on the basis of the expression of cortical maps of orientation and direction preference (Figure 12.3): (a) ferrets with normal visual experience, reared under 12-hour day-night cycles, that develop normal functional maps; (b) ferrets with abnormal visual experience, reared with their eyelids kept shut, that display marked impairments in cortical orientation and direction selectivity; (c) dark-reared ferrets without visual experience, reared in absolute darkness, that show modest orientation selectivity, but severe impairments in direction selectivity; and (d) dark-reared ferrets that are given normal experience two weeks after eye-opening and are normal in every respect, except for the persistence of severe impairments in cortical direction selectivity. In addition, animals that were reared in absolute darkness before eye-opening and then were provided normal visual experience after the time of natural

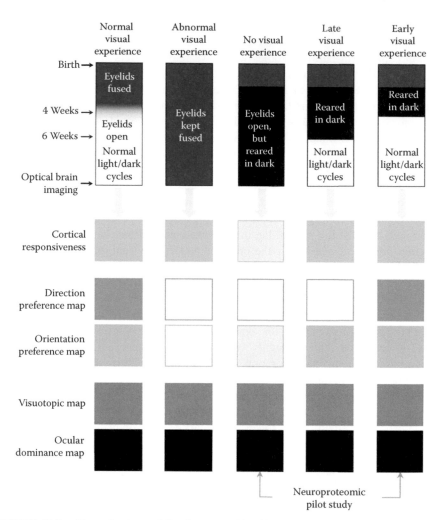

FIGURE 12.3 (See color insert following page 172.) Illustration of experimental protocol for generating physiological phenotypes that vary in the development of functional maps in ferret visual cortex. Normal ferrets (*far left column*) are characterized by cortical maps of direction and orientation preference, visuotopy, and ocular dominance. The former two maps are "computational," representing the cortical distribution of orientation and direction preference computed by thalamocortical and intracortical circuits, and the latter two maps are "topological," representing the near neighbor relations established in the retina and conveyed to the visual cortex by input from the thalamic lateral geniculate nucleus. Rearing ferrets with abnormal visual experience (*second column from left*) prevents the development of the maps of direction and orientation preference (see Figure 12.2). Rearing animals without visual experience (*middle column*) reduces cortical responsiveness by about half and prevents the development of the map of direction preference, while the map of orientation preference is present but weak (saturation of color in corresponding blocks indicates status of map development). Providing late visual experience after early dark-rearing through the second week after eye-opening (*second column from right*) restores all visual cortical maps, except for the map of direction preference. The normal expression of all functional maps in ferrets that were dark-reared only until eye-opening (*far right column*) highlights the importance of early visual experience, especially for the development of the map of direction preference [see (57)].

eye-opening are normal with respect to the development of functional maps in the visual cortex (see Figure 12.3, far right column). This group demonstrates that early visual experience within two weeks of eye-opening is crucial for the development of direction selectivity [see (57)]. Further work is necessary to better characterize these various phenotypes neurophysiologically, neuroanatomically, and in terms of their general physiology and motoric behavior. Nevertheless, these models provide the foundation for neuroproteomic studies aimed at understanding the identity and function of the protein networks that instantiate the benefits of normal visual experience and the disturbances of such networks that must mediate the maladaptive consequences of abnormal experience.

12.6 UNITING SYSTEMS NEUROSCIENCE AND NEUROPROTEOMICS FOR UNDERSTANDING DEVELOPMENT OF ORIENTATION AND DIRECTION SELECTIVITY IN VISUAL CORTEX

In the last several years, a few groups have begun to address the impact of sensory experience on early brain development using large-scale genomic or proteomic approaches [e.g., (65–68)]. Much of this effort has been directed at identifying systems of proteins that are regulated by developmental stage and differentiating the expression of such proteins from those whose expression is more strictly regulated by the acute onset of sensory experience. With few exceptions, these studies have been undertaken in rodent models of the mammalian visual system, with the principal focus being the regulation of the critical period for ocular dominance plasticity (65,69). The outcome has been the identification of age-specific genes and gene products that are modulated by the presence or absence of visual experience, and another set that is subject to experience-dependent regulation at any age (65). These efforts have demonstrated the power of such high-throughput studies to identify putative molecular pathways and networks that instantiate endogenous and experience-dependent programs of cortical maturation. However, because of limitations in the functional organization of the visual system in rats and mice, it is difficult to understand the significance of such molecular evidence for visual function, especially as it applies to mammals with more advanced visual abilities (e.g., carnivores and primates).

Indeed, very little is known about patterns of gene expression and protein interactions that mediate the initial establishment and subsequent maturation of the functional circuits responsible for receptive field properties that underlie the selective evoked responses of visual cortical neurons, and no studies to date have taken a proteomics approach as a means of understanding the development of functional maps.

One pioneering group has, however, initiated proteomic studies in the cat visual system; their studies are likely to further refine the network of proteins that regulate the onset and close of the critical period for ocular dominance plasticity (67,68). A current account of this lab's research approach and experimental findings is presented in Chapter 11. Our approach has been focused on events that unfold in an earlier phase of cortical development, when receptive field properties such as orientation and direction selectivity are first becoming established and organized into functional maps prior to the peak of any known critical periods for functional map plasticity. By exploiting the developmental phenotypes available in ferret visual cortex that vary in the expression of functional maps, we are aiming to link proteomic data on the expression patterns and regulation of protein networks to neurophysiological and neuroanatomical assessments of defined functional circuits in developing brain.

An intriguing target of this multifaceted approach is the neural mechanisms responsible for the development of direction selectivity in the visual cortex. Using the ferret models outlined above (Figure 12.3), we are focusing this effort on the cellular and molecular mechanisms that underlie the unique visual dependence of direction selectivity by identifying proteins that are both regulated by visual experience and critical for the development of cortical direction selectivity. Perhaps the most instructive contrast is a comparison of protein expression in the visual cortex of juvenile ferrets that are specifically deficient in cortical direction selectivity (late visual experience group in Figure 12.3) against age-matched ferrets that express a normal map of direction preference (early visual experience group in Figure 12.3). The results should bring to light candidate proteins that are likely to play a role in the neural events that shape the functional circuits from which computations of direction selectivity arise. Before aiming at this particular comparison, we considered it prudent to verify that positive findings would result from this general approach starting with a more robust, albeit a less insightful contrast, a comparison between juvenile dark-reared ferrets and age-matched control ferrets from the "early vision" group (Figure 12.3). Thus, we first sought to determine if we could isolate and quantify proteins that were differentially expressed in the visual cortex of two groups of ferrets: one experimental group that was dark-reared until the time of sacrifice and severely impaired in cortical direction selectivity, but also modestly impaired in tuning for other receptive field properties (57), and an age-matched control group that was subjected to early dark-rearing, but permitted visual experience during the critical period for the formation of direction selectivity. This general approach is outlined schematically in Figure 12.4 and the details of these general neuroproteomic methods are discussed in more depth in Chapters 3, 4, 5, and 6.

We began by dark-rearing both groups of ferrets from postnatal day 17, which is prior to the onset of cortical responsiveness to visual stimulation (40,70); this allowed us to ensure equivalent conditions between groups during the early stages of corticogenesis and thalamocortical circuit construction (71,72). On postnatal day 35, one group of animals was placed in standard vivarium conditions with a 12-hour light/dark cycle, while the other group remained in the dark. Based on recent evidence (57), we know that ferrets given 2–3 weeks of visual experience beginning on postnatal day 35 (after early dark-rearing) proceed to develop mature maps of direction preference. Thus, a comparison of tissue from these two groups should reveal the proteomes of animals that express direction selectivity and those that are severely impaired (Figure 12.3). On postnatal day 50, both groups of animals were sacrificed, their brains were rapidly removed and the visual cortex was blocked, flash-frozen in isopentane cooled on liquid nitrogen, and stored at −80°C for further processing.

Visual cortical tissue was prepared for two-dimensional, difference-in-gel electrophoresis (2D-DIGE) analysis using the general methods that are detailed in Chapter 4.

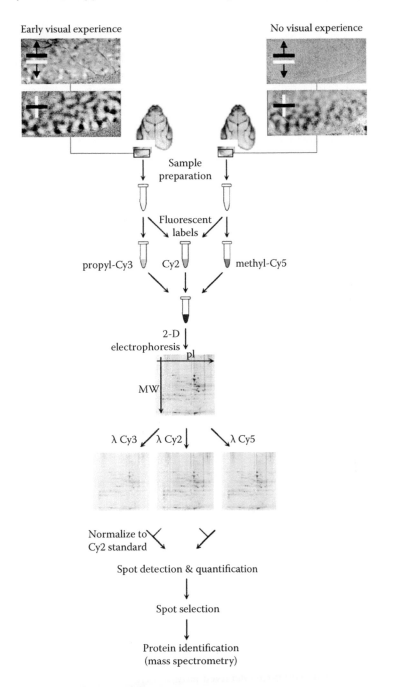

FIGURE 12.4 (See color insert following page 172.) Schematic overview of 2D-DIGE/MS analysis of visual cortical tissue from juvenile ferrets with early visual experience (and cortical maps of direction preference) and without visual experience (lacking cortical maps of direction preference). See Chapters 3, 4, 5, and 6 for further methodological details.

A sample prepared from a continuously dark-reared ferret was labeled with methyl-Cy5 and a control sample from the light-reared ferret was labeled with propyl-Cy3; additional samples from each ferret were mixed together and labeled with Cy3. The samples were then combined and proteins isolated by two-dimensional gel electrophoresis. To facilitate statistical validation of the results, sample preparation and 2D-DIGE were performed in triplicate, with one of the three runs reversing the dye-labeling scheme between dark-reared and control tissue. Separately, samples were run to create master gels, which were used later for picking protein spots for identification by mass spectrometry.

Each individual set of proteins was visualized as a converged image in a mixture based on the spectral properties of the dyes (Figure 12.5). The three 2D-DIGE gels were then imported into the DeCyder Biological Variation Analysis (GE Healthcare) workstation. An automated analysis of individual protein spots was then performed and the results displayed as peaks, characterized by amplitude and shape. The peaks were normalized to the Cy2 labels as a reference in each gel, which minimized the impact of gel-to-gel variation and maximized the accuracy of protein quantification across gels. We then performed a pair-wise comparison of sample ratios, in which effects are measured by running gels that compare internal variations in the combined samples, allowing for the determination of changes between experimental samples with statistical meaning. T-tests were performed comparing the expression levels of isolated proteins between rearing conditions, using a significance threshold of two standard deviations.

More than 2000 protein spots were detected across three biological replicates. Following conservative statistical analysis, 16 spots were significantly different in each of the three replications (Table 12.1). Of these 16, three proteins (including lower inset in Figure 12.5) showed increased expression in the visual cortex of the

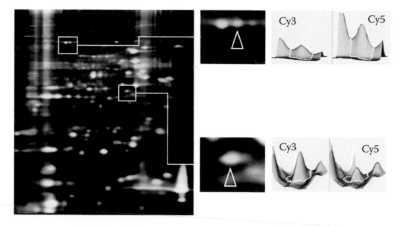

FIGURE 12.5 (See color insert following page 172.) 2D-DIGE analysis of visual cortical samples from normal and visually deprived ferrets. (*Left*) Combined image of 2-D gel generated by the Typhoon 9410 Variable Mode Imager; normal tissue is labeled with Cy3 (*green*) and deprived tissue is labeled with Cy5 (*red*). High magnification insets highlight spots (*arrowheads*) that are expressed more in the dark-reared visual cortex (*upper panel*) or the control cortex (*lower panel*); these two spots were analyzed quantitatively with DeCyder software (*right surface plots*).

TABLE 12.1
2D-DIGE and Mass Spectrometry for Differential Protein Expression in Visual Cortex from Ferrets Reared Normally and Ferrets Reared in Absolute Darkness

	2D-DIGE			Mass Spectrometry			
Spot #	t Value	Expression Ratio	Group with Greater Expression (Normal or Dark-Reared)	Database Accession	Ion Score[a]	Protein	Proposed Function
227	0.008	−1.46	Dark-reared	n.a.			
306	0.048	−1.36	Dark-reared	n.a.			
311	0.045	−1.68	Dark-reared	n.a.			
312	0.027	−1.5	Dark-reared	AAH61999	28	Aconitate hydratase	Mitochondrial tricarboxylic acid cycle enzyme
316	0.025	−1.86	Dark-reared	n.a.			
317	0.009	−2.05	Dark-reared	n.a.			
322	0.011	−1.94	Dark-reared	AAH04645	231	Aconitate hydratase	Mitochondrial tricarboxylic acid cycle enzyme
348	0.005	−1.64	Dark-reared	CAA71370	49	Unc-33–like protein (ULIP) 2	Axonal growth (semaphorin signaling)
367	0.021	1.57	Normal	Q9UK02 HUMAN	456	GRP78	Facilitates multimeric protein assembly in ER
465	0.042	−1.49	Dark-reared	CAD28503	188	Unc-33–like protein (ULIP) 6	Axonal growth (semaphorin signaling)
466	0.030	−1.43	Dark-reared	CAD28503	63	Unc-33–like protein (ULIP) 6	Axonal growth (semaphorin signaling)
473	0.030	−1.30	Dark-reared	A38093	64	Stress-induced phosphoprotein 1	Mediates association of molecular chaperones
750	0.049	−1.49	Dark-reared	n.a.			
771	0.017	−1.17	Dark-reared	BAB18142	273	Fructose-bisphosphate aldolase C	Glycolysis enzyme
776	0.013	1.94	Normal	AAF31668	134	Neuronal tropomodulin	Regulates actin polymerization
854	0.049	1.39	Normal	AAH11429	98	Chromatin-modifying protein 4b	Mediates delivery of proteins to lysosomes

[a] Score of the quality of MS/MS peptide fragment ion matches; scores of 20 or greater are significant (86).

n.a., not analyzed.

control ferret (with direction selectivity), while the remaining 13 proteins (including upper inset in Figure 12.5) were expressed to a greater degree in the dark-reared ferret that was severely impaired in direction selectivity. This pilot result immediately demonstrates that differences in protein expression related to the development of direction selectivity are likely to entail both up- and down-regulation of protein networks within visual cortical tissue. Closer inspection of the gels suggests that some of these significant "hits" could represent physiologically relevant modification of primary proteins. For example, proteins numbered 306–322 (highlighted in upper inset of Figure 12.5) are adjacent to one another, separated by charge but not molecular weight. This is consistent with a single primary protein (aconitate hydratase, in this case) that may be present in multiple states of phosphorylation, a possibility that could be explored by antibodies specific for phosphorylated states and/or phospho-specific dyes, or with mass spectrometry following the methods described in Chapter 5.

From the master gel, we identified a subset of the statistically significant proteins for identification by mass spectrometry (MS). Peptide mass fingerprinting was performed on 10 proteins and the resulting spectra were identified by reference to non-redundant databases maintained at the NIH National Center for Biotechnology Information (http://www.ncbi.nlm.nih.gov/entrez/) and/or the Swiss data bank (http://us.expasy.org/). This analysis positively identified each of the 10 proteins with a high degree of certainty, based upon the conventional standards of the number of peptides that matched the theoretical digest of the primary protein, the score of the peptide-mass fingerprint or MS/MS peptide fragment, and the score of the MS/MS peptide fragment ion match (Table 12.1).

Three proteins were expressed at higher levels in the age-matched, normally reared ferret relative to the continuously dark-reared ferret. One was GRP78, which is a chaperone protein localized to the endoplasmic reticulum (73,74). Another was neuronal tropomodulin, which is a protein that regulates the elongation of actin filaments and contributes to the structure of the neuronal cytoskeleton (75,76). The third protein in this set was chromatin modifying protein 4b, a component of the ESCRT-III complex, which is involved in the delivery of transmembrane proteins into the lumen of lysosomes for degradation (77,78). The up-regulation of these proteins in ferrets provided with the benefit of visual experience could reflect the maturation and consolidation of neural circuits in the visual cortex that underlie direction selectivity (and other aspects of visual cortical function).

The remaining proteins identified by MS were all expressed at higher levels in the visual cortex of the continuously dark-reared ferret that was severely impaired in direction selectivity. These proteins included enzymes involved in glucose metabolism (aconitate hydratase, aldolase C), the association of molecular chaperones (stress-induced phosphoprotein 1), and, most interestingly, Unc-33–like proteins (ULIP2/6). The ULIP2/6 proteins are believed to play a role in the semaphorin signaling pathway and the regulation of axonal growth (79,80). The higher expression of at least some of these proteins in dark-reared visual cortex is consistent with the notion that neural development is dysregulated (if not delayed) in dark-reared animals (62,64). This feasibility study demonstrates the robust potential of this experimental design to isolate and identify proteins that are regulated by visual experience and implicated in the development of cortical direction selectivity, setting the stage for the next phase of study targeting the differential expression of proteins in ferrets that were reared to develop severe impairments in cortical direction selectivity, while all other receptive field properties of the visual cortex achieve normal bandwidths of tuning. Such studies should shed light on which proteins are implicated in the construction of the cortical circuits that compute direction of motion and/or the cellular and synaptic mechanisms of direction selectivity itself.

12.7   TOWARD A COMPREHENSIVE VIEW OF EXPERIENCE-DEPENDENT BRAIN DEVELOPMENT

Much effort is currently devoted to teasing apart the intricate molecular mechanisms responsible for establishing the basic layout of functional maps in visual cortex and subcortical visual centers. The most influential studies of this genre are those that are aimed at understanding the development and subsequent plasticity of the functional map of ocular dominance, which reflects the topological distribution of monocular inputs from the lateral geniculate nucleus of the thalamus to layer 4 of the primary division of the visual cortex. The results of such efforts have led to the now conventional view that functional maps are established early in development under the principal governance of molecular recognition mechanisms that are subject to activity-dependent modulation as nascent neural circuits become functional (54,55). This view considers sensory experience as influential only in later stages of map maintenance when previously established map structures are sustained and fine-tuned according to the demands imposed by the functional ecology of the organism interacting with its environment. However, this broad framework for understanding the development of functional maps in visual cortex is difficult to reconcile with recent evidence obtained from studies of orientation and direction map development. A somewhat different view of cortical maturation and the role of sensory experience is needed to more fully account for the development of functional maps in visual cortex that emerge from computations performed by thalamocortical and intracortical circuits. Rather than serving only as a means of maintaining prior map structure, studies of developing orientation and direction maps in normal and visual deprived ferrets indicate that experience can exert a profound influence over the formation and maturation of these cortical maps, which normally proceeds during a phase of rapid circuit construction in visual cortex. Moreover, the impact of experience can be either beneficial or deleterious: normal patterns of visual experience can interact synergistically with endogenous programs of development to promote the full maturation of columnar structure in the visual cortex, or—if patterns of activity are rendered abnormal, as in the case of rearing animals with the eyelids kept closed—this synergy may be abrogated and early columnar structures nearly lost. However, not all functional maps are equally sensitive to the impact of experience: the map of orientation preference (and ocular dominance) forms in the absence of vision, but the map of direction preference cannot. The formation and maturation of the map of direction preference in ferret visual cortex requires normal visual experience in a remarkably brief period of time (perhaps less than two weeks) beginning at about the time of natural eye-opening. This period of sensitivity concludes just as the so-called critical period for ocular dominance plasticity is nearing its peak (81,82). Apparently, lessons learned from studies of ocular dominance plasticity, as instructive as they may be for understanding competition-based rules of synaptic plasticity, may not readily apply to other functional maps in the visual cortex.

To achieve a comprehensive description of the development of functional circuitry in the cerebral cortex, including the circuitry from which direction selective responses emerge, it is essential to build an integrated understanding of the cellular, synaptic, and molecular mechanisms that must transduce sensory-driven neural activity into patterns of gene expression, protein modification, synaptic plasticity, and directed axonal growth and synaptogenesis. Most importantly, such data must be interpretable in the context of experimental phenotypes that target functional properties of cortical networks. Unfortunately, we are still far from achieving this long-term goal. Nevertheless, with the advent of a wide variety of molecular probes that may be introduced into developing cortical systems in vivo and the ability to track neuronal preferences and selectivity with two-photon imaging of functional signals (21,83,84), it is now becoming possible to explore the full spectrum of neurobiological mechanisms of map formation at the level of populations of identified neurons sampled from columnar structures within functional maps. When combined with neuroproteomics, as outlined in this chapter and in Chapters 1, 11, and 14, such systems-level, neurophysiological studies of neurons and neural circuits can be extended to the level of network cell biology.

The functional significance of neuroproteomic data, such as those generated in our pilot studies, are best understood in the light shed by the emerging interdisciplinary field of network biology, where the identified proteins are considered as key elements of complex molecular pathways and interacting protein–protein modules (85) (see Chapter 8 for additional discussion of protein interaction networks). One future challenge, therefore, is to understand how these elements interact within neurons (and non-neuronal cells) in the visual cortex, how their interactions are regulated by experience, and why the orchestration of such networks by experience is essential for the development of direction selectivity. These questions can be addressed experimentally using interference RNA technologies in vivo to suppress the expression of target proteins in developing visual cortex followed by functional screens to assay the consequences of such interventions for functional map development. Such studies will ask how physiological and proteomic phenotypes are altered by manipulations of targeted proteins. Current evidence suggests that the protein interaction networks we and others are seeking to identify are "scale-free networks" characterized by a high degree of interconnectivity but with a small number of interaction nodes [(85), and Chapter 8]. Thus, changes in the expression of key nodal proteins— as might occur with visual deprivation and as would be possible in such siRNA experiments—could impact several modular structures in the network and impair its overall function. Clearly, to apply these emerging theories to the study of visual cortical development, there must be interdisciplinary collaboration involving experimentalists, theoreticians, and bio-informaticians. Such experimental possibilities provide unprecedented opportunity for new insights into the early events that are responsible for the development of functional maps in sensory cortex and the mechanisms by which experience shapes the ongoing construction of cortical circuits.

ACKNOWLEDGMENTS

The author gratefully acknowledges the work of the Duke University Neuroproteomics Laboratory in the analysis of the cortical tissue, Oscar Alzate for his collaboration throughout the proteomic experiments described in this chapter, and David Fitzpatrick and Ye Li for critical discussions of this work. This effort was supported by a research grant from The Whitehall Foundation, Inc.

REFERENCES

1. Bourgeois, J. P. and Rakic, P. (1993) Changes of synaptic density in the primary visual cortex of the macaque monkey from fetal to adult stage, *J Neurosci* 13:280120.
2. Cragg, B. G. (1975) The development of synapses in the visual system of the cat, *J Comp Neurol* 160:147–66.
3. Erisir, A. and Harris, J. L. (2003) Decline of the critical period of visual plasticity is concurrent with the reduction of NR2B subunit of the synaptic NMDA receptor in layer 4, *J Neurosci* 23:5208–18.
4. Bosking, W. H., Zhang, Y., Schofield, B. and Fitzpatrick, D. (1997) Orientation selectivity and the arrangement of horizontal connections in tree shrew striate cortex, *J Neurosci* 17:2112–27.
5. Gilbert, C. D. and Wiesel, T. N. (1989) Columnar specificity of intrinsic horizontal and corticocortical connections in cat visual cortex, *J Neurosci* 9:2432–42.
6. Malach, R., Amir, Y., Harel, M. and Grinvald, A. (1993) Relationship between intrinsic connections and functional architecture revealed by optical imaging and in vivo targeted biocytin injections in primate striate cortex, *Proc Natl Acad Sci U S A* 90:10469–73.
7. Burkhalter, A., Bernardo, K. L. and Charles, V. (1993) Development of local circuits in human visual cortex, *J Neurosci* 13:1916–31.
8. Callaway, E. M. and Katz, L. C. (1990) Emergence and refinement of clustered horizontal connections in cat striate cortex, *J Neurosci* 10:1134–53.
9. Callaway, E. M. and Katz, L. C. (1992) Development of axonal arbors of layer 4 spiny neurons in cat striate cortex, *J Neurosci* 12:570–82.
10. Durack, J. C. and Katz, L. C. (1996) Development of horizontal projections in layer 2/3 of ferret visual cortex, *Cereb Cortex* 6:178–83.
11. Galuske, R. A. and Singer, W. (1996) The origin and topography of long-range intrinsic projections in cat visual cortex: A developmental study, *Cereb Cortex* 6:417–30.
12. Lubke, J. and Albus, K. (1992) Rapid rearrangement of intrinsic tangential connections in the striate cortex of normal and dark-reared kittens: Lack of exuberance beyond the second postnatal week, *J Comp Neurol* 323:42–58.
13. Luhmann, H. J., Singer, W. and Martinez-Millan, L. (1990) Horizontal interactions in cat striate cortex: I. Anatomical substrate and postnatal development, *Eur J Neurosci* 2:344–57.
14. Ruthazer, E. S. and Stryker, M. P. (1996) The role of activity in the development of long-range horizontal connections in area 17 of the ferret, *J Neurosci* 16:7253–69.
15. White, L. E. and Fitzpatrick, D. (2007) Vision and cortical map development, *Neuron* 56:327–38.
16. Katz, L. C. and Callaway, E. M. (1992) Development of local circuits in mammalian visual cortex, *Annu Rev Neurosci* 15:31–56.

17. Purves, D., White, L. E. and Riddle, D. R. (1996) Is neural development Darwinian? *Trends Neurosci* 19:460–4.
18. Hubel, D. H. and Wiesel, T. (2005) *Brain and visual perception* (New York: Oxford University Press).
19. Blasdel, G. G. and Salama, G. (1986) Voltage-sensitive dyes reveal a modular organization in monkey striate cortex, *Nature* 321:579–85.
20. Bonhoeffer, T. and Grinvald, A. (1996) Optical imaging based on intrinsic signals: The methodology. In *Brain mapping: The methods*, eds. Toga, A. W. and Mazziotta, J. C., 55–97 (San Diego: Academic Press).
21. Ohki, K., Chung, S., Ch'ng, Y. H., Kara, P. and Reid, R. C. (2005) Functional imaging with cellular resolution reveals precise micro-architecture in visual cortex, *Nature* 433:597–603.
22. Blasdel, G. G. (1992) Orientation selectivity, preference, and continuity in monkey striate cortex, *J Neurosci* 12:3139–61.
23. Bonhoeffer, T. and Grinvald, A. (1991) Iso-orientation domains in cat visual cortex are arranged in pinwheel-like patterns, *Nature* 353:429–31.
24. Rao, S. C., Toth, L. J. and Sur, M. (1997) Optically imaged maps of orientation preference in primary visual cortex of cats and ferrets, *J Comp Neurol* 387:358–70.
25. Albright, T. D. (1984) Direction and orientation selectivity of neurons in visual area MT of the macaque, *J Neurophysiol* 52:1106–30.
26. Diogo, A. C., Soares, J. G., Koulakov, A., Albright, T. D. and Gattass, R. (2003) Electrophysiological imaging of functional architecture in the cortical middle temporal visual area of *Cebus apella* monkey, *J Neurosci* 23:3881–98.
27. Malonek, D., Tootell, R. B. and Grinvald, A. (1994) Optical imaging reveals the functional architecture of neurons processing shape and motion in owl monkey area MT, *Proc Biol Sci* 258:109–19.
28. Shmuel, A. and Grinvald, A. (1996) Functional organization for direction of motion and its relationship to orientation maps in cat area 18, *J Neurosci* 16:6945–64.
29. Swindale, N. V., Grinvald, A. and Shmuel, A. (2003) The spatial pattern of response magnitude and selectivity for orientation and direction in cat visual cortex, *Cereb Cortex* 13:225–38.
30. Weliky, M., Bosking, W. H. and Fitzpatrick, D. (1996) A systematic map of direction preference in primary visual cortex, *Nature* 379:725–8.
31. Xu, X., Collins, C. E., Kaskan, P. M. et al. (2004) Optical imaging of visually evoked responses in prosimian primates reveals conserved features of the middle temporal visual area, *Proc Natl Acad Sci U S A* 101:2566–71.
32. Kisvarday, Z. F., Buzas, P. and Eysel, U. T. (2001) Calculating direction maps from intrinsic signals revealed by optical imaging, *Cereb Cortex* 11:636–47.
33. Basole, A., Kreft-Kerekes, V., White, L. E. and Fitzpatrick, D. (2006) Cortical cartography revisited: A frequency perspective on the functional architecture of visual cortex, *Prog Brain Res* 154:121–34.
34. Basole, A., White, L. E. and Fitzpatrick, D. (2003) Mapping multiple features in the population response of visual cortex, *Nature* 423:986–90.
35. Mante, V. and Carandini, M. (2005) Mapping of stimulus energy in primary visual cortex, *J Neurophysiol* 94:788–98.
36. Baker, T. I. and Issa, N. P. (2005) Cortical maps of separable tuning properties predict population responses to complex visual stimuli, *J Neurophysiol* 94:775–87.
37. Chklovskii, D. B. and Koulakov, A. A. (2004) Maps in the brain: What can we learn from them? *Annu Rev Neurosci* 27:369–92.
38. Swindale, N. V. (2000) How many maps are there in visual cortex? *Cereb Cortex* 10:633–43.

39. Horton, J. C. and Adams, D. L. (2005) The cortical column: A structure without a function, *Philos Trans R Soc Lond B Biol Sci* 360:837–62.
40. Chapman, B. and Stryker, M. P. (1993) Development of orientation selectivity in ferret visual cortex and effects of deprivation, *J Neurosci* 13:5251–62.
41. Crair, M. C., Gillespie, D. C. and Stryker, M. P. (1998) The role of visual experience in the development of columns in cat visual cortex, *Science* 279:566–70.
42. Horton, J. C. and Hocking, D. R. (1996) An adult-like pattern of ocular dominance columns in striate cortex of newborn monkeys prior to visual experience, *J Neurosci* 16:1791–807.
43. Hubel, D. H. and Wiesel, T. N. (1963) Receptive fields of cells in striate cortex of very young, visually inexperienced kittens, *J Neurophysiol* 26:994–1002.
44. Katz, L. C. and Crowley, J. C. (2002) Development of cortical circuits: Lessons from ocular dominance columns, *Nat Rev Neurosci* 3:34–42.
45. Sengpiel, F. and Kind, P. C. (2002) The role of activity in development of the visual system, *Curr Biol* 12:R818–26.
46. White, L. E., Coppola, D. M. and Fitzpatrick, D. (2001) The contribution of sensory experience to the maturation of orientation selectivity in ferret visual cortex, *Nature* 411:1049–52.
47. Chapman, B., Stryker, M. P. and Bonhoeffer, T. (1996) Development of orientation preference maps in ferret primary visual cortex, *J Neurosci* 16:6443–53.
48. Wiesel, T. N. and Hubel, D. H. (1974) Ordered arrangement of orientation columns in monkeys lacking visual experience, *J Comp Neurol* 158:307–18.
49. Fregnac, Y. and Imbert, M. (1978) Early development of visual cortical cells in normal and dark-reared kittens: Relationship between orientation selectivity and ocular dominance, *J Physiol* 278:27–44.
50. Goodhill, G. J. and Xu, J. (2005) The development of retinotectal maps: A review of models based on molecular gradients, *Network* 16:5–34.
51. Lemke, G. and Reber, M. (2005) Retinotectal mapping: New insights from molecular genetics, *Annu Rev Cell Dev Biol* 21:551–80.
52. O'Leary, D. D. and McLaughlin, T. (2005) Mechanisms of retinotopic map development: Ephs, ephrins, and spontaneous correlated retinal activity, *Prog Brain Res* 147:43–65.
53. Cang, J., Kaneko, M., Yamada, J. et al. (2005) Ephrin-as guide the formation of functional maps in the visual cortex, *Neuron* 48:577–89.
54. Crowley, J. C. and Katz, L. C. (2002) Ocular dominance development revisited, *Curr Opin Neurobiol* 12:104–9.
55. Huberman, A. D., Speer, C. M. and Chapman, B. (2006) Spontaneous retinal activity mediates development of ocular dominance columns and binocular receptive fields in v1, *Neuron* 52:247–54.
56. Coppola, D. M. and White, L. E. (2004) Visual experience promotes the isotropic representation of orientation preference, *Vis Neurosci* 21:39–51.
57. Li, Y., Fitzpatrick, D. and White, L. E. (2006) The development of direction selectivity in ferret visual cortex requires early visual experience, *Nat Neurosci* 9:676–81.
58. Godecke, I., Kim, D. S., Bonhoeffer, T. and Singer, W. (1997) Development of orientation preference maps in area 18 of kitten visual cortex, *Eur J Neurosci* 9:1754–62.
59. Hatta, S., Kumagami, T., Qian, J. et al. (1998) Nasotemporal directional bias of V1 neurons in young infant monkeys, *Invest Ophthalmol Vis Sci* 39:2259–67.
60. Blakemore, C. and Van Sluyters, R. C. (1975) Innate and environmental factors in the development of the kitten's visual cortex, *J Physiol* 248:663–716.
61. Imbert, M. and Buisseret, P. (1975) Receptive field characteristics and plastic properties of visual cortical cells in kittens reared with or without visual experience, *Exp Brain Res* 22:25–36.

62. Cynader, M. and Mitchell, D. E. (1980) Prolonged sensitivity to monocular deprivation in dark-reared cats, *J Neurophysiol* 43:1026–40.

63. Hensch, T. K. (2005) Critical period plasticity in local cortical circuits, *Nat Rev Neurosci* 6:877–88.

64. Mower, G. D., Berry, D., Burchfiel, J. L. and Duffy, F. H. (1981) Comparison of the effects of dark rearing and binocular suture on development and plasticity of cat visual cortex, *Brain Res* 220:255–67.

65. Majdan, M. and Shatz, C. J. (2006) Effects of visual experience on activity-dependent gene regulation in cortex, *Nat Neurosci* 9:650–9.

66. Tropea, D., Kreiman, G., Lyckman, A. et al. (2006) Gene expression changes and molecular pathways mediating activity-dependent plasticity in visual cortex, *Nat Neurosci* 9:660–8.

67. Van den Bergh, G., Clerens, S., Firestein, B. L., Burnat, K. and Arckens, L. (2006) Development and plasticity-related changes in protein expression patterns in cat visual cortex: A fluorescent two-dimensional difference gel electrophoresis approach, *Proteomics* 6:3821–32.

68. Van den Bergh, G., Clerens, S., Vandesande, F. and Arckens, L. (2003) Reversed-phase high-performance liquid chromatography prefractionation prior to two-dimensional difference gel electrophoresis and mass spectrometry identifies new differentially expressed proteins between striate cortex of kitten and adult cat, *Electrophoresis* 24:1471–81.

69. Mataga, N., Nagai, N. and Hensch, T. K. (2002) Permissive proteolytic activity for visual cortical plasticity, *Proc Natl Acad Sci U S A* 99:7717–21.

70. Krug, K., Akerman, C. J. and Thompson, I. D. (2001) Responses of neurons in neonatal cortex and thalamus to patterned visual stimulation through the naturally closed lids, *J Neurophysiol* 85:1436–43.

71. Crowley, J. C. and Katz, L. C. (2000) Early development of ocular dominance columns, *Science* 290:1321–4.

72. Jackson, C. A. and Hickey, T. L. (1985) Use of ferrets in studies of the visual system, *Lab Anim Sci* 35:211–5.

73. Li, J. and Lee, A. S. (2006) Stress induction of GRP78/BiP and its role in cancer, *Curr Mol Med* 6:45–54.

74. Paschen, W. (2004) Endoplasmic reticulum dysfunction in brain pathology: Critical role of protein synthesis, *Curr Neurovasc Res* 1:173–81.

75. dos Remedios, C. G., Chhabra, D., Kekic, M. et al. (2003) Actin binding proteins: Regulation of cytoskeletal microfilaments, *Physiol Rev* 83:433–73.

76. Schafer, D. A. and Cooper, J. A. (1995) Control of actin assembly at filament ends, *Annu Rev Cell Dev Biol* 11:497–518.

77. Horii, M., Shibata, H., Kobayashi, R. et al. (2006) CHMP7, a novel ESCRT-III-related protein, associates with CHMP4b and functions in the endosomal sorting pathway, *Biochem J* 400:23–32.

78. Katoh, K., Shibata, H., Hatta, K. and Maki, M. (2004) CHMP4b is a major binding partner of the ALG-2-interacting protein Alix among the three CHMP4 isoforms, *Arch Biochem Biophys* 421:159–65.

79. Ellis, R. J. (2006) Molecular chaperones: Assisting assembly in addition to folding, *Trends Biochem Sci* 31:395–401.

80. Ricard, D., Rogemond, V., Charrier, E. et al. (2001) Isolation and expression pattern of human Unc-33-like phosphoprotein 6/collapsin response mediator protein 5 (Ulip6/CRMP5): Coexistence with Ulip2/CRMP2 in Sema3a- sensitive oligodendrocytes, *J Neurosci* 21:7203–14.

81. Issa, N. P., Trachtenberg, J. T., Chapman, B., Zahs, K. R. and Stryker, M. P. (1999) The critical period for ocular dominance plasticity in the ferret's visual cortex, *J Neurosci* 19:6965–78.
82. Liao, D. S., Krahe, T. E., Prusky, G. T., Medina, A. E. and Ramoa, A. S. (2004) Recovery of cortical binocularity and orientation selectivity after the critical period for ocular dominance plasticity, *J Neurophysiol* 92:2113–21.
83. Ohki, K., Chung, S., Kara, P. et al. (2006) Highly ordered arrangement of single neurons in orientation pinwheels, *Nature* 442:925–8.
84. Stosiek, C., Garaschuk, O., Holthoff, K. and Konnerth, A. (2003) In vivo two-photon calcium imaging of neuronal networks, *Proc Natl Acad Sci U S A* 100:7319–24.
85. Barabasi, A. L. and Oltvai, Z. N. (2004) Network biology: Understanding the cell's functional organization, *Nat Rev Genet* 5:101–13.
86. Parker, C. E., Warren, M. R., Loiselle, D. R. et al. (2005) Identification of components of protein complexes, *Methods Mol Biol* 301:117–51.

13 Behaviorally Regulated mRNA and Protein Expression in the Songbird Brain

Miriam V. Rivas and Erich D. Jarvis

CONTENTS

13.1 INTRODUCTION

Most biological processes are the result of the regulated expression of genes into their protein products in a defined temporal and spatial manner. The study of these dynamic protein changes in the brain is called neuroproteomics. A class of proteins expressed in the brain that have generated significant interests are those regulated by neural activity, called activity-dependent genes. The activity of neurons in the brain is important for normal brain function and expression of behavior. One system where the study of these features has been brought together is that of the songbird vocal learning system. Songbirds, and a limited number of animals (i.e., parrots, hummingbirds, bats, cetaceans, seals, elephants, and humans) are capable of vocal learning, the ability to produced imitative and improvisational sounds. Most animals do not have vocal learning but produce species-specific innate sounds used for alarm (e.g., predator) or other communication functions (e.g., alert for food or attracting a mate). The production of learned vocalizations requires the animal to process what

239

it hears in the auditory pathway of the brain and then to produce the sounds heard, as song in songbirds or speech in humans, through specialized motor learning pathways. In songbirds and other vocal learning birds, this song learning motor pathway is organized into anatomically discrete nuclei that are not found in vocal non-learning birds. In this regard, songbirds represent a relatively unique animal model suited to the study of molecular mechanisms of a learned behavior with parallels to human speech. In recent years, a number of genes and their protein products have been identified that are regulated by neural activity during hearing and singing in the auditory and vocal brain pathways of songbirds. This chapter focuses on proteins regulated in the vocal pathway of songbirds performing learned vocalization. A discussion of proteins regulated in the auditory pathway by hearing is presented by Pinaud et al. in Chapter 14.

13.2   AUDITORY AND VOCAL PATHWAYS

All avian species, whether vocal learners or vocal non-learners, share similar brain auditory pathways (Figure 13.1a) (1). Sounds activate ear cells, which synapse onto sensory neurons that project to cochlear and lemniscal nuclei in the brainstem. These neurons in turn project to the midbrain auditory mesencephalic lateral dorsal nucleus (MLd) and to the thalamic nucleus uvaeformis (Uva). The MLd projects to the thalamic auditory nucleus ovoidalis (Ov). Ov projects to primary auditory cell populations in field L2 of the forebrain pallium. L2 neurons then form a complex network to

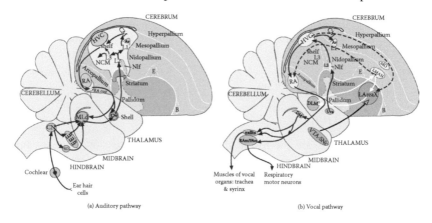

FIGURE 13.1   (See color insert following page 172.) Diagram of auditory (a) and vocal (b) pathways of the songbird brain. Only the most prominent or most studied projections are indicated. For the auditory pathway, NCM actually lies in a parasagittal plane medial to that depicted and reciprocal connections between pallial areas are not indicated. For the vocal pathway, black arrows show connections of the nuclei (in *dark grey*) of the posterior vocal pathway, white arrows show connections of the nuclei (in *white*) of the anterior vocal pathway, and dashed lines indicate connections between the two pathways. Only the lateral part of the anterior vocal pathway is shown, and the connection from Uva to HVC is not depicted. *Source*: Modified from Jarvis, E. D., Gunturkun, O., Bruce, L. et al. (2005) Avian brains and a new understanding of vertebrate brain evolution, *Nat Rev Neurosci* 6:151–9.

higher order auditory neurons in the caudal pallium, to fields L3 and L1 that project to the caudal medial nidopallium (NCM) and the caudal mesopallium (CM), respectively. In addition, L2 also projects to a shelf region below the HVC (used as a letter-based name) song nucleus and to an auditory region of the caudal striatum (CSt). A descending auditory pathway sequentially connects the HVC shelf to the intermediate arcopallium (Ai), which is adjacent to the song nucleus called the robust nucleus of the arcopallium (RA), known as the RA cup. Then the RA cup connects to the shell regions around Ov and MLd. This descending pathway is thought to modulate ascending auditory information (2–6).

Unliker the auditory brain pathway that is found in all avian species whether a vocal learner or non-learner, a forebrain vocal pathway is found only in species that learn their vocalizations. This pathway consists of seven brain regions, distributed into two sub-pathways, the posterior vocal pathway and the anterior vocal pathway (Figure 13.1b) (1). In the posterior pathway, the interfacial nucleus of the nidopallium (NIf) projects to HVC that in turn projects to RA; RA projects out of the forebrain to the dorsal medial nucleus of the midbrain (DM) and to the vocal motor neurons in the tracheosyringeal subdivision of the hypoglossal nucleus, nXIIts. The nXIIts projects to the muscles of syrinx, the avian vocal organ. The posterior vocal pathway is required for the production of learned vocalizations, as lesions to posterior vocal pathway nuclei HVC and RA either prevent singing or for HVC result in variable random–like singing, called subsong (7–9). The anterior vocal pathway is composed of the magnocellular nucleus of the anterior nidopallium (MAN; separated into lateral and medial parts, LMAN and MMAN) that projects to the striatal song nucleus Area X (also separated into lateral and medial parts, LArea X and MArea X). Area X projects to nuclei within the dorsal thalamus (DLM, the medial nucleus of the dorsalateral thalamus and DIP, dorsal intermediate posterior nucleus), which projects back to MAN forming closed loops (10–12). A mesopallium oval (MO) nucleus is part of this pathway in parrots, but its connectivity in songbirds is not known (13–15). The anterior vocal pathway is necessary for song learning, as lesions to anterior vocal pathway nuclei prevent juvenile song learning and adult song plasticity (16,17). Connections from the posterior to the anterior vocal pathway occur via HVC to Area X, and from the anterior to the posterior occur via LMAN to RA and MMAN to HVC (12).

The source of auditory pathway input into the vocal pathway is an unresolved question. It has been proposed that input from the auditory pathway to the posterior vocal pathway can be by way of the HVC shelf to HVC, the RA cup to RA, or L2 into NIf (3). Other alternatives are that CM sends projections to NIf, and then NIf sends the auditory information to HVC and that CM sends a direct auditory projection to HVC (18). The thalamic nucleus Uva also sends projections to NIf, which sends a projection to HVC as well a direct projection from Uva to HVC (Figure 13.1a). It is conceivable that there are multiple contributing inputs. With this anatomical background, we now describe behaviorally regulated mRNA and protein expression in the song system.

13.3   EXPRESSION OF SINGING–REGULATED GENES AND THEIR PROTEINS

13.3.1   INITIAL DISCOVERIES WITH THE ZENK GENE

Increased neuronal firing in the auditory pathway of birds hearing song and in the vocal pathway of birds singing is associated with high increases in the synthesis of mRNA and protein products of the immediate early gene (IEG) zenk (19,20). ZENK protein (also known as Zif268, Egr-1, NGFIA, and Krox 24) is a transcription factor that regulates the expression of other genes. In the brain, IEG, its mRNA expression, is activity dependent; zenk requires increased neuronal firing for its increased expression (21). The study of Zenk has led to a sub-field of research on the study of mRNA and protein regulation in the brain by natural behaviors, using songbirds. The regulated expression that occurs in the auditory pathway is discussed in Chapter 14. In this chapter, we discuss the behaviorally regulated expression found in the vocal pathways during singing.

ZENK expression increases in vocal nuclei (also known as song nuclei) when birds sing (Figure 13.2). The increase occurs in the absence of hearing, i.e., in deafened birds that sing, and in the absence of somatosensory feedback from the syrinx muscle, i.e., in muted birds from nXIIts nerve cuts that attempt to sing but cannot move their syrinx muscles (silent song) (19). The amount of increased zenk mRNA and protein expression per 30 minutes correlates with the amount of song the bird produces per 30-minute period. Increased expression of zenk mRNA peaks at 30 minutes after singing starts, and then partly habituates to a steady state lower level in the song nuclei, as a bird continues to sing (19). A similar profile appears to occur with the protein with a delay of about 30 minutes, i.e., 60 minutes for peak expression (22).

Within the song nucleus HVC, depending on the amount of singing, up to 40%–80% of the neurons that project to RA, called RA-projecting neurons, and a comparable percentage of neurons that project to Area X, called X-projecting neurons, show singing-induced ZENK protein expression (14). These findings suggest that a large portion of HVC neurons participate in singing behavior.

Some of the initial discoveries of behaviorally regulated gene expression in songbirds were also made with another IEG transcription factor, c-fos (23). Expression of the c-Fos protein was found to be induced in the RA-projecting neurons of HVC and in the RA nucleus itself, but very little in the X-projecting neurons of HVC or in Area X itself. However, this apparent difference in ZENK expression could be due to social context (see Social Context Regulation of Gene Expression). As seen for ZENK, c-Fos expression in song nuclei depends on the act of singing and the amount of protein product correlates with the amount of song produced (23).

13.3.2   SOCIAL CONTEXT REGULATION OF GENE EXPRESSION

Expression of ZENK in specific song nuclei was also found to depend on the social context in which a songbird sings. When a zebra finch male sings directed song facing another bird, ZENK expression is lower in the singing male's song nuclei RA,

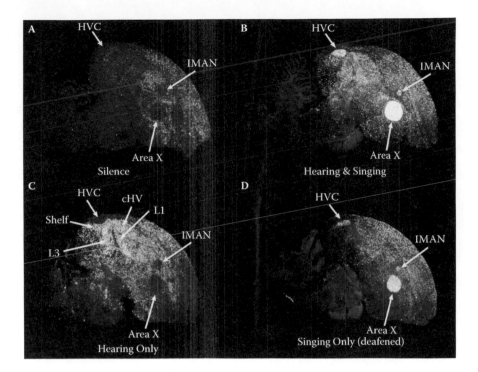

FIGURE 13.2   (See color insert following page 172.) ZENK mRNA expression in canary brain. Shown are dark field views of cresyl violet stained sagittal brain sections reacted in situ hybridization experiments with a ^{35}S-zenk riboprobe (*white grains*). One can see a subset of structures (*arrows*), as outlined in Figure 13.1, which showed singing or hearing song-induced zenk expression. (a) Bird exposed to silence and that did not sing; (b) bird that sang 61 songs during a 30-minute period; (c) bird in an adjacent cage that did not sing during the same 30-minute period; and (d) deafened bird that sang 30 songs during 30 minutes. *Source*: Figure modified from Jarvis, E. D. and Nottebohm, F. (1997) Motor-driven gene expression, *Proc Natl Acad Sci U S A* 94:4097–4102.

LMAN, and LArea X than when he sings undirected song not facing another bird or by himself (Figure 13.3) (14). This differential IEG expression is associated with similar electrophysiological changes in LArea X and LMAN, where the neural firing is less and more distinct during directed singing relative to undirected singing (24). The source of this modulatory effect on both gene expression and electrical activity in LArea X and LMAN may come from outside the song system, in particular dopaminergic modulation from the ventral tegmental area (VTA) and substantia nigra pars compacta (SNc) of the midbrain into Area X and RA (Figure 13.1b) (25–27) and from neuroadrenergic modulation of the locus ceruleus (LoC) also possibly into Area X (26–28). The VTA-SNc neurons have higher levels of firing with associated higher levels of dopamine release in Area X during directed singing relative to undirected singing (25,26). Dopamine is a modulatory neurotransmitter that reinforces motivation behaviors such as sexual motivation (31). Within the VTA-SNc, the GABAergic inhibitory interneurons showed increased ZENK expression during

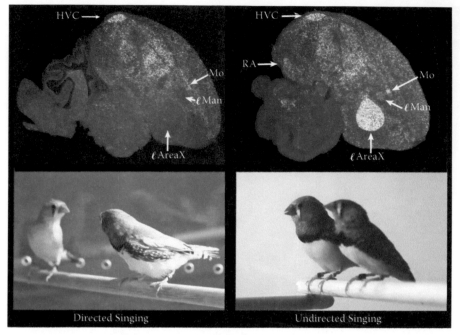

(a) Zenk

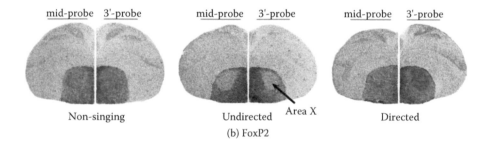

(b) FoxP2

FIGURE 13.3 Social-context-dependent gene expression in the song system of adult male zebra finches. (a) Top panels: Sagittal brain sections (*dark field images*) showing high singing-driven zenk mRNA expression (*white*) in all song nuclei during undirected singing (*right panel*) and low expression in lateral AreaX (LAreaX), LMAN, and RA during directed singing (*left panel*). Bottom panels: Examples of birds singing to a female (*left*) or singing undirected song (*right*). (b) Coronal brain sections (autoradiograph film images) showing decreased singing-driven Foxp2 mRNA expression (*white*) in Area X during undirected singing (*middle panel*) with no change in Area X during directed singing (*right panel*). The left hemispheres were hybridized with a mid-probe made to the coding region of the mRNA; the right hemispheres were hybridized to the 3′ probe. *Sources*: Panel (a) is modified from Jarvis, E. D., Scharff, C., Grossman, M. R., Ramos, J. A. and Nottebohm, F. (1998) For whom the bird sings: Context-dependent gene expression, *Neuron* 21:775–88; and panel (b) from Teramitsu, I. and White, S. A. (2006) FoxP2 regulation during undirected singing in adult songbirds, *J Neurosci* 26:7390–4.

directed singing (27), suggesting that the depressed amount of ZENK expression in Area X during directed singing could be modulated by the inhibitory activities of the GABAergic interneurons. However, although lesions of VTA-SNc prevent high levels of singing-induced ZENK expression in Area X, they do not prevent the social context differences (27).

The song syllables produced during undirected singing are slightly more variable than during directed singing and this variability is controlled by activity from LMAN into RA (17,32). This variability is associated with vocal exploration, and as such, the increased ZENK protein expression in LMAN and RA could be associated with consolidating newly explored song syllables. Female zebra finches also detect this variability, preferring to perch next to speakers with playbacks of the more stereotyped directed song versus undirected song; this preference is associated with selective increased ZENK protein expression in the CM auditory area of the female brain (33). Not only are auditory neurons sensitive to these two different song types, but they also have differences of ZENK expression depending on whether the listening birds know other birds are present (in the dark); hearing song-induced expression is higher in birds with experience of being housed together than when housed alone (34). Thus, social context regulation of ZENK mRNA and protein expression can occur in both vocal and auditory pathways.

Later studies found that social context regulation occurs for other genes, but with different patterns than those found for ZENK. One such gene is FoxP2, a transcription factor associated with spoken language function in humans and song learning in songbirds (35–37). Starting at high basal levels, FoxP2 mRNA decreases in expression in Area X when zebra finch males sing undirected song, and no change occurs when they sing directed song (Figure 13.3) (38). Further, even in the absence of singing, when juvenile zebra finches are in their developmental, plastic to stereotype phase of song learning or canaries are in their seasonal stereotyped song phase, basal levels of FoxP2 expression are higher in Area X than during other stages (35). These correlations suggest that down-regulation of FoxP2 in Area X may be linked with exploratory singing behavior and song learning. This idea was tested in manipulation experiments using RNAi to knock down FoxP2 protein in Area X of zebra finch juveniles learning to sing. Compared to RNAi controls, juvenile zebra finches with knocked down FoxP2 protein could still sing, but they were unable to properly imitate the song syllables and song sequences of their tutors (Figure 13.3) (39). If FoxP2 has a similar function during adult undirected singing, then it might also be involved in modulating song exploration. FoxP2 is thought to suppress the expression of most of its target genes (40). Thus, when FoxP2 is down-regulated, its target genes are presumably up-regulated. The regulation during undirected singing suggests that those target genes could be involved in synaptic plasticity.

Another gene, synaptotagmin IV, which produces a synapse-associated protein, was found to be regulated in a social-context-dependent manner. During undirected singing, synaptotagmin IV expression is up-regulated in LMAN, HVC, and RA, and down-regulated in Area X (41,42). During directed singing, there are no changes in synaptotagmin IV expression in any of the song nuclei. Synaptotagmin IV is a presynaptic protein that modulates neurotransmitter vesicle release at nerve terminals, but there is no consensus on whether it suppresses or enhances the vesicle release

(41). Synaptotagmin IV null mutant mice show enhanced short-term synaptic plasticity as well as deficits in associative passive avoidance memory (43,44). Thus, one possible consequence of increased synaptotagmin IV in pallial song nuclei (LMAN, HVC, and RA) during undirected singing would be to subsequently increase the ability to release synaptic vesicles for future singing events and maintain song memories. Another would be to simply replace synaptotagmin IV protein that was used up in the act of singing.

Overall, the social context experiments demonstrate that the song system is highly engaged during undirected singing when the birds are producing more variable song, that each activity-dependent gene tested to date has its own regulatory pattern in different social context, and that there is not a simple link between neural activity and gene regulation in a behaving animal. The findings also indicate that there are other behaviorally regulated genes to be discovered, as discussed in the next section.

13.3.3 An Apparent Cascade of Genes

Since the initial studies with zenk and c-fos, a total of 36 genes regulated by singing have been discovered (Figures 13.4 and 13.5a). The discoveries were based upon either prior knowledge of activity-regulated genes in other organisms (41,42,45,46), serendipitous findings (38,47), or by non-biased screenings with cDNA microarrays hybridized to mRNA probes from song nuclei of birds that sang (42,48). The discovered genes of these and past studies generate known proteins that span a diverse set of functions. These include signal transduction proteins (egr-1, c-fos, FoxP2, c-jun, similar to junB, Atf4, Hspb1, UbE2v1, HnrpH3, Shfdg1, and Madh2), chromosome scaffold proteins (H3f3B and H2AfX), actin-interacting cytoskeletal proteins (Arc, sim Fmnl, Tagln2, ARHGEF9, and β-actin), a Ca2+-regulating protein (Cacyb), cytoplasmic proteins with enzymatic activity (Prkar1a, Atp6v1b2, Ndufa5, and GADPH), protein kinase (Gadd45b), folding (Hsp70-8), binding, and transporting functions (Hsp40, Hsp90a, Hsp25, UCHL1), and membrane (Stard7, Syt4, and Ebag9) and synaptically released proteins (JSC, BDNF, and Penk; Figure 13.5a-c). Like FoxP2, several of these (ARHGEF9 and similar to NPD014) are down-regulated in Area X as a result of singing. This diversity of recruited genes suggests that large signal transduction networks are activated during production of a learned behavior. However, the networks may be song nuclei specific.

The study of Wada et al. (2006) (42) was able to group the singing-regulated genes into four anatomical categories: (a) those regulated in all four major song nuclei; (b) a combination of one or two pallial song nuclei and in the striatal song nucleus Area X; (c) those regulated in one or more pallial song nuclei only (HVC, RA, and LMAN); and (d) those regulated in the striatal nucleus only (Figure 13.4). Of these, Area X and HVC had the highest percentages (94% and 76%, respectively) of genes regulated by singing, and LMAN and RA the lowest (30% and 33%, respectively). This suggests that Area X and HVC are more molecularly dynamic than the other song nuclei, and thus potentially open to greater neural plasticity. It also indicates that in each song nucleus, overlapping or different gene networks are activated during production of song.

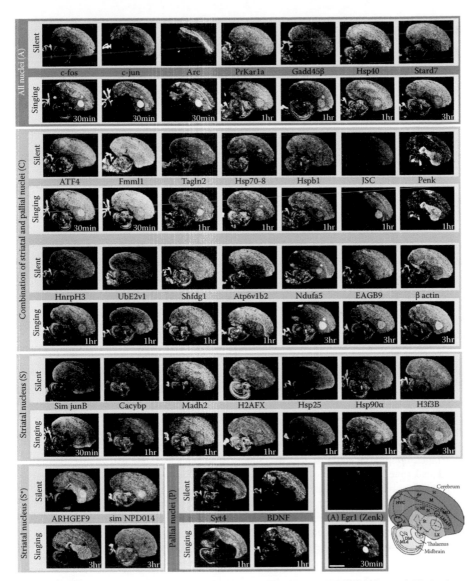

FIGURE 13.4   (See color insert following page 172.) In situ hybridizations of 33 singing-driven genes in song nuclei. Shown are inverse images of autoradiographs; white is mRNA expression. Images are ordered from top to bottom according to four overall expression patterns and from left to right in temporal order of peak expression (time shown in *lower right corner*). Zenk expression is shown to the left of the brain diagram (*bottom right*) for anatomical reference. *Abbreviations*: A, arcopallium; Av, avalanche; DM, dorsal medial nucleus; LX, lateral AreaX of the striatum; MO, oval nucleus of the mesopallium; N, nidopallium; NIf, interfacial nucleus of the nidopallium; P, pallidum; RA robust nucleus of the arcopallium; St, striatum. (Scale bar, 2 mm.). Gene abbreviations and further description in Figure 13.5. *Source*: Figure from Wada, K., Howard, J. T., McConnell, P. et al. (2006) A molecular neuroethological approach for identifying and characterizing a cascade of behaviorally regulated genes, *Proc Natl Acad Sci U S A* 103:15212–7, with permission.

#	Gene	Annotated name	Cellular Location	Molecular Function	Biological Process	HVC	RA	MAN	X	Peak
1	zenk	zif268, egr-1, NGFIA, Krox24	nucleus	transcription factor	signal transduction	+	+	+	+	0.5
2	c-fos	cellular-fos	nucleus	transcription factor-pol II	signal transduction	+	+	+	+	0.5
3	c-jun	cellular-jun	nucleus	transcription factor-pol II	signal transduction	+	+	+	+	0.5
4	sim junB	similar to jun B	nucleus	transcription factor	signal transduction				+	0.5
5	Atf4	activating transcription factor 4	nucleus	transcription factor-pol II	signal transduction				+	0.5
6	FoxP2	Forkhead box 2	nucleus	transcription factor	transcription suppression	+				
7	Hspb1	heat shock protein binding protein 1	nucleus	trans co-repressor of Hsf 1	transcription suppression				+	0.5
8	UbE2v1	ubiquitin-conjugating E2 variant 1	nucleus	transcription act & enzyme	transcription regulation	+			+	1
9	HnrpH3	heterogeneous nuclear ribonucleoprotein H3	nucleus	RNA binding	RNA splicing arrest	+			+	1
10	H3F3B	H3 histone, family 3B	chromosome	DNA binding	nucleosome assembly				+	1
11	H2AFX	H2A histone, family X	chromosome	DNA binding	nucleosome assembly				+	1
12	Shfdg1	split hand/foot deleted gene 1, short clone	cytos to nucleus	transcription factor	cell differentiation	+		+	+	1
13	Madh2	mothers against DPP homolog	cytos to nucleus	transcription mediator	signal transduction				+	1
14	Arc	activity regulated cytoskeletal-associated protein	cytoskeleton	actin binding	synaptic plasticity	+	+	+	+	0.5
15	sim Fmnl1	similar to formin-like protein	cytoskeleton	actin binding protein	actin filament elongation				+	0.5
16	Tagln2	transgelin	cytoskeleton	actin-associated protein	cell differentiation	+		+	+	0.5
17	β-actin	β-actin	cytoskeleton	structural molecule	axongenesis/cell motility	+		+	+	3
18	ARHGEF9	Cdc42 guanine nucleotide exchange factor 9	cytoskeleton	actin GTPase	signal transduction				−	3
19	Prkar1α	protein kinase, cAMP-dependent regulatoryIα	cytoplasm	sig to cAMP	sig trans/cell proliferation	+	+	+	+	1
20	Gadd45	growth arrest & DNA-damage-inducible 45β	cytoplasm	protein binding	protein synthesis & other	+	+	+	+	1
21	Hsp70-8	heat shock 70kDa protein 8c	cytoplasm	ATPase/chaperone	prot folding/cell cycle	+	+	+	+	0.5
22	Hsp40	heat shock protein 40 (aka DnaJa1)	cytoplasm	protein binding/chaperone	protien transport-PR recep	+	+	+	+	1
23	Hsp90α	heat shock protein 90α	cytoplasm	protein binding/chaperone					+	1
24	Hsp25	heat shock protein 25	cytoplasm	aggresome	protein aggregation				+	1
25	UCHL1	ubiquitin carboxy-terminal hydrolase L1	cytoplasm/axon	ubiquiton binding	axonogenesis	+	n.d.	n.d.	n.d.	n.d.
26	GAPDH	glyceraldehyde-3-phosphate dehydrogenase	cyto/mitochondria	oxidoreductase	glucose metabolism	−	n.d.	n.d.	n.d.	n.d.
27	Cacybp	calcyclin binding protein	cytopl/nucl mem	protein binding	unknown				+	1
28	Stard7	START domain containing 7	membrane	phosphatidylcholine transfer?	cell signalling?	+	+	+	+	1
29	Syt4	synaptotagmin IV	synp vesicle	vesicular trafficking	synaptic vesicle regulation	+	+	+		1
30	Atp6v1b2	ATPase, H+ transporting, V1 subunit B, isoform 2	cytopl/mem	hydrolase-ATPsynthesis	H+ transport/cell motility	+			+	1
31	Ndufa5	NADH dehydrogenase (ubiquione) α subcomplex 5	mito inner mem	B13 sub. NADH dehydrogenase	e− transport to ubiquione	+			+	3
32	EAGB9	estrogen receptor binding site associated antigen 9	golgi membrane	protease-ligand?	cell growth	+			+	1
33	JSC	jun-suppressed chemokine	extracellular?	cytokine	chemoattractant	+			+	0.5
34	BDNF	brain derived neurotrophic factor	extracellular	ligand	neuron differentiation	+	+	+		1
35	Penk	proenkephalin	extracellular	neuropeptide	cell-cell signaling	+			+	1
36	sim NPD014	similar to NPD014 protein	unknown	unknown	unknown				−	3

(a) Table of singing regulated genes

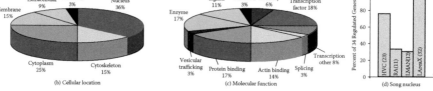

(b) Cellular location

(c) Molecular function

(d) Song nucleus

FIGURE 13.5 (See color insert following page 172.) Summary of singing-driven genes in songbird song nuclei. (a) Table of 36 singing-regulated genes verified to date. Shown are the inferred cellular locations, molecular functions, and biological processes based on ontology definitions of homologous genes in other species. The list is organized according to cellular location (nucleus-to-extracellular space), the number of song nuclei showing regulated expression, and peak time (0.5–3 hours) of expression. Sim, similar to the named gene at 60%–74% protein identity. (b and c) Pie-chart quantifications of cellular location (b) and molecular function (c) of all genes (except the one gene [#36] of unknown function). (d) Percentage of genes regulated by singing in each song nucleus. Numbers in parentheses ( ) show number of genes for each nucleus. Only 34 genes are included in the calculation, as two of the remaining genes were not examined in all song nuclei. n.d., not determined.

These gene networks appear to involve molecules of various functions and cellular locations. The largest proportion (36%) are involved in the nucleus with transcription factors being the largest functional group (18%) (Figures 13.5b and c). The second largest are involved in cytoplasmic (25%) and cytoskelatal (15%) functions, with smaller percentages being membrane and synaptically released proteins. These proportional relationships suggest a trajectory of activation from the nucleus to released proteins at the membrane. If this trajectory were a cascade, though, then a question is why the proportional representation of the cellular categories does not increase from the nucleus to the membrane. One reason could be that the screening for singing-regulated genes focused on early time points (within one hour) after singing (42,48), where a focus on all time points after singing could reveal a higher proportion of genes involved in extra-nuclear functions. A time course analysis, however, of the currently identified genes does suggest a cascade from the nucleus out (Figures 13.4-13.6).

As shown by Wada et al. (2006) (42), the first genes that show the first peak of expression from the start of singing, at 30 minutes, are mainly transcription factors (ZENK, c-fos, c-jun, sim junB, Atf4) (Figure 13.4). The only structural molecules with peak expression to date at 30 minutes are Arc and a gene similar to formin-like protein, involved in actin filament elongation (Figure 13.5a). Thereafter, expression of these genes either remains high or decreases as singing continues. Many of the cytoplasmic, cytoskeleton, and other non-nuclear genes show peak expression at one hour, again with some remaining high and some decreasing as singing continues. Besides basic cytoplasmic proteins, these include RNA binding (HnrpH3) and protein binding/heat shock proteins (Hsp25, Hsp40, Hsp70-8, Hsp90a) (Figures 13.4–13.6). By three hours after singing, mainly structural molecules (β-actin, Stard7, H3F3B) show peak expression.

Some of the non-transcription factor genes have been studied in further detail during singing. A discussion of several of them may help with understanding the functional consequences of behaviorally regulated gene expression. One gene that has received a lot of attention is BDNF, i.e., brain derived neurotrophic factor. BDNF is a synaptically released protein at axon endings, and has a variety of functions including enhancement of neuronal survival. Singing drives increased BDNF expression in the pallial song nuclei (HVC, RA, and LMAN) (42,45). Within HVC, BDNF shows a preference of increased expression in the RA-projecting neurons, with very little to no induction in the X-projecting neurons. Like ZENK, the amount of BDNF in the RA-projecting neurons correlates with the amount of singing. Intriguingly, the RA-projecting neurons in HVC are renewed throughout the animals' life, whereas the X-projecting neurons are not renewed and remain stable. In this regard, it was found that more singing leads to enhanced survival of new RA-projecting neurons that enter HVC (45,49). These findings suggest that singing induces BDNF in the RA-projecting neurons, which in turn enhances the survival of those same or adjacent RA-projecting neurons. The survival role of BDNF may be explained by BDNF acting in local networks with several other IEGs. In the mouse regulatory region of BDNF, the CCAAT enhancer binding protein (C/EBP), AP-1 (composed of a dimer of c-Jun and c-Fos), and ZENK binding sites were identified (50). A study by Calella et al. (51), using microarray analysis, showed that the binding of BDNF to its receptor TrkB initiates the transcription of a number of transcription factor such as c-fos, zenk 1, and zenk 2. Thus, it is possible that BDNF, ZENK, and c-fos may interact in a molecular pathway that completes a circular loop whereby the neurotrophin stimulates transcription of immediate early genes whose protein products regulate the expression of the same neurotrophic factors and other proteins required for neuron maintenance and/or strengthening of neuronal connections during and after the act of singing.

A related but not identical finding was seen with the ubiquitin carboxy-terminal hydrolase 1 (UCHL1) gene (47). UCHL1 is part of a ubiquitin-proteasome system necessary for protein degradation, and has been implicated in the death of neurons in neurodegenerative diseases, such as Parkinson's, Alzheimer's, and Huntington's.

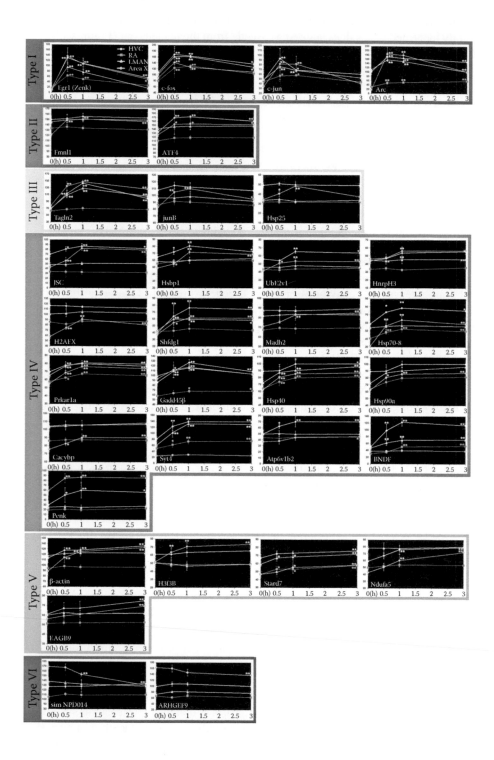

With RNA from laser captured microdissected neurons of HVC hybridized to cDNA microarrays, Lombardino et al. (47,48) found that UCHL1 was expressed at high levels in the X-projecting neurons and at low levels in the RA-projecting neurons. When the birds sang, UCHL1 levels increased in the RA-projecting neurons, but there was no change in the already high levels of the X-projecting neurons. The singing-induced expression in the RA-projecting neurons was still lower than the expression in the X-projecting neurons. A mouse mutant for UCHL1 shows axonal loss and cell death (52). Thus, it has been suggested that naturally low UCHL1 levels in replaceable neurons allow for these neurons to undergo natural cell death, and that the singing up-regulation in these neurons extends their life (47).

The above two examples are of genes that function in neuron survival and death. As such, the turnover of the RA-projecting neurons has been proposed to be involved in memory functions (53). Another type of change one would expect to be involved in memory formation is whether expression of any of the singing-regulated genes differs when birds are producing well learned song versus when they are practicing singing in the early juvenile stages of life. Such a difference has been found for proenkephalin (Penk). Penk shows higher singing-induced expression in HVC when juveniles sing plastic song than in adults singing stable stereotyped song (42). Proenkephalin is a precursor protein that contains several copies of enkephalins. Enkephalins are pentapeptides, released at synapses, and are best known for their ability to compete with and mimic the effect of opiate drugs. They are also involved in memory and emotional functions (54–56). How high singing-regulated enkephalin expression in song nuclei during juvenile development functions in the development of song is a question that remains to be answered. Most other singing-regulated genes that have been tested differ little or at all during singing in juveniles versus adults (14,19,42,57). ZENK is higher in song nuclei when juveniles sing, but basal levels are already higher.

Future experiments will be necessary to determine the functional, cellular, and behavioral consequences of individual genes regulated during production of a learned behavior, or during the actual learning of the behavior. The experimental designs will have to consider how faithful the behaviorally regulated protein changes reflect the mRNA changes, the subject of the next section (14,57).

13.4   RNA AND PROTEIN CONCORDANCE

Most genes show concordant singing-regulated mRNA and protein expression, in which protein and mRNA have been examined. Singing-induced expression of mRNA

FIGURE 13.6   (See color insert following page 172.) Time course of singing-regulated expression of 33 genes in four song nuclei. The genes are categorized into one of six types of temporal expression patterns by Wada et al. (2006) based on expression peak time (0.5, 1, or 3 hours) and expression profile. Expression levels were measured as pixel density on x-ray film minus background on the film. The * indicates significant differences at one or more time points relative to silent controls at the 0 hour (0 h) (ANOVA by post hoc probable least-squares difference test. *, $P < 0.05$; **, $P < 0.01$). *Source*: Figure from Wada, K., Howard, J. T., McConnell, P. et al. (2006) A molecular neuroethological approach for identifying and characterizing a cascade of behaviorally regulated genes, *Proc Natl Acad Sci U S A* 103:15212–7.

and protein has been shown for the transcription factor ZENK in all forebrain song nuclei, with the protein localized in the nucleus (14,19,58) as expected for a transcription factor. The peak of protein expression lags that of mRNA expression by 30 minutes (3). Not only the mRNA, but ZENK protein expression in Area X, LMAN, and RA depends on the social context in which singing occurs, with increased expression in birds that produce undirected song and less of an increase when they sing directed song to females (14). However, in the song nucleus RA, translation of zenk mRNA and into ZENK protein depends on whether the birds sing in the presence of other birds and on hearing (59). In the absence of other birds, zenk mRNA synthesis is induced in RA, but only a proportion of the mRNA is translated into protein. When other birds are present, then singing-driven zenk mRNA is faithfully translated into ZENK protein (Figure 13.7a and b). This effect of translation into protein requires that the singing male hear the others birds while singing, because when he is deaf the lower translation levels are found (57). This finding indicates that in RA, another layer of regulation, potentially on the translational level from mRNA to protein, is at work. This regulation could involve microRNAs or other repression mechanisms of mRNA translation by the binding of a repressor to a specific sequence in the 5′ end of the mRNA. An alternative mechanism would be rapid ZENK post-translational protein degradation, such as through the ubiquitin pathway, when birds sing alone. The effect is not observed in HVC where relative amounts of zenk mRNA matched the protein levels (Figure 13.7b).

Comparable spatial and temporal regulation of singing-induced mRNA and protein was confirmed for several other genes by in situ hybridization and immunocytochemistry: c-fos, BDNF, c-jun, and Penk (42). For c-fos and BDNF, protein expression changes follow the mRNA expression changes, although not all song nuclei have been tested (23,42,45). It is also difficult to assess BDNF protein expression, as it is synaptically released and difficult to detect. The transcription factor c-jun had the expected nuclear localization within neuronal cells and enkephalin, a protein product of proenkephalin, was localized to neuronal processes (42). However, for β-actin, no changes in protein expression were detected in song nuclei one to two hours after singing. It is possible that the increase in the actin mRNA is compensated by a decrease in the protein expression, or that an increase in protein expression occurs at a later time point. In general, five of six genes tested show singing upregulation at both the mRNA and protein levels (39). However, the mRNA and protein changes do not faithfully follow each other in all instances, indicating that it is necessary to study regulation at both levels of expression.

13.5 DIFFERENCES BETWEEN VOCAL AND AUDITORY PATHWAYS

There has been some overlap of genes regulated by hearing in the auditory pathway and singing in the vocal pathway. These include ZENK, ARC, c-Fos, and c-Jun. The activity-dependent regulation of these genes was discovered first either by hearing or singing, and then checked for the other condition. At first it was thought that c-Fos protein was not regulated in NCM by hearing song (23), but this negative result has since not been confirmed, as hearing song can induce c-Fos in NCM (46).

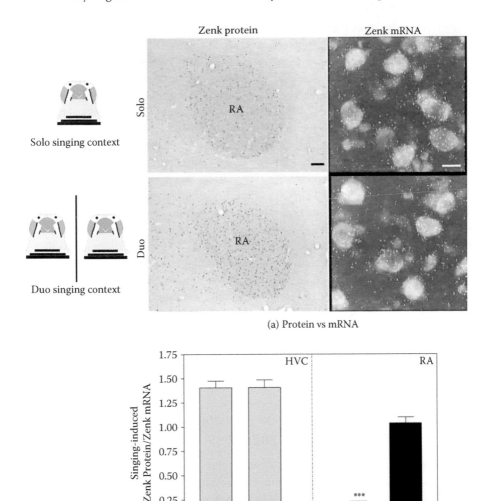

(a) Protein vs mRNA

(b) Quantitation

FIGURE 13.7   In RA singing drives zenk mRNA, while ZENK protein varies as a function of social context. (a) (*Left*) ZENK protein-labeled cells are shown at low magnification in RA for birds that sang alone (solo) or in the presence of another singing male (duo), for 45 minutes. The number of ZENK-protein-expressing neurons in RA is low after singing solo compared to birds that sang duo or birds that heard another male conspecific sing while singing. (*Right*) Zenk mRNA-labeled (*silver grains*) over neurons in RA at high power in dark field images from emulsion-dipped slides, of adjacent sections from the same animals shown for ZENK protein. Scale bar 10 μm. Singing-induced zenk mRNA occurs regardless of social context. (b) Quantification of the number of neurons that show ZENK protein versus mRNA expression in HVC and RA in the different social context. No difference is found for HVC, but a difference is found for RA. Source: Figure modified from Whitney, O. and Johnson, F. (2005) Motor-induced transcription but sensory-regulated translation of ZENK in socially interactive songbirds, *J Neurobiol* 65:251–9, and constructed by Dr. Whitney.

The discrepancy between the studies could be due to an apparent higher threshold of activity needed to induce c-Fos, where the threshold may have been reached in most but not all studies (15). When using unbiased gene screening approaches, there has been little overlap between the genes discovered that are regulated by hearing in the auditory pathway and singing in the vocal pathway. The reasons for this non-overlap may be due to differences in approaches used: microarray analysis for the vocal pathway (42) and proteomics for the auditory pathway (60) (also see Pinaud et al., Chapter 14). It may also be due to real differences in expression of proteins as seen in the expression of 19 glutamate receptor subtypes in the zebra finch brain (61). Glutamate receptors, upon binding the neurotransmitter glutamate, regulate the expression of activity-dependent genes such as ZENK, as determined in the mammalian brain (62). Of the 21 glutamate receptor subtypes studied in songbirds, 19 showed differential expression in song nuclei relative to the respective surrounding brain subdivisions, whereas the auditory pathways had expression with few differences relative to the surrounding areas (60). In this regard, it is possible that the differential expression of neurotransmitter receptors may result in different, but overlapping sets of genes regulated by singing in song nuclei and hearing in the auditory nuclei. A resolution to this difference may be resolved in future studies that test both the vocal and auditory pathway using the same methods, and that test singing-regulated genes discovered in the vocal pathway for their potential regulation in the auditory pathway and vice versa. Such an initial comparison was recently made while this chaper was in press, where more overlap was in the hearing and singing related genes (63).

13.6   PROPOSED HYPOTHESIS OF SINGING-REGULATED GENES AND PROTEINS

It has been proposed that activity-dependent regulatory cascades are the basic contributory mechanisms for long-term memory formation in the nervous system of a wide range of organisms from sea mollusks to mammals (64–66). The exact causal relationship between neuronal activation and gene expression response is unknown. One hypothesis (46,67) has been that presynaptic neurotransmitter release leads to postsynaptic neurotransmitter receptor activation, which in turn induces calcium-dependent entry through voltage-dependent channels (Figure 13.8) (68). This will then lead to the activation of calcium-dependent signal transduction pathways (e.g., kinases) and activation of temporally inactive transcription factor proteins such as CREB or C/EBP (20). CREB, C/EBP, and other factors, some already bound to the promoter regions, then initiate the transcription of IEG mRNAs such as ZENK, c-Fos, c-Jun, and ARC. The extent to which electrophysiological and gene activation are coupled in neurons depends on the location and type of neuron affected (69,70). This coupling of action potentials to gene activation has been referred to as the genomic potential (71). In this regard, it has been argued that the rate of neuronal firing of the presynaptic neuron controls the amount of behaviorally regulated gene expression in the postsynaptic neuron (67). In support of a presynaptic-postsynaptic relationship, lesion of a presynaptic song nucleus either prevents or disrupts the

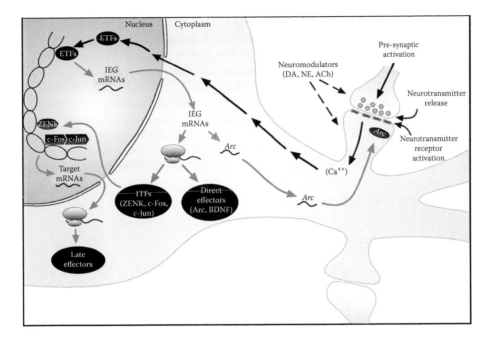

FIGURE 13.8 Hypothesis of synaptic mechanism for immediate early gene induction by neural activity associated with behavior. Shown is a schematic representation of the cascade (*arrows*) triggered upon synaptically activated presynaptic neurotransmitter release onto postsynaptic receptors, followed by calcium entry via the receptors into the postsynaptic neuron. DNA is drawn as a double-stranded helix inside the nucleus; mRNAs are drawn as squiggly lines, which are translated into proteins by ribosomes (*gray ovals*). This cascade of events occurs simultaneously while the animals perform behaviors, and is thought to have long-term consequences on subsequent perception and behavior. *Abbreviations*: Ach, acetylcholine; DA, dopamine; ETF, early transcription factors; ITF, inducible transcription factors; NE, norepinephrine. *Source*: Figure from Mello, C. V. and Jarvis, E. D. (2008) Behavior-dependent expression of inducible genes in vocal learning birds. In *Neuroscience of birdsong*, eds. Zeigler, H. P. and Marler, P., 381–97 (Cambridge University Press); modified from Velho, T. A., Pinaud, R., Rodrigues, P. V. and Mello, C. V. (2005) Co-induction of activity-dependent genes in songbirds, *Eur J Neurosci* 22:1667–78.

modulation of singing-regulated ZENK expression in its downstream postsynaptic targets (12).

Another open question is what is the role of the singing-regulated mRNA and protein expression in cellular physiology and behavior? Two possibilities include neural plasticity (i.e., learning) and neural homeostasis (i.e., behavior maintenance). The genome as it concerns the neuron must express genes that are necessary for the homeostasis of the cell. Homeostasis presumably would include genes that have singing-regulated expression levels in the song brain nuclei at relatively the same levels in adult and juvenile animals. Examples are NADH dehydrogenase, glyceraldehyde-3-phosphate dehydrogenase, β-actin, and actin-associated molecules (42). These gene products may need to be replenished on a regular basis as they get used up or damaged during the active process of the genomic potential. Consistent with this idea is the fact that a number of the singing-regulated genes are heat shock proteins (Figure 13.5a). As such, one can imagine that neural activity causes some cellular fatigue. The heat shock proteins, known to be involved in neuroprotection (72), could protect the cells from further fatigue and death. In contrast, genes involved in plasticity may be expected to be regulated differentially in juveniles and adults. Examples are Penk, zenk, and foxp2. All are expressed at higher levels in specific juvenile song nuclei relative to adults (14,35,42). Thus, in juveniles actively learning how to sing, the neuroprotective mechanisms may be at work, overlaid with synaptic plasticity mechanisms to form new motor memories of the songs the birds imitate. Future experiments are necessary to test these hypotheses, using gene manipulation tools and assessing the cellular and behavioral consequences.

13.7 FUTURE NEEDS

In future experiments, it will be necessary to link the behaviorally regulated genes into gene regulatory networks. This can be accomplished with a combination of gene manipulation and computational studies. With computational analyses, these molecules can be placed into inferred gene and/or protein networks to gain insight into how these interactions affect hearing and singing behavior (73,74). For example, one can generate protein–DNA, protein–protein, and other interaction networks of the singing–regulated genes using an interaction algorithm that utilizes known molecular interactions in the literature (Ingenuity System, Path Designer Network software, 2000–2008). We applied the Ingenuity algorithm to the 36 singing-regulated genes and found that the highest scoring network was a neural molecular network that included 18 of the genes, with the immediate early gene transcription factors (JUN, JUNB, FOS, and EGR-1 [aka ZENK]) as central hubs with direct or indirect interactions to the other genes (Figure 13.9). In this network, for example, the binding of JUN to proenkephalin (Penk) DNA (PD [1], direct protein-DNA interaction Figure 13.9) is consistent with the temporal (c-jun on first) and anatomical regulation found in singing animals. Although this is a small-scale analysis, it serves as a testable hypotheses of how the cascade of singing-regulated genes may interact with each other and other genes. Once inferred, these networks can to be used to perform gene manipulation experiments in live animals via either RNAi or over-expression to either inhibit or enhance a gene product in the song nuclei of birds and observe a result on cellular physiology and singing behavior. These challenges are within the grasp of the foreseeable future and promise to unlock the mystery of behaviorally regulated gene expression.

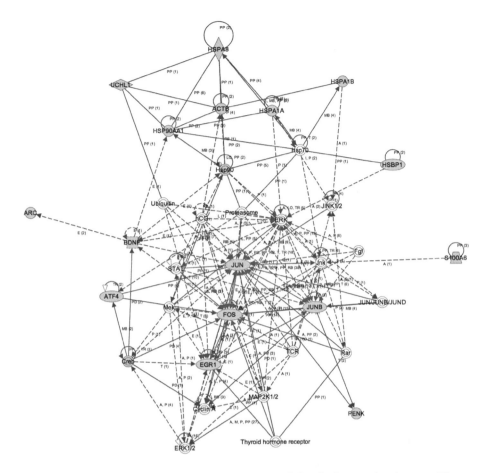

FIGURE 13.9 Hypothesized interaction pathway of the singing-regulated genes (Figure 13.5), according to pathway analyses interactions of the Ingenuity Systems software (Sept. 2008 version). Filled objects (nodes) are the gene/protein-products-regulated singing. Open objects (also nodes) are other interacting gene/protein products based on the ingenuity analysis. The shape of the objects indicates difference types of molecules, with oval indicating transcription factors. Solid lines (edges) indicate direct interactions. Dashed lines (also edges) indicate indirect interactions via intervening genes or proteins. Numbers along edges indicate the number of intervening interacting molecules (either 1 for none, or greater than 1 for more molecules). The abbreviations along the edges indicate the type of interactions: A, activation/deactivation; RB, regulation of binding; PR, protein-mRNA binding; PP, protein-protein binding; PD, protein-DNA binding; B, binding; E, expression; I, inhibition; L, proteolysis; M, biochemical modification; O, other; P, phosphorylation/dephosphorylation; T, transcription; and LO, localization. See http://www.ingenuity.com/products/pathways_analysis.html for software information.

REFERENCES

1. Jarvis, E. D., Gunturkun, O., Bruce, L. et al. (2005) Avian brains and a new understanding of vertebrate brain evolution, *Nat Rev Neurosci*, 6, 151–9.
2. Vates, G. E., Broome, B. M., Mello, C. V. & Nottebohm, F. (1996) Auditory pathways of caudal telencephalon and their relation to the song system of adult male zebra finches, *J Comp Neurol*, 366, 613–42.
3. Mello, C. V., Vates, G. E., Okuhata, S. & Nottebohm, F. (1998) Descending auditory pathways in the adult male zebra finch (*Taeniopygia guttata*), *J Comp Neurol*, 395, 137–60.
4. Gentner, T. Q. & Margoliash, D. (2003) Neuronal populations and single cells representing learned auditory objects, *Nature*, 424, 669–74.
5. Theunissen, F. E., Amin, N., Shaevitz, S. S. et al. (2004) Song selectivity in the song system and in the auditory forebrain, *Ann N Y Acad Sci*, 1016, 222–45.
6. Coleman, M. J., Roy, A., Wild, J. M. & Mooney, R. (2007) Thalamic gating of auditory responses in telencephalic song control nuclei, *J Neurosci*, 27, 10024–36.
7. Nottebohm, F., Stokes, T. M. & Leonard, C. M. (1976) Central control of song in the canary, Serinus canarius, *J Comp Neurol*, 165, 457–86.
8. Simpson, H. B. & Vicario, D. S. (1990) Brain pathways for learned and unlearned vocalizations differ in zebra finches, *J Neurosci*, 10, 1541–56.
9. Aronov, D., Andalman, A. S. & Fee, M. S. (2008) A specialized forebrain circuit for vocal babbling in the juvenile songbird, *Science*, 320, 630–4.
10. Vates, G. E. & Nottebohm, F. (1995) Feedback circuitry within a song–learning pathway, *Proc Natl Acad Sci USA*, 92, 5139–43.
11. Luo, M., Ding, L. & Perkel, D. J. (2001) An avian basal ganglia pathway essential for vocal learning forms a closed topographic loop, *J Neurosci*, 21, 6836–45.
12. Kubikova, L., Turner, E. A. & Jarvis, E. D. (2007) The pallial basal ganglia pathway modulates the behaviorally driven gene expression of the motor pathway, *Eur J Neurosci*, 25, 2145–60.
13. Durand, S. E., Heaton, J. T., Amateau, S. K. & Brauth, S. E. (1997) Vocal control pathways through the anterior forebrain of a parrot (*Melopsittacus undulatus*), *J Comp Neurol*, 377, 179–206.
14. Jarvis, E. D., Scharff, C., Grossman, M. R., Ramos, J. A. & Nottebohm, F. (1998) For whom the bird sings: context–dependent gene expression, *Neuron*, 21, 775–88.
15. Feenders, G., Liedvogel, M., Rivas, M. et al. (2008) Molecular mapping of movement–associated areas in the avian brain: a motor theory for vocal learning origin, *PLoS ONE*, 3, e1768.
16. Scharff, C. & Nottebohm, F. (1991) A comparative study of the behavioral deficits following lesions of various parts of the zebra finch song system: implications for vocal learning, *J Neurosci*, 11, 2896–913.
17. Kao, M. H., Doupe, A. J. & Brainard, M. S. (2005) Contributions of an avian basal ganglia–forebrain circuit to real-time modulation of song, *Nature*, 433, 638–43.
18. Bauer, E. E., Coleman, M. J., Roberts, T. F. et al. (2008) A synaptic basis for auditory–vocal integration in the songbird, *J Neurosci,* 28, 1509–22.
19. Jarvis, E. D. & Nottebohm, F. (1997) Motor-driven gene expression, *Proc Natl Acad Sci USA*, 94, 4097–102.
20. Mello, C. V., Vicario, D. S. & Clayton, D. F. (1992) Song presentation induces gene expression in the songbird forebrain, *Proc Natl Acad Sci USA*, 89, 6818–22.
21. Sheng, M., McFadden, G. & Greenberg, M. E. (1990) Membrane depolarization and calcium induce c–fos transcription via phosphorylation of transcription factor CREB, *Neuron*, 4, 571–82.

22. Mello, C. V. & Ribeiro, S. (1998) ZENK protein regulation by song in the brain of song-birds, *J Comp Neurol*, 393, 426–38.

23. Kimpo, R. R. & Doupe, A. J. (1997) FOS is induced by singing in distinct neuronal populations in a motor network, *Neuron*, 18, 315–25.

24. Hessler, N. A. & Doupe, A. J. (1999) Social context modulates singing-related neural activity in the songbird forebrain, *Nat Neurosci*, 2, 209–11.

25. Sasaki, A., Sotnikova, T. D., Gainetdinov, R. R. & Jarvis, E. D. (2006) Social context–dependent singing–regulated dopamine, *J Neurosci*, 26, 9010–4.

26. Yanagihara, S. & Hessler, N. A. (2006) Modulation of singing-related activity in the songbird ventral tegmental area by social context, *Eur J Neurosci*, 24, 3619–27.

27. Hara, E., Kubikova, L., Hessler, N. A. & Jarvis, E. D. (2007) Role of the midbrain dopaminergic system in modulation of vocal brain activation by social context, *Eur J Neurosci*, 25, 3406–16.

28. Castelino, C. B. & Ball, G. F. (2005) A role for norepinephrine in the regulation of context–dependent ZENK expression in male zebra finches (*Taeniopygia guttata*), *Eur J Neurosci*, 21, 1962–72.

29. Castelino, C. B., Diekamp, B. & Ball, G. F. (2007) Noradrenergic projections to the song control nucleus area X of the medial striatum in male zebra finches (*Taeniopygia guttata*), *J Comp Neurol*, 502, 544–62.

30. Cornil, C. A., Castelino, C. B. & Ball, G. F. (2008) Dopamine binds to alpha(2)-adrener-gic receptors in the song control system of zebra finches (*Taeniopygia guttata*), *J Chem Neuroanat*, 35, 202–15.

31. Young, L. J. & Wang, Z. (2004) The neurobiology of pair bonding, *Nat Neurosci*, 7, 1048–54.

32. Olveczky, B. P., Andalman, A. S. & Fee, M. S. (2005) Vocal experimentation in the juvenile songbird requires a basal ganglia circuit, *PLoS Biol*, 3, e153.

33. Woolley, S. C. & Doupe, A. J. (2008) Social context-induced song variation affects female behavior and gene expression, *PLoS Biol*, 6, e62.

34. Vignal, C., Andru, J. & Mathevon, N. (2005) Social context modulates behavioural and brain immediate early gene responses to sound in male songbird, *Eur J Neurosci*, 22, 949–55.

35. Haesler, S., Wada, K., Nshdejan, A. et al. (2004) FoxP2 expression in avian vocal learn-ers and non-learners, *J Neurosci*, 24, 3164–75.

36. Marcus, G. F. & Fisher, S. E. (2003) FOXP2 in focus: what can genes tell us about speech and language? *Trends Cogn Sci*, 7, 257–262.

37. Teramitsu, I., Kudo, L. C., London, S. E., Geschwind, D. H. & White, S. A. (2004) Parallel FoxP1 and FoxP2 expression in songbird and human brain predicts functional interaction, *J Neurosci*, 24, 3152–63.

38. Teramitsu, I. & White, S. A. (2006) FoxP2 regulation during undirected singing in adult songbirds, *J Neurosci*, 26, 7390–4.

39. Haesler, S., Rochefort, C., Georgi, B. et al. (2007) Incomplete and inaccurate vocal imi-tation after knockdown of FoxP2 in songbird basal ganglia nucleus Area X, *PLoS Biol*, 5, e321.

40. Spiteri, E., Konopka, G., Coppola, G. et al. (2007) Identification of the transcriptional targets of FOXP2, a gene linked to speech and language, in developing human brain, *Am J Hum Genet*, 81, 1144–57.

41. Poopatanapong, A., Teramitsu, I., Byun, J. S. et al. (2006) Singing, but not seizure, induces synaptotagmin IV in zebra finch song circuit nuclei, *J Neurobiol*, 66, 1613–29.

42. Wada, K., Howard, J. T., McConnell, P. et al. (2006) A molecular neuroethological approach for identifying and characterizing a cascade of behaviorally regulated genes, *Proc Natl Acad Sci USA*, 103, 15212–7.

43. Ferguson, G. D., Anagnostaras, S. G., Silva, A. J. & Herschman, H. R. (2000) Deficits in memory and motor performance in synaptotagmin IV mutant mice, *Proc Natl Acad Sci USA*, 97, 5598–603.

44. Ferguson, G. D., Wang, H., Herschman, H. R. & Storm, D. R. (2004) Altered hippocampal short-term plasticity and associative memory in synaptotagmin IV (–/–) mice, *Hippocampus*, 14, 964–74.

45. Li, X. C., Jarvis, E. D., Alvarez–Borda, B., Lim, D. A. & Nottebohm, F. (2000) A relationship between behavior, neurotrophin expression, and new neuron survival, *Proc Natl Acad Sci USA*, 97, 8584–9.

46. Velho, T. A., Pinaud, R., Rodrigues, P. V. & Mello, C. V. (2005) Co-induction of activity-dependent genes in songbirds, *Eur J Neurosci*, 22, 1667–78.

47. Lombardino, A. J., Li, X. C., Hertel, M. & Nottebohm, F. (2005) Replaceable neurons and neurodegenerative disease share depressed UCHL1 levels, *Proc Natl Acad Sci USA*, 102, 8036–41.

48. Lombardino, A. J., Hertel, M., Li, X. C. et al. (2006) Expression profiling of intermingled long-range projection neurons harvested by laser capture microdissection, *J Neurosci Methods*, 157, 195–207.

49. Alvarez-Borda, B. & Nottebohm, F. (2002) Gonads and singing play separate, additive roles in new neuron recruitment in adult canary brain, *J Neurosci*, 22, 8684–90.

50. Hayes, V. Y., Towner, M. D. & Isackson, P. J. (1997) Organization, sequence and functional analysis of a mouse BDNF promoter, *Brain Res Mol Brain Res*, 45, 189–98.

51. alella, A. M., Nerlov, C., Lopez, R. G. et al. (2007) Neurotrophin/Trk receptor signaling mediates C/EBPalpha, -beta and NeuroD recruitment to immediate-early gene promoters in neuronal cells and requires C/EBPs to induce immediate-early gene transcription, *Neural Develop*, 2, 4.

52. Mukoyama, M., Yamazaki, K., Kikuchi, T. & Tomita, T. (1989) Neuropathology of gracile axonal dystrophy (GAD) mouse. An animal model of central distal axonopathy in primary sensory neurons, *Acta Neuropathol*, 79, 294–9.

53. Nottebohm, F. (2004) The road we travelled: discovery, choreography, and significance of brain replaceable neurons, *Ann N Y Acad Sci*, 1016, 628–58.

54. Dudas, B. & Merchenthaler, I. (2003) Topography and associations of leu-enkephalin and luteinizing hormone-releasing hormone neuronal systems in the human diencephalon, *J Clin Endocrinol Metab*, 88, 1842–8.

55. Owczarek, D., Garlicka, M., Pierzchala-Koziec, K., Skulina, D. & Szulewski, P. (2003) [Met-enkephalin plasma concentration and content in liver tissue in patients with primary biliary cirrhosis], *Przegl Lek*, 60, 461–6.

56. Zhang, M., Balmadrid, C. & Kelley, A. E. (2003) Nucleus accumbens opioid, GABAergic, and dopaminergic modulation of palatable food motivation: contrasting effects revealed by a progressive ratio study in the rat, *Behav Neurosci*, 117, 202–11.

57. Jin, H. & Clayton, D. F. (1997) Localized changes in immediate-early gene regulation during sensory and motor learning in zebra finches, *Neuron*, 19, 1049–59.

58. Ribeiro, S., Cecchi, G. A., Magnasco, M. O. & Mello, C. V. (1998) Toward a song code: evidence for a syllabic representation in the canary brain, *Neuron*, 21, 359–71.

59. Whitney, O. & Johnson, F. (2005) Motor-induced transcription but sensory-regulated translation of ZENK in socially interactive songbirds, *J Neurobiol*, 65, 251–259.

60. Pinaud, R., Osorio, C., Alzate, O. & Jarvis, E. D. (2008) Profiling of experience-regulated proteins in the songbird auditory forebrain using quantitative proteomics, *Eur J Neurosci*, 27, 1409–22.

61. Wada, K., Sakaguchi, H., Jarvis, E. D. & Hagiwara, M. (2004) Differential expression of glutamate receptors in avian neural pathways for learned vocalization, *J Comp Neurol*, 476, 44–64.

62. Lerea, L. (1997) Glutamate receptors and gene induction: signalling from receptor to nucleus, *Cell Signal*, 9, 219–226.
63. Dong, S., Replogle, K. L., Hasadsri, L. et al. (2009) Discrete molecular states in the brain accompany changing responses to a vocal signal, *Proc Natl Acad Sci USA*, 106, 11364–9.
64. Goelet, P., Castellucci, V. F., Schacher, S. & Kandel, E. R. (1986) The long and the short of long-term memory—a molecular framework, *Nature*, 322, 419–22.
65. Robertson, H. A. (1992) Immediate-early genes, neuronal plasticity, and memory, *Biochem Cell Biol*, 70, 729–37.
66. Knapska, E. & Kaczmarek, L. (2004) A gene for neuronal plasticity in the mammalian brain: Zif268/Egr-1/NGFI-A/Krox-24/TIS8/ZENK? *Prog Neurobiol*, 74, 183–211.
67. Jarvis, E. D. (2004) Brains and birdsong, in: Marler, P. & Slabbekoorn, H. (Eds.) Nature's Music: The Science of Bird Song, pp. 226–271 (San Diego, Elsevier–Academic Press).
68. Mello, C. V. & Jarvis, E. D. (2008) Behavior-dependent expression of inducible genes in vocal learning birds, in: Zeigler, H. P. & Marler, P. (Eds.) *Neuroscience of birdsong*, pp. 381–397 (Cambridge University Press).
69. Mello, C. V. & Clayton, D. F. (1995) Differential induction of the ZENK gene in the avian forebrain and song control circuit after metrazole-induced depolarization, *J Neurobiol*, 26, 145–61.
70. Jarvis, E. D. & Mello, C. V. (2000) Molecular mapping of brain areas involved in parrot vocal communication, *J Comp Neurol*, 419, 1–31.
71. Clayton, D. F. (2000) The genomic action potential, *Neurobiology of Learning & Memory*, 74, 185–216.
72. Yenari, M. A. (2002) Heat shock proteins and neuroprotection, *Adv Exp Med Biol*, 513, 281–99.
73. Smith, V. A., Jarvis, E. D. & Hartemink, A. J. (2002) Evaluating functional network inference using simulations of complex biological systems, *Bioinformatics*, 18 Suppl 1, S216–24.
74. Yu, J., Smith, V. A., Wang, P. P., Hartemink, A. J. & Jarvis, E. D. (2004) Advances to Bayesian network inference for generating causal networks from observational biological data, *Bioinformatics*, 20, 3594–603.

14 Proteomics of Experience-Dependent Plasticity in the Songbird Auditory Forebrain

Raphael Pinaud, Oscar Alzate, and Liisa A. Tremere

CONTENTS

14.1 INTRODUCTION

The brain processing of complex auditory information and the formation of auditory memories are critical to the biology of a number of social organisms that use vocal communication, as in the case of humans and songbirds [for reviews, see (1,2)]. In mammals, the auditory cortex is considered to play a central role in the processing of complex acoustic stimuli, including those that involve communication signals, such as vocalizations in animals and speech in humans (3,4). Songbirds have emerged as a powerful model to study the brain processing of complex, behaviorally relevant auditory communication signals, due to their complex communication relative to many other animal groups. For instance, songbirds are capable of vocal learning—the ability to modify their vocalizations based on auditory input—a rare behavior that is necessary for the acquisition of spoken language in humans, and found only in few animal groups (humans, cetaceans, elephants, some species of bats, parrots, and hummingbirds, in addition to songbirds) (5–8). It is thought that all other animal groups, including non-human primates, do not exhibit vocal learning behavior and, therefore, only display innate species-specific vocalizations (6). Songbirds also rely in the perceptual processing of songs for other important behaviors including those that involve individual identification, territorial defense, and mate selection (9–14).

14.2   BRAIN SUBSTRATES OF SONG LEARNING AND PRODUCTION

The brains of songbirds contain a system of interconnected areas, known as the song-control system, that controls the learning and production of vocalizations (8,15–21) (Figure 14.1). This brain system can be sub-divided into two pathways: (a) a direct vocal-motor pathway, or posterior forebrain pathway (PFP) that controls production of learned vocalizations through projections from telencephalic stations to brainstem centers involved in vocal and respiratory function (17,20,22), and (b) an anterior forebrain pathway (AFP) required for learning and maintenance of the bird's own song production (21,23–27) (Figure 14.1). Projections of the PFP originate from nucleus HVC of the nidopallium (a letter-based name) and target the robust nucleus of the arcopallium (RA). Descending projections from RA then target the dorsomedial nucleus of the intercollicular complex (DM) and the tracheosyringeal component of the hypoglossal nerve nucleus (nXIIts), which innervates muscles of the syrinx, the vocal organ in songbirds, in addition to respiratory centers in the medulla. The AFP encompasses a loop of topographically organized projections from area X of the medial striatum to the medial part of the dorsolateral thalamic nucleus (DLM), from DLM to the lateral magnocellular nucleus of the anterior nidopallium (LMAN), and from LMAN back to area X (16,28–31). This projection system is comparable to mammalian cortico-basal ganglia-thalamic loops that are thought to be involved in learning and/or performance of movements requiring fine sequential sensorimotor integration [for reviews, see (32–36)].

In juvenile songbirds, the acquisition and development of songs (vocal learning behavior) is directly dependent on intact auditory processing. For instance, deafening and interference with appropriate auditory feedback prevents song learning (7,8,37–39). Interestingly, complete or selective impairment of auditory feedback in adult songbirds leads to a gradual deterioration of learned song structure (39–41), a phenomenon that is also observed in human speech (42–45). These findings

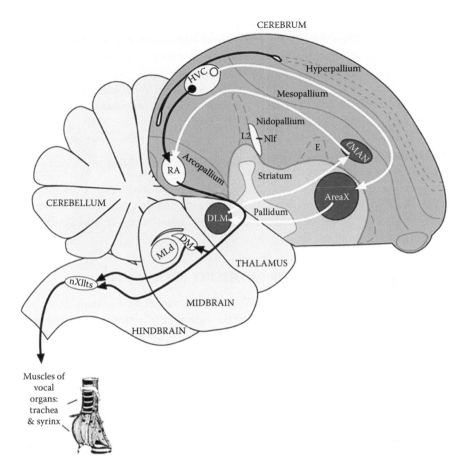

FIGURE 14.1   Schematic diagram of a parasagittal section through a zebra finch brain detailing the connectivity of the main stations of song-control system. For clarity, only the main nuclei and projections are shown in each diagram. Projection systems participating in the posterior forebrain pathway (PFP) are indicated by black arrows, while projections that compose the anterior forebrain pathway (AFP) are detailed by white arrows. Anatomical abbreviations not mentioned in text: DM, dorsal medial mesencephalic nucleus; E, entopallium; H, hyperpalium; Hp, hippocampus; M, mesopallium; MLd, dorsal lateral mesencephalic nucleus; N, nidopallium; NIf, interface nucleus; Ov, ovoidalis; St, striatum; v, ventricle. *Source*: Figure provided by E. D. Jarvis.

indicate that intact hearing is required for both the acquisition and maintenance of the structure of learned vocal communication signals.

The song-control system is clearly involved in vocal learning, as lesions in this circuit during a critical period early in life prevent juvenile birds from appropriately acquiring songs from adult tutors (23,26); for reviews, see (8). Furthermore, as indicated above, auditory information is necessary for vocal learning; it is thought that songbirds rely on hearing to generate auditory memories that are used as templates for the normal development of vocal behavior (7); for reviews, see (5,8,46,47).

However, the role of the song-control system in the auditory processing of songs is unclear and, perhaps, paradoxical. In anesthetized male songbirds, auditory-driven responses have been documented for all nuclei that constitute the song-control system, and have been shown to exhibit selectivity to the bird's own song (48–51), with such selectivity emerging during the critical period for vocal learning (52–55). These findings have contributed fundamental information on how auditory information shapes neuronal activity within the AFP and PFP. However, hearing-evoked responses in nuclei of the song-control system are primarily seen under anesthesia or during sleep in songbirds. In awake animals, hearing-driven electrophysiological responses are substantially reduced or even absent (56–58); but see (59–61), suggesting that the contributions of the song-control system to the auditory processing of songs may be limited. Moreover, perceptual processing and discrimination of songs plays a central role in the biology of female songbirds whose song-control system is inexistent or largely atrophied (62).

14.3   BRAIN SUBSTRATES FOR AUDITORY PROCESSING

Even though songbirds have been extensively used as a model for the study of vocal learning behavior, auditory discrimination and the formation of auditory memories are necessary for other central biological processes, especially those involving individual recognition. For instance, territorial demarcation is achieved through the auditory processing of songs; it is through discrimination of the acoustic features of songs that songbirds are able to differentiate the presence of a neighbor versus that of a trespasser (9,10,12,14). The recognition and discrimination of individuals in a pool of conspecifics is also mediated through the processing of auditory cues (63,64). Moreover, the selection of breeding mates by female songbirds, who largely lack a song control system, also centers around discrimination and preference for certain spectral and temporal parameters associated with the song of potential partners (11–13).

Given that auditory processing of complex natural communication signals is required for a number of key behaviors in the biology of songbirds, a significant effort has been directed at understanding the anatomical and functional organization of brain areas involved in the auditory processing of natural vocal communication signals. As with other avian species and vertebrate groups, songbirds possess an ascending auditory pathway that consists of a series of medullary, pontine, mesencephalic and thalamic nuclei that convey auditory information from the cochlea to telencephalic cerebral centers, although the detailed connectivity of this pathway up to the midbrain has not yet been determined in songbirds (Figure 14.2). At the level of the telencephalon, several auditory areas are located within a prominent caudomedial lobule (31,65). These areas include the auditory thalamo-recipient zone field L2, as well as major field L2 targets, namely the caudomedial nidopallium (NCM) and the caudomedial mesopallium (CM) (Figure 14.2). From field L and these primary targets, auditory inputs reach other telencephalic areas and originate descending auditory projections that target both thalamic and mesencephalic nuclei (65). Although the specific role of individual constituent nuclei or areas has not been clearly established, these central auditory pathways are thought to be crucial for song

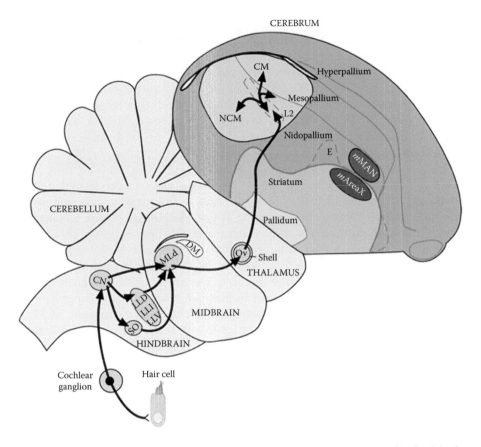

FIGURE 14.2   Schematic representation of a parasagittal section through a zebra finch brain illustrating the main stations of the songbird ascending auditory pathway. For clarity, only the main nuclei and projections are shown in each diagram. The focus of this chapter, nucleus NCM, receives input from the thalamo-recipient layer field L2, and has reciprocal connectivity with the caudal mesopallium (CM). Anatomical abbreviations not mentioned in text: CN, cochlear nuclei; LL, lateral lemniscal nuclei; SO, superior olive. *Source*: Figure provided by E. D. Jarvis.

auditory processing, perception, and possibly the song memorization required for perceptual discrimination and vocal learning [for reviews, see (66–68)].

14.4   NCM SELECTIVITY, PLASTICITY, AND POTENTIAL ROLES IN MEMORY FORMATION

A brain area that has received significant attention with respect to the neural basis of auditory processing, discrimination, and auditory memory formation has been NCM, a telencephalic area considered analogous to the supragranular layers of the mammalian primary auditory cortex (31,65,69,70). NCM displays strong electrophysiological responses to song stimulation in the awake animal, with greater selectivity for complex stimuli, as compared to responses obtained at earlier stations in the ascending auditory pathway, such as the thalamo-recipient field L2 (71–76).

Our current understanding of NCM's anatomo-functional organization was significantly advanced by the use of activity-regulated markers, especially the expression of immediate early genes (IEGs), a class of genes that are expressed rapidly and transiently in response to a variety of cellular stimuli including neuronal activation [for reviews, see (77–80)]. Within this scope, zenk (aka *egr-1*, NGFI-A, *krox-24*, *zif268*) has been the most extensively used IEG to map neuronal activity in the songbird brain associated with auditory processing and song production (68,81,82). The *zenk* gene encodes a zinc-finger transcription factor (ZENK) that is highly sensitive to neuronal depolarization and that has been implicated as a central molecule putatively involved in the induction of neuronal plasticity in a variety of preparations [for reviews, see (68,77–80)]. The monitoring of both mRNA and protein products that result from *zenk* expression provides a valuable tool to evaluate global patterns of neuronal ensemble activation, with cellular resolution (Figure 14.3). Using *zenk* expression as a tool for neuronal activity, Mello et al. (1992) (68) demonstrated that NCM is more robustly activated when songbirds hear conspecific songs, as compared to heterospecific songs or artificial stimuli such as pure tones. These findings have also been demonstrated on the basis of electrophysiological activity, whereby NCM neurons fire at a higher rate to conspecific, relative to heterospecific or artificial auditory stimuli, suggesting that NCM not only participates in auditory processing but may also contribute to auditory discrimination of behaviorally relevant acoustic cues (71,72). Consistent with this concept is the finding that repeated presentations of the same song leads to a marked decrease, or "habituation," in the number of *zenk*-expressing cells in NCM; this decrease in the *zenk*-positive cell population in NCM can be reinstated following presentation of a novel song (83). Likewise, electrophysiological responses to successive presentations of the same song lead to a rapid and marked decrease in the responsiveness of NCM neurons to that particular stimulus (71). This song-specific electrophysiological habituation is also recovered upon stimulation of animals with novel, previously unheard songs, suggesting that NCM neurons are capable of discriminating across auditory stimuli and, for such, may retain a memory trace of previously "heard" songs (71,72). Brain areas that provide input to NCM, such as field L2, do not appear to exhibit song-specific habituation, indicating that this network property likely arises from within NCM circuitry. Importantly, song-specific habituated responses in NCM are long-lasting and also display conspecific song selectivity; habituated responses last significantly longer (>40 hours) for conspecific songs, as compared to heterospecific songs (~5 hours) (71,72). Furthermore, the long-term maintenance of song-specific habituated responses depends on de novo protein synthesis, as this phenomenon is blocked by either RNA or protein synthesis inhibitors (71). Together, these findings suggest that NCM exhibits experience-dependent plasticity that may underlie long-term changes in neuronal performance and be a substrate for auditory discrimination. Importantly, such plasticity appears to depend on the engagement of protein regulatory cascades.

Although there is currently no available evidence suggesting a causal relationship between song-specific habituation of electrophysiological responses and habituation

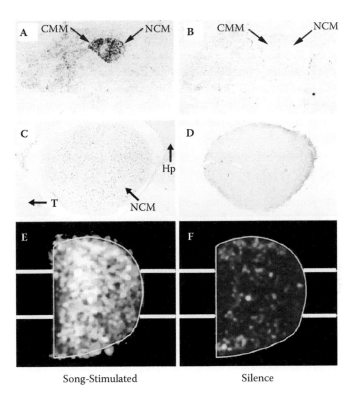

Song-Stimulated   Silence

FIGURE 14.3 (See color insert following page 172.) Induction of zenk mRNA and protein levels in the songbird forebrain following exposure to song. A: In situ hybridization auto-radiogram of a section of an adult male zebra finch exposed to 45 minutes of playbacks of conspecific songs. B: Unstimulated control. These sections correspond to parasagittal plane 250 μm lateral to the medial surface of the brain. C–D: ZENK expression, revealed by immu-nocytochemistry, in the caudomedial telencephalon. Notice the presence of numerous ZENK-labeled cell nuclei after song stimulation (C) in the caudomedial nidopallium (NCM) of a zebra finch from the hearing-only group. D: Unstimulated control. E: Density/intensity map of ZENK expression in the NCM of a canary resulting from presentation of a playback of con-specific songs. F: Unstimulated control. Colors and brightness correspond to relative density and intensity of labeling of ZENK positive cells. Anatomical abbreviations not mentioned above or in text: Hp, hippocampus; T, telencephalon. Sources: Adapted from Mello, C. V. and Pinaud, R. (2006) Immediate early gene regulation in the auditory system. In Immediate early genes in sensory processing, cognitive performance and neurological disorders, eds. Pinaud, R. and Tremere, L. A., 35–56 (New York: Springer-Verlag); Mello, C. V., Vicario, D. S. and Clayton, D. F. (1992) Song presentation induces gene expression in the songbird fore-brain, *Proc Natl Acad Sci USA* 89:6818–22; and Mello, C. V. and Ribeiro, S. (1998) ZENK protein regulation by song in the brain of songbirds, *J Comp Neurol* 393:426–38.

of *zenk* expression, these findings indicate that NCM undergoes experience-depen-dent plastic changes that may underlie higher order brain functions. In accordance with this notion, it has been shown that neuronal activation in NCM, as revealed by the expression of IEGs (*zenk* and *c-fos*) is strongly correlated with the number of song elements that birds copy from tutor songs (84–86), a finding that is also corroborated by electrophysiological findings (87). In addition, recent experiments involving restricted NCM lesions in adult animals showed a significant impact in some aspects of the behavioral discrimination of songs (88). These data suggest that NCM, a central auditory forebrain area, may be a key site involved in auditory discrimination and, possibly, the formation of auditory memories. Moreover, the findings discussed above indicate that experience-dependent modifications in the functional organization of NCM depend on a protein regulatory network, the identity of which is largely unknown.

14.5   KNOWN PROTEIN REGULATORY EVENTS IN NCM

Before large-scale proteomic screening, within a 15-year period of extensive studies in NCM, our knowledge of the identity experience-regulated molecules in NCM was limited to the IEGs *zenk*, *c-jun*, *c-fos*, and *arc* (68,81,82,89–91), the protease caspase-3 and its endogenous inhibitor (BIRC4) (92), and the extracellular-signal regulated kinase (ERK), that is phosphorylated in NCM as a result of auditory stimulation (93). These studies contributed substantially to our understanding of the biochemical and gene expression pathways activated by behaviorally relevant auditory stimuli in NCM. However, a complete, and unbiased, picture of the molecular and cellular processes affected by auditory experience is currently lacking. Uncovering these processes will be fundamental to understanding how the physiology of NCM neurons is affected by sensory experience, and may provide valuable information on the specific roles of NCM in the auditory and perceptual processing of songs and, perhaps, the formation of auditory memories.

14.6   QUANTITATIVE PROTEOMICS SCREENING IN NCM: A METHODOLOGICAL SUMMARY

We recently initiated efforts to elucidate on a more global level how the NCM's proteome is impacted by sensory experience. Using two-dimensional differential in-gel electrophoresis (2D-DIGE)-based proteomics, coupled with extensive data analyses for protein quantification and tandem mass spectrometry, we conducted the first large-scale quantitative proteomics screening in the songbird brain, focusing on high-abundance proteins (94). In these initial efforts we asked how the NCM's high-abundance proteome is dynamically regulated over time as a result of the length of auditory experience (exposure of animals to playbacks of a medley of novel conspecific songs), a paradigm known to induce experience-dependent plasticity in NCM circuitry. To achieve this goal, we placed animals in soundproof chambers overnight prior to experimentation to minimize hearing-regulated protein activity. We subsequently divided the animals into four groups for a time course experiment: group 1—controls, which were sacrificed in the absence of auditory experience, and three groups of animals that were exposed to different lengths of auditory stimulation; group 2—five minutes; group 3—one hour; and group 4—three hours of continuous auditory stimulation (Figure 14.4A). The choice of these time points was based in our interest of studying fast changes, such as post-translational modifications (five

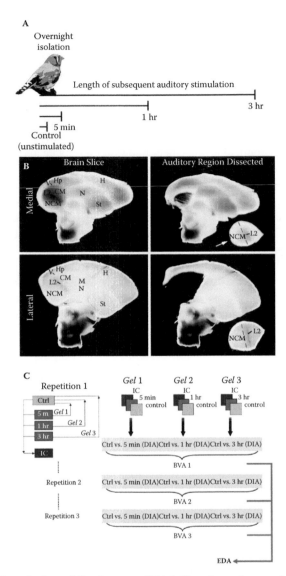

FIGURE 14.4 (See color insert following paage 172.) A: Experimental approach used to investigate experience-regulated proteins in the songbird NCM. All animals were isolated overnight in sound-proof chambers and randomly included into one of four experimental groups. Control animals were sacrificed after isolation without stimulation. The remaining animals were stimulated with playbacks of conspecific songs for 5 minutes, 1 hour, or 3 hours. B: Representative brain sections were obtained with a vibratome for dissection of NCM. The red dotted lines indicate the approximate locations where sections were performed to isolate NCM. Note the darker color of the thalamo-recipient field L2 in these preparations, which significantly facilitates the isolation of NCM. C: Representation of the 2D-DIGE experiments conducted by Pinaud et al. (94). Gels were run for individual comparisons across experimental groups (control vs. 5 minutes; control vs. 1 hour; and control vs. 3 hours). Each experiment was repeated three times with different animals. Intra-gel (DIA), inter-gel (BVA), as well as across-repetition (EDA) comparisons were conducted in a quantitative manner in order to identify statistically significant differences across these variables.

minutes), the first wave of immediate early gene responses (one hour) and a second wave proposed to consist of late responses gene (three hours) (71,80). We conducted these experiments in females given that they do not sing in response to hearing the song playbacks, avoiding potential confounding effects associated with hearing oneself.

NCM is the most caudomedial forebrain structure in the songbird brain, and its boundaries are distinctly delineated by the ventricular zone in its caudal, dorsal, and ventral aspect, and by field L2 and the lamina mesopallialis (LM) in its rostral aspect (Figure 14.2). This anatomical arrangement allows for a reliable and speedy dissection of NCM for proteomics processing (Figure 14.4B).

We used the internal standard methodology to establish the potential differential expression of sensory-regulated proteins (95–97); see also Chapter 4. Briefly, total protein was extracted from NCM samples of each animal group; control samples were labeled with Cy3 (green dye) while each of the experimental samples (five minutes, one hour, and three hours) were independently labeled with Cy5 (red dye). An internal standard (IS) consisting of protein samples from all groups were pooled and labeled with Cy2 (blue dye) (Figure 14.4C). The IS is central for the normalization of sample concentrations across gels, and across repetitions of each experiment, enabling for a highly quantitative assessment of protein levels in each studied condition. In addition, the IS enables the correction of distortions in the pattern in which proteins run across gels, which is also required for reliable protein-level comparisons.

In each 2D gel we included samples from control (Cy3-labeled), an experimental group (Cy5-labeled), and the IS (Cy2-labeled). Thus, a complete experimental set included three gels (control vs. five minutes, control vs. one hour, control vs. three hours) (Figure 14.4C). Each comparison, however, was replicated a minimum of three times (n = 3 animals) so that statistically meaningful comparisons could be performed. The protein combined samples (control, experimental, and IS) were fractionated in two dimensions: first by isoelectric focusing, utilizing an immobilized pH gradient, and second by molecular weight using standard polyacrylamide gel electrophoresis. The resulting tri-chromatic gel was subsequently imaged with a Typhoon 9410 scanning system (GE Healthcare) that allows for signal associated with each channel to be acquired independently. The resulting images were fed into an analytical software (DeCyder 6.5) that is designed to perform all pair-wise, in-gel comparisons, as well as across-gel comparisons through individual specialized algorithms (for a more detailed description of the methodology and comparison parameters see (94) and Chapter 4). The identity of differentially regulated proteins was achieved through tandem mass spectrometry, using standard approaches (94); see Chapters 4, 5, and 6 for detailed description of the techniques used.

14.7   EXPERIENCE-REGULATED PROTEINS IN NCM

We used 2D-DIGE–based proteomics to screen approximately 2500 protein spots (Figure 14.5). Principal component analysis conducted in all protein spots indicated that each experimental group was significantly different from the other (within a 95% confidence interval) (Figure 14.6), suggesting that the NCM proteome dynamically

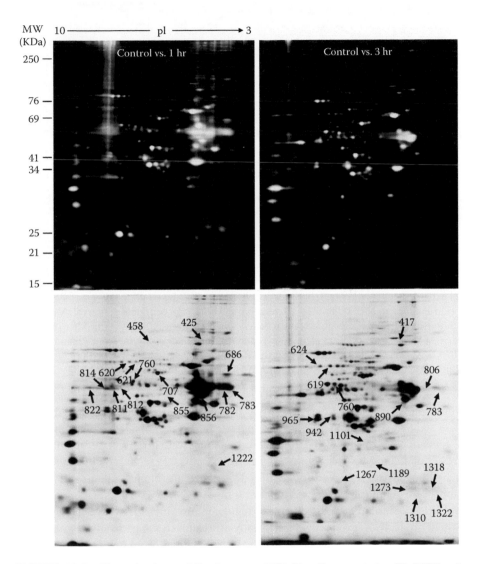

FIGURE 14.5 (See color insert following page 172.) Top: Representative 2D-DIGE gels illustrating fractionated proteins from NCM in a comparison for the 1-hour (control vs. 1 hour; *left*) and 3-hour (control vs. 3 hours; *right*) groups. Control samples were labeled with Cy3 (*green*) while experimental samples were labeled with Cy5 (*red*). Internal control samples were labeled with Cy2 (blue, not shown). Bottom: Coomassie blue-stained gels illustrating differentially regulated spots (*arrows*) in the 1-hour (*left*) and 3-hour (*right*) conditions, as revealed by quantitative and statistical analyses with DeCyder software. All differentially regulated spots underwent protein fingerprinting by mass-spectrometry. *Source*: Figure from Pinaud, R., Osorio, C., Alzate, O. and Jarvis, E. D. (2008) Profiling of experience-regulated proteins in the songbird auditory forebrain using quantitative proteomics, *Eur J Neurosci* 27:1409–22.

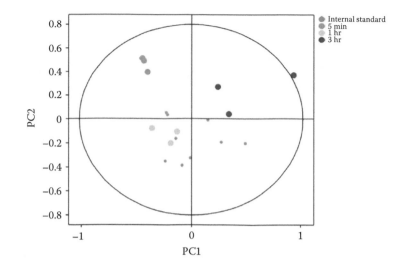

FIGURE 14.6 (See color insert following page 172.) Principal component analysis (PCA) of ~2500 protein spots per condition reveals that the NCM's proteome is significantly different across experimental conditions. Independent gels for each of the conditions are not significantly different from each other. The circle represents the 95% confidence interval. Values that cluster in a quadrant are significantly different from values falling within other quadrants. See Chapter 4 description of internal standard (IS). *Source*: Figure from Pinaud, R., Osorio, C., Alzate, O. and Jarvis, E. D. (2008) Profiling of experience-regulated proteins in the songbird auditory forebrain using quantitative proteomics, *Eur J Neurosci* 27:1409–22.

changes as a result of sensory experience (94). When compared to controls, although no significant changes were detected in the five-minute group, one hour and three hours of auditory stimulation significantly impacted the expression levels of a number of proteins in NCM. We detected 16 protein spots that were differentially regulated as a result of auditory experience in the 1-hour group. Of these 16 protein spots, 12 were significantly up-regulated (range 9%–37% above unstimulated control levels) and four were significantly down-regulated (range 37%–117% below unstimulated control levels). Similarly, 16 differentially regulated protein spots were detected in the 3-hour group, with seven spots significantly up-regulated (range 13%–40% above unstimulated control levels) and nine significantly down-regulated (range 10%–44% below unstimulated control levels) (Table 14.1).

Using matrix-assisted laser desorption ionization–time of flight (MALDI-TOF)/ TOF-mass spectrometry (Chapter 5), we explored the identity of these 32 differentially regulated protein spots from 1-hour and 3-hour groups. We were unable to reliably (>95% confidence) identify four proteins (of 16 spots) in the 1-hour group, and seven proteins (of 16 spots) in the 3-hour group. The lack of confidence in reliably identifying certain protein spots was directly related to low protein concentration in those particular spots.

In the 1-hour group, the 12 reliably identified protein spots consisted of nine proteins, while the nine spots detected in the 3-hour group consisted of seven proteins. The presence of repeated protein spots in multiple gel locations suggest that sensory

TABLE 14.1

Quantitative Analysis of Differentially Regulated Spots in NCM, as Revealed by 2D-DIGE–Based Proteomics

Control versus 1 hr		Control versus 3 hr	
Spot Number	Average Ratio	Spot Number	Average Ratio
425	1.27	417	−1.35
458	1.09	619	1.14
620	1.20	624	1.13
621	1.16	760	1.19
686	−2.35	783	−1.16
707	1.11	806	−1.31
760	1.19	890	−1.10
782	−1.36	942	1.19
783	−1.31	965	1.16
811	1.20	1101	1.40
812	1.19	1189	1.20
814	1.24	1267	−1.17
822	1.37	1273	−1.44
855	1.21	1310	−1.28
856	1.35	1318	−1.39
1222	−1.37	1322	−1.40

Note: Indicated are spot numbers of differentially regulated proteins ($p <$ 0.05) and the average expression difference between control and experimental samples. Positive ratios indicate spots where experimental samples have higher expression levels compared to control samples (up-regulation as a result of auditory experience) and negative ratios indicate spots where the control samples have higher expression levels compared to experimental samples (down-regulation as a result of auditory experience).

experience may modify the conformational, charge, or structural properties of each of these proteins.

The identified proteins belonged to multiple functional categories that were putatively distributed across multiple cellular organelles (Figure 14.7A; see Table 14.2). Many proteins were primarily cytoplasmic proteins, whereas others are found in the endoplasmic reticulum, cell nucleus, and pre-synaptic terminals (Figure 14.7B). The large majority of differentially regulated proteins were involved in metabolic and catabolic functions, most of which have participation in the synthesis of adenosine tri-phosphate (ATP). Perhaps not surprisingly, these findings suggest that auditory input drives protein changes in NCM that are associated with cellular respiration. The proteins whose expression levels are altered by sensory experience and that participate in ATP synthesis are well positioned to trigger long-lasting modifications in the metabolic processing in NCM.

TABLE 14.2

Hearing Regulated Proteins in NCM at 1 Hour and 3 Hours as Determined from 2D-DIGE Experiments and MS/MS Analyses

				1-hr Group						
Spot No.	Protein Name	Abbrev.	Accession ID	Species	Pred. MW	Obs. MW/pI	Protein Score	Ion Score	Coverage (%)	# of Peptides
425	Heat shock protein 108	hsp108	Q90WA6	G. gallus	91.1	91.2/4.86	83	42	17	13
620	Synapsin II	SYN2	Q6RSD2	T. guttata	23.8	71.3/7.35	126	106	19	4
621	Synapsin II	SYN2	Q6RSD2	T. guttata	23.8	71.3/7.18	68	40	20	5
707	Collapsin response mediator protein-2B	CRMP-2B	Q71SG1	G. gaius	62.2	62.2/6.05	679	515	53	21
760	Pyruvate kinase	PKM2	P00548	G. gallus	57.8	58.0/7.29	178	121	31	13
812	Glutamate dehydrogenase	GLUD	P10860	R. norvegicus	61.3	61.4/8.05	133	92	25	11
811	Glutamate dehydrogenase	GLUD	P00368	G. gallus	55.6	59.0/7.84	181	119	30	13
814	ATP synthase alpha subunit	ATP5A1	Q8UVX3	G. gallus	60.1	59.0/7.97	300	213	34	15
822	ATP synthase alpha subunit	ATP5A1	Q8UVX3	G. gallus	60.1	58.4/8.47	735	608	39	19
855	Similar to NK-3 transcription factor	Nkx3-1	P97436	M. musculus	26.8	57.0/6.24	39	23	13	3
856	ATP synthase beta subunit	ATPB	Q5ZLC5	G. gallus	56.5	56.7/5.23	914	758	47	19
1222	Tropomyosin 3 gamma	TRM3	Q5ZLJ7	G. gallus	28.7	34.3/4.58	285	151	42	19

3-hr Group

Spot No.	Protein Name	Abbrev.	Accession ID	Species	Pred. MW	Obs. MW/pI	Protein Score	Ion Score	Coverage (%)	# of Peptides
619	Synapsin II	SYN2	Q6RSD2	*T. guttata*	23.8	71.2/7.58	140	85	39	7
624	Synapsin II	SYN2	Q6RSD2	*T. guttata*	23.8	71.1/7.89	52	32	19	4
760	Pyruvate kinase	PKM2	P00548	*G. gallus*	57.9	58.0/7.29	233	184	29	12
806	Tubulin beta chain	TBB	P07437	*H. sapiens*	49.7	59.9/4.5	145	64	26	13
942[a]	Phosphoglycerate kinase	PGK	Q76BF6	*O. latipes*	41.2	50.7/7.5	104	50	28	10
965	Phosphoglycerate kinase	PGK	Q76BF6	*O. latipes*	41.2	49.4/7.97	323	287	26	8
1273	14-3-3 protein gamma (PKC inhibitor)	1433G	Q5F3W6	*O. aries*	17.9	31.8/5.23	195	80	58	11
1310	Calretinin	CALB2	Q08331	*M. musculus*	22	31.3/4.94	125	73	37	8
1318	Synaptosomal-associated protein-25	SNP25	Q5R505	*P. pygmmaeus*	23.1	30.3/4.28	90	78	20	3

Note: Ion score is a measure of how well MS/MS spectra match the observed peptide. The "expect" value indicates the probability that the observed match between MS/MS spectra and peptide sequence would be found by chance. The protein score [S = −10 log(P)] is a measure of the probability (P) that a found peptide sequence matches the observed spectrum. The score for an MS/MS match is based on the absolute P that the match between the data and the corresponding match from the database is a random event. The lower the score, the lower the probability that the data represent a real match (see Chapter 5 for a complete discussion of these parameters). Spot numbers are labeled according to Figure 14.5. Accession IDs are those from NCBI. Pred, predicted molecular weight (MW) in kDa from MASCOT; Obs, observed MW and isoelectric point calculated from 2D-DIGE gels.

[a] Oxidized.

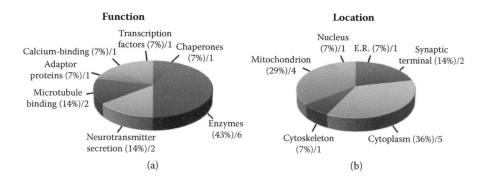

FIGURE 14.7   Pie charts illustrating the cellular location (a) and molecular function (b) of the proteins identified in the proteomics screening conducted by Pinaud et al. [(94); detailed in Table 14.2]. Data are expressed as percentage of the total number of differentially expressed proteins (*in brackets*). Numbers adjacent to the brackets indicate the number of proteins identified in that category. Categories were determined based upon known gene ontologies. No protein was grouped into more than one category for (a) and (b). The protein synapsin II, which is involved in both calcium binding and neurotransmitter secretion, was placed in the latter category as its main function.

Our 2D-DIGE screenings also successfully identified differentially regulated proteins involved in neurotransmitter release. For instance, we found that auditory experience drives the regulation of the synaptosomal-associated protein-25 (SNAP-25) (94), a target SNARE involved in neurotransmitter release (98,99), and synapsin II, a protein involved in the regulation of the size of the readily releasable pool of vesicles (100,101). The up-regulation of synapsin II that we detected with our 2D-DIGE proteomics screening may provide a substrate for forms of experience-dependent plasticity in NCM circuitry, namely the song-specific habituation of electrophysiological responses.

As indicated above, repeated presentation of the same song leads to a long-lasting, song-specific decrease (or habituation) in the responsiveness of NCM neurons. The mechanisms underlying NCM's habituation are completely unknown. It is possible that the up-regulation of synapsin II in our 3-hour group (a stimulation protocol that leads to robust habituation in NCM) may be part of the neural basis underlying this form of experience-dependent plasticity. Synapsin II anchors vesicles to actin filaments. In response to calcium influx that results from action potential generation, synapsin II is phosphorylated, thereby releasing vesicles from the actin filaments and replenishing the readily releasable pool of vesicles for subsequent action potential bouts. Given that we identified an up-regulation of synapsin II, we may be able to propose that higher synapsin II levels may lead to greater binding of vesicles to actin filaments and, consequently, fewer vesicles in the readily releasable pool. Such molecular modification could, in theory, provide a mechanism by which successive song presentations would lead to decreased electrophysiological responses over time. A causal link between song-specific habituation and synapsin II expression, however, remains to be investigated. Overall, our results indicate that proteins that play a direct role in synaptic transmission are significantly impacted by brief and acute episodes of auditory experience in NCM.

14.8   CALCIUM-DRIVEN PLASTICITY: ACTIVATION OF THE EXTRACELLULAR-SIGNAL REGULATED KINASE PATHWAY

As indicated above, the most carefully characterized protein change in NCM is the up-regulation of ZENK as a result of auditory stimulation [for reviews, see (68,81)]. The expression of *zenk* is dependent on calcium influx through glutamatergic NMDA receptor activation and it is believed that this IEG plays a central role in integrating electrophysiological changes that occur at the cell membrane with the genomic machinery in the cell nucleus (77,87,101–105). Furthermore, it has been proposed that *zenk* may play a role in triggering experience-dependent plastic changes in NCM circuitry that underlie the formation of auditory memories required for song learning (68,81). Calcium influx, either through NMDA receptor activation, voltage-gated calcium channels, or mobilization of intracellular stores, has been shown to be central to various types of synaptic plasticity, including those that underlie phenomena proposed to be the basis of learning and memory formation (79,103,106–110). Remarkably, our findings obtained with unbiased, quantitative 2D-DIGE-based proteomics, and the findings of other research groups utilizing biased screening approaches, have consistently demonstrated that auditory experience drives calcium-regulated proteins that participate primarily in a single biochemical pathway: the extracellular-signal regulated kinase (ERK)/mitogen-activated protein kinase (MAPK) pathway (Figure 14.8) (91,93,94,111). Thus the remainder of this review focuses on the ERK pathway, which is known to mediate several forms of plasticity in neural circuits (111). Activation of the ERK pathway in NCM may be central to experience-regulated changes in the functional organization of NCM, providing a possible substrate for higher order network properties such as auditory discrimination and memory formation.

A series of biochemical studies, conducted in different models and preparations, have detailed key steps associated with the induction of the ERK pathway (Figure 14.8). The influx of calcium through the mechanisms outlined above directly activates Ras, a small guanine triphosphatase (GTPase). In its inactive state, Ras is bound to guanine diphosphate (GDP). Upon activation, a number of guanine nucleotide exchange factors (GEFs), for example RasGRF (112), exchange GDP by guanine triphosphate (GTP). When GTP is bound, Ras becomes active and recruits a kinase known as Raf (Raf proto-oncogene serine/threonine protein kinase), which contains two domains in its structure: a Ras binding domain and a kinase domain (111,113). These events trigger a cascade of kinase activation: recruitment of Raf activates its kinase domain, which consequently phosphorylates a domain that is rich in cysteine residues within a second kinase, the MAPK/ERK kinase (MEK) (111,114,115). Following phosphorylation, MEK becomes active and phosphorylates ERK either on tyrosine or threonine residues. Phosphorylated ERK dimerizes, dissociates from MEK, and migrates to the cell nucleus, where it regulates transcriptional events (111) (Figure 14.8). For clarity, only the main constituents of this pathway

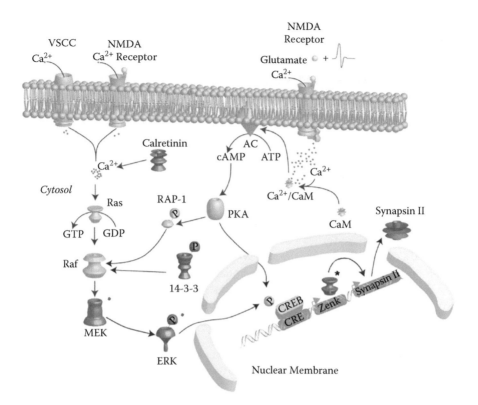

FIGURE 14.8 Schematic representation of known interacting biochemical pathways of proteins detected in the study by Pinaud et al. (94) mapped onto previously known pathways. Proteins identified to be regulated by auditory stimulation in Pinaud et al. (94) and previous studies are displayed in dark grey, and identified without and with asterisks, respectively. Most of the interactions of the molecular signaling pathways shown here have been determined in non-songbird species, including the ZENK binding site in the synapsin II promoter. For a detailed discussion of potential mechanisms see text and (103). *Source*: Figure modified from Pinaud, R., Osorio, C., Alzate, O. and Jarvis, E. D. (2008) Profiling of experience-regulated proteins in the songbird auditory forebrain using quantitative proteomics, *Eur J Neurosci* 27:1409–22.

have been detailed here; however, multiple isoforms have been reported for selected molecules that participate in this pathway including, but not limited to, three isoforms of Raf (A-Raf, B-Raf, and Raf-1 or c-Raf) (116), two MEK isoforms (MEK 1 and 2), and two isoforms of ERK (ERK 1 and 2) (117,118).

In the songbird NCM, phosphorylation of ERK occurs in response to auditory experience (93). Moreover, activation of MEK and, consequently, ERK is necessary for song-induced expression of several IEGs in NCM, namely *zenk*, *c-fos*, and *arc* (91,93), suggesting that activation of the ERK pathway and its downstream targets are markedly affected by auditory experience in NCM.

Our proteomics screenings also uncovered the up-regulation of the synapsin II protein by song presentation—a protein that is downstream of the ERK activation

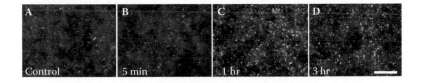

FIGURE 14.9 High power photomicrographs illustrating the hearing-induced up-regulation of synapsin II in NCM. Synapsin II is expressed in neuropils, as suggested by the punctate staining, which is consistent with presynaptic terminals. Scale bar = 25 μm. *Source*: Figure modified from Pinaud, R., Osorio, C., Alzate, O. and Jarvis, E. D. (2008) Profiling of experience-regulated proteins in the songbird auditory forebrain using quantitative proteomics, *Eur J Neurosci* 27:1409–22.

pathway (Figure 14.8)—a finding that was validated by immunocytochemical approaches (94) (Figure 14.9). Proteins belonging to the synapsin family, including synapsin II, have been shown to play a central role in neurotransmitter release, more specifically by controlling the size of the readily releasable pool of vesicles via calcium-regulated mechanisms [for review, see (100)]. Importantly, synapsin II expression has been shown to be directly regulated by *zenk* in cell cultures (119). Our proteomics findings on the regulation of synapsin II by song in NCM, coupled with the large body of evidence on song-driven *zenk* expression, suggest that synapsin II may be regulated by ZENK in NCM in vivo, as a function of auditory experience. In fact, we showed that ZENK and synapsin II are co-induced in the NCM of the same animals (94). However, evidence for a direct synapsin II regulation by ZENK in NCM is still lacking. Regardless of the identity of the transcriptional machinery controlling the expression of synapsin II, the up-regulation of this protein as a result of hearing behavioral-meaningful auditory stimuli may be directly related to the long-term maintenance of the song-specific "habituation" observed in NCM following repeated presentations of the same song (71,72,120).

An additional level of control over calcium-sensitive biochemical and protein regulatory cascades is through calcium-binding proteins. Our proteomics screening uncovered a significant song-induced down-regulation of calbindin-2 (aka calretinin) in NCM. Given that calbindin-2 is involved in the binding of calcium, our findings suggest that the down-regulation of this protein may allow for an experience-dependent increase in the concentrations of intracellular calcium levels that could, in theory, enhance the activity of calcium-regulated pathways, including the ERK pathway and its downstream targets (Figure 14.8).

Together, the findings to date provide conclusive evidence that auditory stimulation drives the activation of calcium-regulated pathways in NCM, including the ERK pathway and some of its downstream targets, including the IEGs *zenk*, *c-fos*, and *arc*. The regulation of late-response proteins (e.g., synapsin II), which are thought to mediate long-lasting changes in the anatomical and functional organization of neural circuits, were also impacted by auditory experience in NCM. These processes may be central for multiple forms of experience-dependent plasticity in NCM that may provide the physical basis for higher order network functions that appear to be performed by this auditory area, such as auditory discrimination and memory formation.

14.9 CONCLUDING REMARKS

NCM, an auditory forebrain area that exhibits selective responses to natural vocal communication signals, undergoes forms of experience-dependent neural plasticity that may underlie auditory memory formation required for important behaviors such as vocal learning, mate selection, and individual recognition. These experience-dependent changes appear to involve a complex network of protein regulatory events, the identities of which are just now starting to be uncovered. Detailing the identities of the proteins that compose these networks, and understanding how they are dynamically regulated, will be central to elucidating how NCM's anatomical and functional organization is impacted by sensory experience and the precise roles this auditory area plays in higher order brain functions, as in the case of auditory discrimination and the formation of auditory memories. Future studies can now be focused at characterizing in-depth the identity of the NCM proteome, further studying how it is impacted by sensory experience, and establishing the precise roles of experience-regulated proteins to the physiology of NCM neurons. Quantitative proteomics approaches, such as 2D-DIGE, hold the promise of becoming central tools in the quest for the sensory-, learning-, and memory-regulated proteome in the songbird brain, and are expected to shed significant light into potential molecular and cellular mechanisms regulated by the auditory processing of behaviorally relevant vocal communication signals.

REFERENCES

1. Hauser, M. D. and Konishi, M. (1999) *The design of animal communication* (Cambridge: MIT Press).
2. Kroodsma, D. E. and Miller, E. H. (1982) *Acoustic communication in birds* (New York: Academic Press).
3. Fitch, R. H., Miller, S. and Tallal, P. (1997) Neurobiology of speech perception, *Annu Rev Neurosci* 20:331–53.
4. Rauschecker, J. P. (1998) Cortical processing of complex sounds, *Curr Opin Neurobiol* 8:516–21.
5. Doupe, A. J. and Kuhl, P. K. (1999) Birdsong and human speech: Common themes and mechanisms, *Annu Rev Neurosci* 22:567–631.
6. Jarvis, E. D. (2004) Learned birdsong and the neurobiology of human language, *Ann N Y Acad Sci* 1016:749–77.
7. Konishi, M. (1965) The role of auditory feedback in the control of vocalization in the white-crowned sparrow, *Z Tierpsychol* 22:770–83.
8. Zeigler, H. P. and Marler, P. (2004) *Behavioral neurobiology of birdsong* (New York: New York Academy of Sciences).
9. Brooks, R. J. and Falls, J. B. (1975) Individual recognition by song in white-throated sparrows. I. Discrimination of songs of neighbors and strangers, *Can J Zool* 53:879–88.
10. Brooks, R. J. and Falls, J. B. (1975) Individual recognition by song in white-throated sparrows. II. Song features used in individual recognition, *Can J Zool* 53:1749–61.

11. Gentner, T. Q. and Hulse, S. H. (2000) Female European starling preference and choice for variation in conspecific male song, *Anim Behav* 59:443–58.
12. Nowicki, S. and Searcy, W. A. (2004) Song function and the evolution of female preferences: Why birds sing, why brains matter, *Ann N Y Acad Sci* 1016:704–23.
13. Ratcliffe, L. and Otter, K. (1996) Sex differences in song recognition. In *Ecology and evolution of acoustic communication in songbirds*, eds. Kroodsma, D. E. and Miller, E. H., 340–55. (Ithaca: Cornell University Press).
14. Searcy, W. A., McArthur, D., Peters, S. and Marler, P. (1981) Response of male song and swamp sparrows to neighbour, stranger and self songs, *Behaviour* 77:152–63.
15. Bottjer, S. W., Brady, J. D. and Cribbs, B. (2000) Connections of a motor cortical region in zebra finches: Relation to pathways for vocal learning, *J Comp Neurol* 420:244–60.
16. Bottjer, S. W., Halsema, K. A., Brown, S. A. and Miesner, E. A. (1989) Axonal connections of a forebrain nucleus involved with vocal learning in zebra finches, *J Comp Neurol* 279:312–26.
17. Nottebohm, F. and Arnold, A. P. (1976) Sexual dimorphism in vocal control areas of the songbird brain, *Science* 194:211–3.
18. Nottebohm, F., Kelley, D. B. and Paton, J. A. (1982) Connections of vocal control nuclei in the canary telencephalon, *J Comp Neurol* 207:344–57.
19. Vates, G. E. and Nottebohm, F. (1995) Feedback circuitry within a song-learning pathway, *Proc Natl Acad Sci U S A* 92:5139–43.
20. Vicario, D. S. (1991) Neural mechanisms of vocal production in songbirds, *Curr Opin Neurobiol* 1:595–600.
21. Wild, J. M. (2004) Functional neuroanatomy of the sensorimotor control of singing, *Ann N Y Acad Sci* 1016:438–62.
22. Wild, J. M. (1997) Neural pathways for the control of birdsong production, *J Neurobiol* 33:653–70.
23. Bottjer, S. W., Miesner, E. A. and Arnold, A. P. (1984) Forebrain lesions disrupt development but not maintenance of song in passerine birds, *Science* 224:901–3.
24. Brainard, M. S. (2004) Contributions of the anterior forebrain pathway to vocal plasticity, *Ann N Y Acad Sci* 1016:377–94.
25. Brainard, M. S. and Doupe, A. J. (2000) Auditory feedback in learning and maintenance of vocal behaviour, *Nat Rev Neurosci* 1:31–40.
26. Scharff, C. and Nottebohm, F. (1991) A comparative study of the behavioral deficits following lesions of various parts of the zebra finch song system: Implications for vocal learning, *J Neurosci* 11:2896–2913.
27. Sohrabji, F., Nordeen, E. J. and Nordeen, K. W. (1990) Selective impairment of song learning following lesions of a forebrain nucleus in the juvenile zebra finch, *Behav Neural Biol* 53:51–63.
28. Johnson, F., Sablan, M. M. and Bottjer, S. W. (1995) Topographic organization of a forebrain pathway involved with vocal learning in zebra finches, *J Comp Neurol* 358:260–78.
29. Luo, M., Ding, L. and Perkel, D. J. (2001) An avian basal ganglia pathway essential for vocal learning forms a closed topographic loop, *J Neurosci* 21:6836–45.
30. Luo, M. and Perkel, D. J. (1999) Long-range GABAergic projection in a circuit essential for vocal learning, *J Comp Neurol* 403:68–84.
31. Vates, G. E., Broome, B. M., Mello, C. V. and Nottebohm, F. (1996) Auditory pathways of caudal telencephalon and their relation to the song system of adult male zebra finches, *J Comp Neurol* 366:613–42.
32. Bottjer, S. W. (2004) Developmental regulation of basal ganglia circuitry during the sensitive period for vocal learning in songbirds, *Ann N Y Acad Sci* 1016:395–415.
33. Bottjer, S. W. and Johnson, F. (1997) Circuits, hormones, and learning: vocal behavior in songbirds, *J Neurobiol* 33:602–18.

34. Farries, M. A. (2004) The avian song system in comparative perspective, *Ann N Y Acad Sci* 1016:61–76.
35. Parent, A. and Hazrati, L. N. (1995) Functional anatomy of the basal ganglia. I. The cortico-basal ganglia-thalamo-cortical loop, *Brain Res Rev* 20:91–127.
36. Perkel, D. J. (2004) Origin of the anterior forebrain pathway, *Ann N Y Acad Sci* 1016:736–48.
37. Marler, P. and Waser, M. S. (1977) Role of auditory feedback in canary song development, *J Comp Physiol Psychol* 91:8–16.
38. Woolley, S. M. (2004) Auditory experience and adult song plasticity, *Ann N Y Acad Sci* 1016:208–21.
39. Woolley, S. M. and Rubel, E. W. (1997) Bengalese finches *Lonchura striata domestica* depend upon auditory feedback for the maintenance of adult song, *J Neurosci* 17:6380–90.
40. Leonardo, A. and Konishi, M. (1999) Decrystallization of adult birdsong by perturbation of auditory feedback, *Nature* 399:466–70.
41. Nordeen, K. W. and Nordeen, E. J. (1992) Auditory feedback is necessary for the maintenance of stereotyped song in adult zebra finches, *Behav Neural Biol* 57:58–66.
42. Cowie, R. and Douglas-Cowie, E. (1983) Speech production in profound postlingual deafness. In *Hearing science and hearing disorders*, eds. Lutman, M. E. and Haggard, M. P., 183–230 (New York: Academic).
43. Cowie, R. and Douglas-Cowie, E. (1992) *Postlingual acquired deafness: Speech deterioration and the wider consequences* (Berlin: Mouton de Gruyter).
44. Cowie, R., Douglas-Cowie, E. and Kerr, A. (1982) A study of speech deterioration in post-lingually deafened adults, *J Laryngol Otol* 96:101–12.
45. Waldstein, R. S. (1990) Effects of postlingual deafness on speech production: Implications for the role of auditory feedback, *J Acoust Soc Am* 88:2099–2114.
46. Koppl, C., Manley, G. A. and Konishi, M. (2000) Auditory processing in birds, *Curr Opin Neurobiol* 10:474–81.
47. Nottebohm, F. (1999) The anatomy and timing of vocal learning in birds. In *The design of animal communication*, ed. Konishi, M., 63–110 (Cambridge: MIT Press).
48. Doupe, A. J. and Konishi, M. (1991) Song-selective auditory circuits in the vocal control system of the zebra finch, *Proc Natl Acad Sci U S A* 88:11339–43.
49. Margoliash, D. (1983) Acoustic parameters underlying the responses of song-specific neurons in the white-crowned sparrow, *J Neurosci* 3:1039–57.
50. Vicario, D. S. and Yohay, K. H. (1993) Song-selective auditory input to a forebrain vocal control nucleus in the zebra finch, *J Neurobiol* 24:488–505.
51. Williams, H. and Nottebohm, F. (1985) Auditory responses in avian vocal motor neurons: A motor theory for song perception in birds, *Science* 229:279–82.
52. Solis, M. M. and Doupe, A. J. (1997) Anterior forebrain neurons develop selectivity by an intermediate stage of birdsong learning, *J Neurosci* 17:6447–62.
53. Solis, M. M. and Doupe, A. J. (1999) Contributions of tutor and bird's own song experience to neural selectivity in the songbird anterior forebrain, *J Neurosci* 19:4559–84.
54. Solis, M. M. and Doupe, A. J. (2000) Compromised neural selectivity for song in birds with impaired sensorimotor learning, *Neuron* 25:109–21.
55. Theunissen, F. E., Amin, N., Shaevitz, S. S. et al. (2004) Song selectivity in the song system and in the auditory forebrain, *Ann N Y Acad Sci* 1016:222–45.
56. Dave, A. S. and Margoliash, D. (2000) Song replay during sleep and computational rules for sensorimotor vocal learning, *Science* 290:812–6.
57. Dave, A. S., Yu, A. C. and Margoliash, D. (1998) Behavioral state modulation of auditory activity in a vocal motor system, *Science* 282:2250–4.
58. Schmidt, M. F. and Konishi, M. (1998) Gating of auditory responses in the vocal control system of awake songbirds, *Nat Neurosci* 1:513–8.

59. Cardin, J. A. and Schmidt, M. F. (2003) Song system auditory responses are stable and highly tuned during sedation, rapidly modulated and unselective during wakefulness, and suppressed by arousal, *J Neurophysiol* 90:2884–99.
60. Prather, J. F., Peters, S., Nowicki, S. and Mooney, R. (2008) Precise auditory-vocal mirroring in neurons for learned vocal communication, *Nature* 451:305–10.
61. Rauske, P. L., Shea, S. D. and Margoliash, D. (2003) State and neuronal class-dependent reconfiguration in the avian song system, *J Neurophysiol* 89:1688–1701.
62. Nottebohm, F. and Arnold, A. P. (1979) Songbirds' brains: Sexual dimorphism, *Science* 206:769.
63. Catchpole, C. K. and Slater, P. J. B. (1995) *Bird song: Biological themes and variations* (Cambridge: Cambridge University Press).
64. Kroodsma, D. E. and Miller, E. H. (1996) *Ecology and evolution of acoustic communication in birds* (Ithaca, NY: Cornell University Press).
65. Mello, C. V., Vates, G. E., Okuhata, S. and Nottebohm, F. (1998) Descending auditory pathways in the adult male zebra finch (*Taeniopygia guttata*), *J Comp Neurol* 395:137–60.
66. Bolhuis, J. J. and Gahr, M. (2006) Neural mechanisms of birdsong memory, *Nat Rev Neurosci* 7:347–57.
67. Gentner, T. Q. (2004) Neural systems for individual song recognition in adult birds, *Ann N Y Acad Sci* 1016:282–302.
68. Mello, C. V., Velho, T. A. and Pinaud, R. (2004) Song-induced gene expression: A window on song auditory processing and perception, *Ann N Y Acad Sci* 1016:263–81.
69. Karten, H. J. and Shimizu, T. (1989) The origins of neocortex: Connections and lamination as distinct events in evolution, *J Cogn Neurosci* 1:291–301.
70. Wild, J. M., Karten, H. J. and Frost, B. J. (1993) Connections of the auditory forebrain in the pigeon (*Columba livia*), *J Comp Neurol* 337:32–62.
71. Chew, S. J., Mello, C., Nottebohm, F., Jarvis, E. and Vicario, D. S. (1995) Decrements in auditory responses to a repeated conspecific song are long-lasting and require two periods of protein synthesis in the songbird forebrain, *Proc Natl Acad Sci U S A* 92:3406–10.
72. Chew, S. J., Vicario, D. S. and Nottebohm, F. (1996) A large-capacity memory system that recognizes the calls and songs of individual birds, *Proc Natl Acad Sci U S A* 93:1950–5.
73. Muller, C. M. and Leppelsack, H. J. (1985) Feature extraction and tonotopic organization in the avian auditory forebrain, *Exp Brain Res* 59:587–99.
74. Sen, K., Theunissen, F. E. and Doupe, A. J. (2001) Feature analysis of natural sounds in the songbird auditory forebrain, *J Neurophysiol* 86:1445–58.
75. Terleph, T. A., Mello, C. V. and Vicario, D. S. (2006) Auditory topography and temporal response dynamics of canary caudal telencephalon, *J Neurobiol* 66:281–92.
76. Terleph, T. A., Mello, C. V. and Vicario, D. S. (2007) Species differences in auditory processing dynamics in songbird auditory telencephalon, *Dev Neurobiol* 67:1498–1510.
77. Herdegen, T. and Leah, J. D. (1998) Inducible and constitutive transcription factors in the mammalian nervous system: Control of gene expression by Jun, Fos and Krox, and CREB/ATF proteins, *Brain Res Brain Res Rev* 28:370–490.
78. Kaczmarek, L. and Robertson, H. A. (2002) *Immediate early genes and inducible transcription factors in mapping of the central nervous system function and dysfunction* (Amsterdam: Elsevier Science B.V.).
79. Pinaud, R. (2004) Experience-dependent immediate early gene expression in the adult central nervous system: Evidence from enriched-environment studies, *Int J Neurosci* 114:321–33.
80. Pinaud, R. and Tremere, L. A. (2006) *Immediate early genes in sensory processing, cognitive performance and neurological disorders* (New York: Springer-Verlag).

81. Mello, C. V. and Pinaud, R. (2006) Immediate early gene regulation in the auditory system. In *Immediate early genes in sensory processing, cognitive performance and neurological disorders*, eds. Pinaud, R. and Tremere, L. A., 35–56 (New York: Springer-Verlag).

82. Mello, C. V., Vicario, D. S. and Clayton, D. F. (1992) Song presentation induces gene expression in the songbird forebrain, *Proc Natl Acad Sci U S A* 89(15):6818–22.

83. Mello, C., Nottebohm, F. and Clayton, D. (1995) Repeated exposure to one song leads to a rapid and persistent decline in an immediate early gene's response to that song in zebra finch telencephalon, *J Neurosci* 15:6919–25.

84. Bolhuis, J. J., Hetebrij, E., Den Boer-Visser, A. M., De Groot, J. H. and Zijlstra, G. G. (2001) Localized immediate early gene expression related to the strength of song learning in socially reared zebra finches, *Eur J Neurosci* 13:2165–70.

85. Bolhuis, J. J., Zijlstra, G. G., den Boer-Visser, A. M. and Van Der Zee, E. A. (2000) Localized neuronal activation in the zebra finch brain is related to the strength of song learning, *Proc Natl Acad Sci U S A* 97:2282–5.

86. Terpstra, N. J., Bolhuis, J. J. and den Boer-Visser, A. M. (2004) An analysis of the neural representation of birdsong memory, *J Neurosci* 24:4971–7.

87. Phan, M. L., Pytte, C. L. and Vicario, D. S. (2006) Early auditory experience generates long-lasting memories that may subserve vocal learning in songbirds, *Proc Natl Acad Sci U S A* 103:1088–93.

88. Gobes, S. M. and Bolhuis, J. J. (2007) Birdsong memory: A neural dissociation between song recognition and production, *Curr Biol* 17:789–93.

89. Mello, C. V. and Ribeiro, S. (1998) ZENK protein regulation by song in the brain of songbirds, *J Comp Neurol* 393:426–38.

90. Nastiuk, K. L., Mello, C. V., George, J. M. and Clayton, D. F. (1994) Immediate-early gene responses in the avian song control system: Cloning and expression analysis of the canary c-jun cDNA, *Brain Res Mol Brain Res* 27:299–309.

91. Velho, T. A., Pinaud, R., Rodrigues, P. V. and Mello, C. V. (2005) Co-induction of activity-dependent genes in songbirds, *Eur J Neurosci* 22:1667–78.

92. Huesmann, G. R. and Clayton, D. F. (2006) Dynamic role of postsynaptic caspase-3 and BIRC4 in zebra finch song-response habituation, *Neuron* 52:1061–72.

93. Cheng, H. Y. and Clayton, D. F. (2004) Activation and habituation of extracellular signal-regulated kinase phosphorylation in zebra finch auditory forebrain during song presentation, *J Neurosci* 24:7503–13.

94. Pinaud, R., Osorio, C., Alzate, O. and Jarvis, E. D. (2008) Profiling of experience-regulated proteins in the songbird auditory forebrain using quantitative proteomics, *Eur J Neurosci* 27:1409–22.

95. Alban, A., David, S. O., Bjorkesten, L. et al. (2003) A novel experimental design for comparative two-dimensional gel analysis: Two-dimensional difference gel electrophoresis incorporating a pooled internal standard, *Proteomics* 3:36–44.

96. Friedman, D. B., Hill, S., Keller, J. W. et al. (2004) Proteome analysis of human colon cancer by two-dimensional difference gel electrophoresis and mass spectrometry, *Proteomics* 4:793–811.

97. Osorio, C., Sullivan, P. M., He, D. N. et al. (2007) Mortalin is regulated by APOE in hippocampus of AD patients and by human APOE in TR mice, *Neurobiol Aging* 28:1853–62.

98. Catterall, W. A. (1999) Interactions of presynaptic Ca2+ channels and snare proteins in neurotransmitter release, *Ann N Y Acad Sci* 868:144–59.

99. Zamponi, G. W. (2003) Regulation of presynaptic calcium channels by synaptic proteins, *J Pharmacol Sci* 92:79–83.

100. Hilfiker, S., Pieribone, V. A., Czernik, A. J. et al. (1999) Synapsins as regulators of neurotransmitter release, *Philos Trans R Soc Lond B Biol Sci* 354:269–79.

101. Pinaud, R., Terleph, T. A. and Tremere, L. A. (2005) Neuromodulatory transmitters in sensory processing and plasticity in the primary visual cortex. In *Plasticity in the visual system: From genes to circuits*, eds. Pinaud, R., Tremere, L. A. and De Weerd, P., 127–51 (New York: Springer-Verlag).

102. Cole, A. J., Saffen, D. W., Baraban, J. M. and Worley, P. F. (1989) Rapid increase of an immediate early gene messenger RNA in hippocampal neurons by synaptic NMDA receptor activation, *Nature* 340:474–6.

103. Pinaud, R. (2005) Critical calcium-regulated biochemical and gene expression programs involved in experience-dependent plasticity. In *Plasticity in the visual system: From genes to circuits*, eds. Pinaud, R., Tremere, L. A. and De Weerd, P., 153–80 (New York: Springer-Verlag).

104. Pinaud, R., Velho, T. A., Jeong, J. K. et al. (2004) GABAergic neurons participate in the brain's response to birdsong auditory stimulation, *Eur J Neurosci* 20:1318–30.

105. Wisden, W., Errington, M. L., Williams, S. et al. (1990) Differential expression of immediate early genes in the hippocampus and spinal cord, *Neuron* 4:603–14.

106. Collingridge, G. L. and Singer, W. (1990) Excitatory amino acid receptors and synaptic plasticity, *Trends Pharmacol Sci* 11:290–6.

107. Karmarkar, U. R. and Buonomano, D. V. (2002) A model of spike-timing dependent plasticity: One or two coincidence detectors? *J Neurophysiol* 88:507–13.

108. Pinaud, R., Tremere, L. A., Penner, M. R. et al. (2002) Complexity of sensory environment drives the expression of candidate-plasticity gene, nerve growth factor induced-A, *Neuroscience* 112:573–82.

109. Tang, Y. P., Shimizu, E., Dube, G. R. et al. (1999) Genetic enhancement of learning and memory in mice, *Nature* 401:63–9.

110. Yuste, R., Majewska, A., Cash, S. S. and Denk, W. (1999) Mechanisms of calcium influx into hippocampal spines: Heterogeneity among spines, coincidence detection by NMDA receptors, and optical quantal analysis, *J Neurosci* 19:1976–87.

111. Grewal, S. S., York, R. D. and Stork, P. J. (1999) Extracellular-signal-regulated kinase signalling in neurons, *Curr Opin Neurobiol* 9:544–53.

112. Farnsworth, C. L., Freshney, N. W., Rosen, L. B. et al. (1995) Calcium activation of Ras mediated by neuronal exchange factor Ras-GRF, *Nature* 376:524–7.

113. Carey, K. D., Watson, R. T., Pessin, J. E. and Stork, P. J. (2003) The requirement of specific membrane domains for Raf-1 phosphorylation and activation, *J Biol Chem* 278:3185–96.

114. Kyosseva, S. V. (2004) Mitogen-activated protein kinase signaling, *Int Rev Neurobiol* 59:201–20.

115. Xia, Z. and Storm, D. R. (2005) The role of calmodulin as a signal integrator for synaptic plasticity, *Nat Rev Neurosci* 6:267–76.

116. Hindley, A. and Kolch, W. (2002) Extracellular signal regulated kinase (ERK)/mitogen activated protein kinase (MAPK)-independent functions of Raf kinases, *J Cell Sci* 115:1575–81.

117. English, J. M. and Cobb, M. H. (2002) Pharmacological inhibitors of MAPK pathways, *Trends Pharmacol Sci* 23:40–5.

118. Fukunaga, K. and Miyamoto, E. (1998) Role of MAP kinase in neurons, *Mol Neurobiol* 16:79–95.

119. Petersohn, D., Schoch, S., Brinkmann, D. R. and Thiel, G. (1995) The human synapsin II gene promoter. Possible role for the transcription factor zif268/egr-1, polyoma enhancer activator 3, and AP2, *J Biol Chem* 270:24361–9.

120. Stripling, R., Volman, S. F. and Clayton, D. F. (1997) Response modulation in the zebra finch neostriatum: Relationship to nuclear gene regulation, *J Neurosci* 17:3883–93.

15 Applications of Proteomics to Nerve Regeneration Research

Mark W. Massing, Grant A. Robinson,
Christine E. Marx, Oscar Alzate, and
Roger D. Madison

CONTENTS

15.1 PERIPHERAL NERVE INJURY: BACKGROUND AND CLINICAL SIGNIFICANCE

Peripheral nerve injury is a major clinical and public health challenge. Although a common and increasingly prevalent wartime condition (1), injury to peripheral nerves, plexuses, and roots is present in 5% of patients seen in civilian trauma centers (2). In one study, almost half of peripheral nerve injuries at trauma centers were due to motor vehicle accidents and about half required surgery (3). Peripheral nerve injuries can substantially impact quality of life through loss of function and increased risk of secondary disabilities from falls, fractures, and other injuries (2).

Neurons are connected in intricate communication networks established during development to convey sensory information from peripheral receptors of sensory neurons to the central nervous system (the brain and spinal cord), and to convey commands from the central nervous system to effector organs such as skeletal muscle innervated by motor neurons. The peripheral nerve environment is quite complex, consisting of axonal projections from neurons, supporting cells such as Schwann cells and fibroblasts, and the blood supply to the nerve.

Connective tissue known as endoneurium surrounds peripheral nerve axons. Within peripheral nerves, axons are grouped into fascicles surrounded by connective tissue known as perineurium. Between and surrounding groups of fascicles is the epineurium. Microvessel plexuses course longitudinally through the epineurium and send branches through the perineurium to form a vascular network of capillaries in the endoneurium (4).

The primary supporting cell for peripheral nerves is the Schwann cell. Schwann cells wrap around axons in a spiral fashion multiple times and their plasma membranes form a lipid-rich tubular cover around the axon known as the myelin sheath or the neurilemma. Schwann cells and the myelin sheath support and maintain axons and help to guide axons during axonal regeneration following nerve injury (5).

It has been known for quite some time that regenerating axons exhibit a strong preference for growing along the inside portion of remaining basal lamina tubes in the distal nerve stump, the well-characterized "bands of Bungner" (6–10). Schwann cells originally associated with myelinated axons form such bands all the way from the transection site to the distal end organ target. The critical concept here is that the eventual distal destination of regenerating axons is largely determined by the Schwann cell tubes as they enter at the nerve transection site (9,11,12).

Recent elegant work with transgenic mice expressing fluorescent proteins in their axons has verified that most (but not all) regenerating axons distal to either a crush or a transection injury remain within a single Schwann cell band as they grow within the distal nerve stump (13,14). Within the motor system it has even been shown that the same axon predominantly reinnervates the same neuromuscular junctions, and that Schwann cell bands act as mechanical barriers to direct axon outgrowth (13).

The neuronal cell body is the site of synthesis of virtually all proteins and organelles in the cell. A complex process known as anterograde transport continuously moves materials from the neuronal cell body via the axon to its terminal synapse. These transported substances include neurotransmitters that facilitate communications between the neuron and end organ tissue across a narrow extracellular space known as the synaptic cleft (5) or, as in the case of the motor neuron innervation of muscle, the neuromuscular junction (15).

Conversely, end organs such as muscle produce substances that act as nerve growth factors. These make their way across the neuromuscular junction to the innervating motor neuron axon (16). Some of these substances, or chemical messengers induced by them, are packaged and conveyed by retrograde transport from the synapse via the axon to the neuronal cell body. In this manner, the neuron and its end organ are continuously informed about the status of the connection between them. It has been suggested that information from end organs takes the form of factors that sustain existing nerve cell connections and promote the regeneration of damaged nerve cells. For instance, it has long been known that muscle exerts a strong influence on developing and regenerating motor neurons, and we have recently shown that even within an individual muscle there are factors that can influence the accuracy of reinnervation (17). Recent work has elegantly shown that if a single muscle fiber is selectively lesioned, the motor neuron axon terminal making up the proximal side of the neuromuscular junction rapidly atrophies and withdraws from the muscle postsynaptic sites within a matter of hours (18).

The clinical significance, prognosis, and treatment of peripheral nerve injury depend on the site and extent of the injury. Despite regeneration, extensive peripheral nerve injuries can result in the effective paralysis of the entire limb or distal portions of the limb. Two peripheral nerve injury classification schemes, the Seddon (19) and the Sunderland (20), are in common use. These classify nerve injury according to whether the injury was confined to demyelination only or a more severe disruption of axons and supporting connective tissue. According to Seddon, the most severe injuries are classified as axonotmesis and neurotmesis. Axonotmesis is a nerve injury characterized by axon disruption rather than destruction of the connective tissue framework. The connective tissue and Schwann tubes are relatively intact. This is typical of stretch injuries common in falls and motor vehicle accidents. In contrast, neurotmesis involves the disruption of the nerve trunk and the connective tissue structure. This would occur in injuries where the nerve has been completely severed or badly crushed.

Prognosis is good in peripheral nerve injuries where endoneurial Schwann cell tubes remain intact. Disruption of the Schwann cell tubes results in the loss of established pathways that regenerating axons follow. For extensive injuries, surgery is usually necessary to remove damaged nerve tissue and join viable nerve ends by direct anastomosis or by a nerve tissue graft (1). Refinement of microsurgical techniques involving the introduction of the surgical microscope and microsutures has increased the accuracy of this mechanical process, yet only 10% of adults will recover normal nerve function using state-of-the-art current techniques (21–23). The limits of microsurgical techniques have been reached; this is not surprising given that the finest suture material and needles (18–22 and 50–75 microns, respectively) are still quite a bit larger than the smallest axons that need to be repaired. The major key to recovery of function following peripheral nerve lesions is the accurate regeneration of axons to their original target end organs. A recognized leader of clinical nerve repair once stated, "The core of the problem is not promoting axon regeneration, but in getting them back to where they belong" (Sunderland, 1991) (23).

At the level of a mixed peripheral nerve where motor and sensory axons are intermixed, correct discriminatory choices for appropriate terminal nerve branches at the lesion site are necessary prerequisites for the subsequent successful reinnervation of appropriate end-organ targets. Motor axons previously innervating muscle may be misdirected to sensory organs, and sensory axons typically innervating skin can be misdirected to muscle. Misdirected regeneration is a major barrier to functional recovery.

In order to understand axonal regeneration and the mechanisms that axons use to navigate to target tissues, our laboratory has conducted a series of studies that are now culminating in proteomic investigations to identify specific biochemical mediators that may be the underlying mechanisms that direct accurate axon regeneration. We describe our work and that of others in the development of a model of axonal regeneration in the rodent femoral nerve and what we have learned from it. Then we will lay out our current research direction illustrating how approaches in proteomics such as two-dimensional differential gel electrophoresis (2D-DIGE) and mass spectrometry can be used to identify the underlying mediators that may lead to new therapies for peripheral nerve injury.

15.2   FEMORAL NERVE REGENERATION MODEL

15.2.1   MOTOR AXON REGENERATION ACCURACY

The rodent femoral nerve is an elegant model to examine motor neuron reinnervation accuracy. Two terminal nerve pathways roughly equal in size, one to the skin and the other to the quadriceps muscle, are intermixed at the level of the parent nerve, but bifurcate distally into distinct terminal nerve branches: a sensory cutaneous branch (continuing as the saphenous nerve and providing local cutaneous innervation) and a muscle branch to the quadriceps.

Weiss and Edds originally introduced the rat femoral nerve in 1945 as a model system to study the fate of axons that originally innervated muscle or skin when they were forcibly misdirected into the inappropriate nerve branch (24). When motor axons were forced into the cutaneous branch (and vice versa) via nerve crosses, axons were able to grow and survive in the foreign territory for extended periods of time. This led these researchers to conclude that neurotropism within the context of distal nerve stumps "has been ruled out conclusively" [(24), p. 173]. The original femoral nerve model has more recently been modified to look at the fate of regenerating axons when they are given equal access to the terminal cutaneous and muscle branches, rather than being forced into foreign nerve stumps. In contrast to the conclusions of Weiss and Edds (1945), the outcome of these equal access experiments showed a preference for regenerating motor axons to reinnervate their original terminal nerve branch, a process that has been termed "preferential" motor reinnervation (PMR) (Figure 15.1) (25,26).

In the normal femoral nerve no motor neurons project into the cutaneous branch. Thus reinnervation of this distal branch by regenerating motor neurons represents a failure of specificity, which can be quantified by retrograde tracing. Previous work with this model system in both rats and mice when muscle and skin contact are maintained has revealed that motor axons initially grow equally into both nerve branches but over time are preferentially retained in the muscle branch, thus resulting in PMR (27–33). Conversely, we have recently shown that when muscle contact is denied but the cutaneous branch remains intact to skin, the cutaneous branch now becomes the preferred terminal nerve branch for motor neuron projections (33,34), suggesting that regeneration accuracy is highly dependent on the accessibility options allowed by a particular surgical preparation.

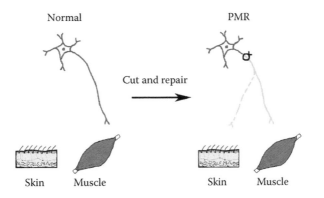

FIGURE 15.1 Axon regeneration from motor neurons illustrating preferential motor reinnervation (PMR) in the femoral nerve. This nerve divides distally into two terminal nerve branches: a cutaneous branch and a motor branch to the quadriceps muscle. Normally, as shown in the left hand panel, motor neurons project their axons exclusively to the quadriceps muscle. After nerve transection and repair motor axons initially grow into both nerve branches, but over time are preferentially retained in the muscle branch. This is partially explained by a process of pruning motor axon collaterals from the inappropriate cutaneous branch (*dashed line* projecting to skin).

15.2.2 INVESTIGATING RELATIVE ROLES OF PATHWAYS AND END ORGANS IN PREFERENTIAL MOTOR REINNERVATION

The above studies indicate that both the end organ and the pathway can influence PMR. We have conducted a series of studies to assess the relative roles of these two sources of influence by developing surgical models that manipulate relative levels of trophic support in the muscle and cutaneous nerve branches. In all of these studies, the measurement of regeneration accuracy is the number of motor neurons that are retrogradely labeled from just one of the two terminal pathways, or simultaneously from both pathways. This is determined as follows. After the predetermined survival period following the initial proximal nerve repair (8 weeks in Figure 15.2), the two terminal nerve branches (which project either to muscle or to skin) are re-exposed and separated from each other by food-grade silicone grease dams, trimmed to ~3 mm distal to the normal parent nerve bifurcation, and randomly assigned to receive crystals of either fluorescein dextran (D-1820, Molecular Probes, Eugene, OR, USA) or tetramethylrhodamine dextran (D-1817, Molecular Probes). After crystal application, each branch is blotted and sealed with silicone grease, and the surgical site is closed in layers. Three days later, animals receive an overdose of anesthetic and are perfused through the heart with 0.1 M phosphate-buffered saline (PBS, pH 7.4) followed by 4% paraformaldehyde in PBS. The lumbar spinal cord is removed, postfixed for several hours, and sucrose protected overnight. The cord is frozen on dry ice and stored at −80°C until being sectioned with a cryostat. Serial 25-μm frozen sections are thawed in PBS, mounted onto glass slides, briefly air-dried, and coverslipped with Prolong according to the manufacturer's instructions (P-7481, Molecular Probes).

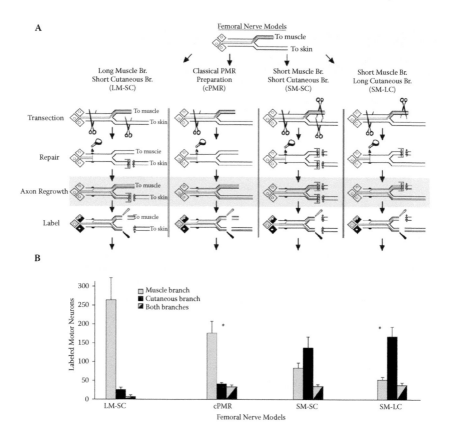

FIGURE 15.2 Data obtained eight weeks after femoral nerve repair in adult rats. (A) All animals received parent femoral nerve transection and repair. In the *long muscle* branch and *short cutaneous* branch model (LM-SC; *left-most column*) the cutaneous branch was also transected, ligated, and capped while the muscle branch was left intact to the quadriceps muscle. In the classical preferential motor reinnervation (PMR), both terminal branches remain intact to their respective end organs of muscle and skin. In the *short muscle* and *short cutaneous* model (SM-SC) all end-organ contact was prevented by transecting, ligating, and capping both terminal branches. In the *short muscle* and *long cutaneous* model (SM-LC; *right-most column*) the muscle branch was also transected, ligated, and capped while the cutaneous branch was left intact to skin. In all models, motor neurons reinnervating each branch were retrogradely labeled eight weeks after the nerve repaired by application of labeled dextran tracers to the two terminal nerve branches just distal to the bifurcation of the nerve. (B) Average motor neuron counts. Labeled motor neurons that contained only one of the labels were counted (indicating projection to a single terminal nerve branch), or both labels, indicating a simultaneous projection to both nerve branches. When muscle contact is allowed (LM-SC and C-PMR) more motor neurons project to the muscle branch. When muscle contact is prevented (SM-SC and SM-LC), more motor neurons project to the cutaneous branch. * = significant differences within each model for number of motor neurons labeled from the muscle and cutaneous branches (ANOVA, and post-hoc Student-Newman-Keuls, $p < 0.05$). Error bars = SEM. *Source*: Adapted from Madison, R. D., Robinson, G. A. and Chadaram, S. R. (2007) The specificity of motor neurone regeneration (preferential reinnervation), *Acta Physiol (Oxf)* 189:201–6.

Retrogradely labeled motor neurons containing a nucleus are identified using a composite filter set that allows simultaneous visualization of both labels (#51006, Chroma Technology, Brattleboro, VT, USA) in a fluorescence-equipped Zeiss Axiophot microscope, at 250× magnification. Labeling with one or both fluorophores is verified by single filters tailored to appropriate absorption and emission spectra (Chroma Technology). Motor neuron counts are carried out by blinded independent observers and scored as either single labeled (fluoroscein or tetramethylrhodamine only) or double labeled (both fluorescein and tetramethylrhodamine). Neurons are tabulated based on staining as projecting to the muscle or cutaneous pathway only, or as projecting to both pathways. Counting variation among observers is ~2%. Neuron counts are corrected for split cells in microtome sections (35). Although there are certainly newer methods to correct for split cell counts, we use the Abercrombie method in order to relate our work to previous works in this field by other laboratories. Previously published work from our laboratory has documented the validity and reliability of these methods (e.g., lack of cross-contamination due to tracer leakage, intra- and inter-counter reliability, etc. (34,36).

One of our working hypotheses regarding the development of PMR in the femoral nerve is that there is a hierarchy of trophic support for regenerating motor neurons, with muscle contact being the highest, followed by the length of the terminal nerve branch and/or contact with skin. Because we hypothesized that muscle contact by regenerating motor axons results in the greatest level of trophic support, we examined the effect on PMR when muscle contact was allowed, but skin contact was prevented by capping the cutaneous branch; a preparation referred to as long-muscle:short cutaneous. This preparation allows trophic support from muscle and the muscle branch without the competing influence of contact with skin or a long cutaneous nerve branch. The results were an extremely robust bias for regenerating motor neurons to project to the muscle branch, with approximately a 9:1 ratio (Figure 15.2, far left column). Conversely, when we allowed skin contact but prevented muscle contact (a preparation referred to as short-muscle:long *cutaneous*) the ratio of regenerating motor neuron projections was reversed, with approximately a 1:3 ratio of motor neurons projecting to the muscle versus the cutaneous branch (far right column of Figure 15.2). The only surgical model that did not result in a significant difference in the number of motor neurons projecting to the two terminal nerve branches was when contact with both end organs was prevented, a preparation expected to make the level of trophic support in both terminal branches the most similar (referred to as short-muscle:short cutaneous).

These findings are consistent with the hypotheses that motor neurons compare relative levels of trophic support from various distal targets, and that there is a hierarchy of trophic support for regenerating motor axons with muscle contact being the highest, followed by the length of the terminal nerve branch and/or contact with skin.

These studies suggest several important aspects of PMR. Since initial axonal growth into the two terminal branches appears to be random, it is not likely that mechanical guidance plays a major role in PMR. The finding that the muscle branch becomes a more hospitable environment for motor axons over time suggests a role for attractive factors in the muscle branch or repulsive factors in the cutaneous branch.

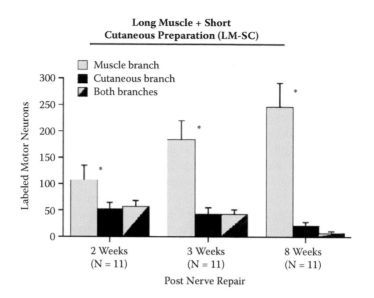

FIGURE 15.3 Development of pathway choice in adult rats after femoral nerve repair. When the influence of the cutaneous pathway/skin was removed, preferential reinnervation of the muscle pathway was more pronounced and was seen beginning at 2 weeks (mean ± SEM, ANOVA, Student-Newman-Keuls, muscle vs. cutaneous branch, * = $p < 0.05$). *Source*: Adapted from Uschold, T., Robinson, G. A., and Madison, R. D. (2007) Motor neuron regeneration accuracy: Balancing trophic influences between pathways and end-organs, *Experimental Neurology* (Elsevier) 205:250–6.

We next focused on the time course of the development of preferential projections in the long-muscle:short cutaneous surgical model, where muscle contact is allowed, but skin contact is prevented by capping the cutaneous branch. We found that preferential projections developed as early as two weeks following the initial surgery [Figure 15.3, from (36)].

Given the rapidity with which preferential projections were established in the muscle branch (i.e., by two weeks), we decided to investigate the importance of maintaining the connection to muscle during this early regeneration time point. Given the fact that retrograde transport is known to continue within denervated rodent nerves for approximately 48 hours (37–42), and that the rate of retrograde transport is at least several millimeters/hour [ibid, see also (43)], muscle-derived signals could easily reach the proximal femoral nerve repair site if they were moved via retrograde transport. We therefore decided to investigate the role of retrograde transport distal to the proximal femoral nerve using the long-muscle:short-cutaneous model that demonstrated PMR at two weeks. The two-week time point is also important because we wanted to minimize the possible influence of regenerating motor neurons physically contacting muscle. By two weeks motor axons are just beginning to reach the quadriceps muscle and motor endplate reinnervation is very rare (27).

Colchicine disrupts microtubule function and has previously been used by many laboratories as a means of disrupting retrograde transport; it is typically used at a concentration of 1–50 mM [e.g., (44–47)]. At the same time as the proximal femoral nerve repair, we injected 3 µL of colchicine (25 mM, in saline) into the distal muscle branch using a pulled glass micro-pipette and a Hamilton syringe. As a control for possible damage due to the injection procedure another group of animals received injection of saline alone. A third group of animals received a crush of the distal muscle branch. The crush was carried out using fine needle holders with a 0.5-mm tip. Pressure was applied for 15 seconds, the orientation of the needle holder reversed, and pressure applied for another 15 seconds. The completeness of the crush procedure was assessed visually; the nerve crush site became completely translucent in all animals. Animals survived for two weeks and then were processed for retrogradely labeled motor neurons as described above.

It is important to keep in mind that the manipulations to the distal muscle branch are carried out immediately after the femoral nerve repair, during the same surgical exposure. Or, in other words, to a nerve that has already had its axons severed more proximally. Both the colchicine and crush groups failed to show PMR at two weeks, whereas the saline injection group did show PMR. The variability of the colchicine group was greater than the crush group as might be expected from a somewhat variable disruption of retrograde transport. The demonstration of PMR in the saline injection group repeats the initial finding above of PMR at two weeks using the long-muscle:short-cutaneous model (Figure 15.4).

Within the confines of this model system, these findings show that when retrograde transport is disrupted between the proximal femoral nerve lesion site and distal muscle, PMR fails to develop. The results of all of the above work suggest

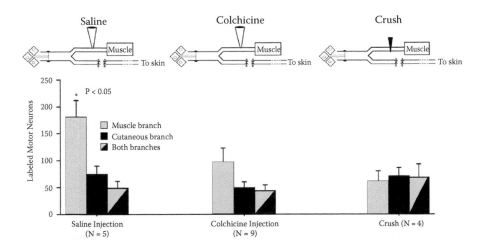

FIGURE 15.4   PMR at two weeks in adult rats is dependent on retrograde transport from the denervated muscle branch. Animals received the long-muscle:short-cutaneous (LM-SC) surgical preparation. In addition, at the same time as the proximal femoral nerve repair the distal muscle branch was injected with saline (control), or colchicine, or received a crush injury. Two weeks later we quantified the number of motor neurons projecting to the two terminal nerve branches. PMR was evident only in the saline group, reproducing the initial finding shown in Figure 15.3 for the two-week time point.

that contact with, or influence from, muscle is a major determinant of motor neuron regeneration accuracy.

15.2.3 Target-Derived Trophic Support and Trophomorphism

As briefly discussed above, a generalized concept of neuronal development that has received much interest over the past several decades has been that neurons require target-derived trophic support for survival (48–50). A related concept is that of *trophomorphism*, which suggests that competition among sibling axonal branches determines which axonal collaterals survive, and this is based on the relative amounts of trophic support to each of the axon collaterals (51,52).

Our findings are consistent with the hypothesis that *trophomorphism* plays a prominent role in PMR. This is especially the case since, during regeneration, axons from the regenerating motor neurons have access to both terminal nerve branches. We have shown that when muscle contact is allowed, significantly more motor neurons project to the distal muscle branch; however, when muscle end-organ contact is prevented, motor axons preferentially reinnervate the foreign cutaneous terminal pathway (Figure 15.2). This suggests that motor neurons assess the level of trophic support from each of the terminal branches and over time become preferentially located in the one that provides the greater amount of trophic support. This explanation is also consistent with the classical work of Campenot showing that neurons retain their axons in compartments with relatively higher levels of trophic support (53,54). Studies from other laboratories using Y-shaped tubes also offer support for the assertion that motor neurons assess relative levels of trophic support and respond accordingly. Regenerating axons have been shown to distinguish between nerve and non-nerve targets, suggesting tissue specificity with respect to attracting and retaining regenerating axons. These studies have also shown that distal nerve volume influences axonal growth (55).

This strong evidence from surgical experiments is consistent with models where trophomorphism plays a prominent role in PMR, especially for motor neurons. Trophic support implies the presence of soluble factors migrating from end organs in a retrograde fashion. Interestingly, it has been known for almost 30 years that the endocytotic activity of skeletal muscle increases after denervation (56). Subsequent work demonstrated a pairing between the increased endocytotic activity with significant increases in exocytotic activity (57). These observations have led to the suggestion that the high exocytotic activity of denervated muscle is due to the increased secretion of neurotrophic factors by the muscle in an attempt to stimulate reinnervation by regenerating axons (58–62). Consistent with such an interpretation is the finding that the expression levels of several neurotrophic factors are indeed increased following either nerve or muscle damage [e.g., hepatocyte growth factor (HGF), fibroblast growth factor (FGF), brain-derived neurotrophic factor (BDNF) (63–69)]. The increased endocytotic/exocytotic activity occurs predominantly at the endplate region of the muscle (70), where it would be ideally situated to pass muscle-derived factors across the neuromuscular junction to the preterminal motor axon.

A major ongoing component of our work focuses on the identification of mediators of trophomorphism within the femoral nerve model. We are taking advantage of novel techniques developed in the field of proteomics. In the remainder of this chapter we consider the various proteomic methodologies and explore how they are used in our laboratory to understand biochemical mediators of axonal regeneration and PMR.

15.3 ROLE OF PROTEOMICS IN IDENTIFYING MEDIATORS OF AXONAL REGENERATION AND PREFERENTIAL MOTOR REINNERVATION

15.3.1 RESEARCH GOALS AND RELEVANCE OF PROTEOMICS

Findings from surgical models and other evidence described previously are consistent with a prominent role for unknown trophic mediators in PMR. It is thought that end organs produce soluble protein trophic factors that either promote or inhibit axonal growth. Levels of these factors are believed to be altered in response to nerve injury, making a terminal nerve branch either more or less attractive to regenerating motor neurons. Trophic factors may be expressed by supporting cells such as Schwann cells, but previously described experimental evidence suggests that muscle tissue has a particularly strong trophic effect.

The objective of our current investigations is to identify the soluble proteins produced by muscle tissue in response to denervation. Specifically, we are looking for proteins that demonstrate changes in expression or are released secondarily to nerve damage. We reason that some subset of these may be involved in axonal regeneration and PMR. This task has only recently become technically feasible with the development of powerful proteomic techniques for biomarker discovery and protein identification.

Biomarker discovery involves the comparison of entire proteomes of sample tissues that differ according to some known process. In our case, we will compare proteomes of normal tissue to those of tissue involved in nerve injury. The objective is to identify protein expression differences that enable us to discover small subsets of proteins somehow involved in nerve regeneration.

15.3.2 POWER AND LIMITATIONS OF PROTEOMICS

Fundamental challenges in biomarker discovery for neuroproteomics research include the immense size and variability of the neuroproteome, phenotypic variations as cells interact with their environment, and the wide range of relative protein abundances. The size of the proteome even within a single cell is immense. Lefkovits et al. (71) have estimated that a single B-lymphoblast cell contains 10^9 protein molecules of 4000 different molecular species. Frequently, complex tissues rather than single cell lines are examined, further increasing the number of protein species. The estimated size of the proteome, all proteins potentially expressed from the genome, is far larger and may exceed 100,000 in humans (72). This implies great potential phenotypic variation between tissues and within cell lines under varying conditions. Add to this the variation introduced by post-translational modification and the magnitude of the challenge becomes truly impressive. Proteomic techniques must also manage the wide range of protein abundances where just a few hundred protein species predominate, and species most likely of interest to researchers are relatively rare. The abundance range comparing the most common to the more rare protein species has been estimated to span seven to eight orders of magnitude (71).

Historically protein chemistry focused on one or a few proteins of interest. Laboratory techniques such as chromatography, electrophoresis, and affinity columns were developed to identify, quantify, and characterize individual proteins. But proteomics required the development of laboratory techniques, information systems, and statistical approaches to process hundreds to thousands of proteins simultaneously. The theoretical advantage of proteomics is in the ability to simultaneously examine all proteins in a biological system so that relations among them may be explored under different conditions. This is a theoretical advantage because, in practice, no proteomic technique has the resolution to examine all proteins, especially those in low abundance. These undetectable low-abundance proteins my make up 70% of the expressed proteome (71).

During the early 1960s, Norman G. Anderson proposed the Molecular Anatomy (MAN) program to catalog the human proteome (73,74). Parts of this program were initiated at Oak Ridge National Laboratory and included many of the objectives of proteomic programs today such as the identification of all cellular proteins, their polymorphisms, their structure and functions, their location, and their chemical properties. Limitations in technology, funding, and a public focus on the Human Genome Project delayed the realization of this dream. In many respects we are only now beginning to realize the ambitious goals set out by Anderson almost a half century ago.

The goals of the MAN project were likely ahead of their time. Their realization awaited the development of techniques that enable large numbers of proteins to be processed simultaneously. In the mid-1970s such a breakthrough occurred in the form of two-dimensional gel electrophoresis (2-DE) (75) (for a discussion of 2-DE see Chapters 3 and 4). By the end of that decade, 2-DE was greatly refined and standardized, leading to proposals for its use in the development of a human protein index (HPI) consistent with the basic concepts of the MAN project (71). Proteomics has been described as the "rebellious child of 2-DE" (73), suggesting a prominent role for 2-DE as a foundation technique in the field. Although automated and refined, the basic concept of 2-DE has changed little since it was first described (75). In his seminal manuscript, O'Farrell reported the ability to resolve a maximum of 5000 proteins with 2-DE (75). This was the first technique to simultaneously evaluate large numbers of proteins, and it led to the advances that make modern protein biomarker discovery possible.

Although a breakthrough, 2-DE has many limitations that make it less than optimal in modern proteomic investigations in biomarker discovery. As proteins migrate across the gel they separate based on molecular weight and chemical properties. The location, shape, and size of the resulting protein spots are greatly influenced by experimental conditions. This is critically important because protein identification and quantification depend on the ability to locate spots on the gel and to measure their exact size. Standardizing experimental protocols and equipment and controlling for variation among gels, experiments, and laboratories have been major challenges. This is especially true when different experimental samples are compared on different gels. This lack of reproducibility among gels is a major limitation of 2-DE that has been solved with the introduction of 2D-DIGE (see Chapter 4, and Section 15.4.2). A number of computational approaches to deal with this problem have been developed with limited success because the sources of gel-to-gel variation are numerous, complex, and difficult to model (76).

The limitations of 2-DE in its ability to resolve or detect low-abundance proteins and to distinguish proteins clustered near high-abundance spots could, in theory, be overcome by mass spectrometry techniques such as surface-enhanced laser desorption/ionization (SELDI) or matrix-assisted laser desorption/ionization (MALDI) mass spectrometry. But in practice the promise of the superiority of mass spectrometry over electrophoresis has yet to be realized without extensive pre-fractionation of protein samples. However, mass spectrometry does offer an alternative to the gels and liquids required for 2-DE (see Chapters 5, 6, and 7 for detailed discussions of mass spectrometry techniques in neuroproteomics).

Although the technical challenges are far from resolved, proteomic methods have evolved rapidly in recent years, making it feasible to simultaneously evaluate expression differences in hundreds of proteins in tissues under different experimental conditions. Our ongoing research is utilizing two well-established, widely accepted, and complementary proteomic techniques: SELDI mass spectrometry and 2D-DIGE. The goal of our research is to use these techniques to discover and identify the soluble protein mediators expressed by muscle tissue that play a role in axonal regeneration and PMR following peripheral nerve injury.

15.3.3   PILOT EXPERIMENTS USING SELDI MASS SPECTROMETRY

We hypothesized that denervated muscle is likely to increase production of specific proteins that are attractive to, and help provide support for, motor neurons and their axons. We have performed pilot experiments examining this hypothesis using SELDI to identify biomarkers associated with denervation of skeletal muscle. Our initial experiments were conducted to demonstrate proof of principle for our proteomics approach.

For our pilot work using SELDI we compared the expression profiles from two groups of rat muscles: 10 normal quadriceps and 10 four-day denervated quadriceps. We refined harvesting and protein-processing protocols to be compatible with the SELDI instrument, focusing first on procedures that simultaneously sample wide categories of proteins (e.g., membrane, cytosolic, etc.). The experiment is outlined in Figure 15.5.

Dilute solutions of the skeletal muscle lysate are applied directly to spots on the protein chip arrays (Ciphergen, Inc., Fremont, CA, USA). The protein chips differ according to the chemical properties of the spot surfaces. The spots are then washed with a series of buffers of different stringencies. Binding of proteins to these spots is a function of the protein affinity to the spot surface, wash stringency/type, and wash parameters (e.g., time and frequency). In this manner, skeletal muscle proteins are fractionated prior to processing in the ProteinChip reader. The presence of chemically active surfaces on the chips distinguishes SELDI from MALDI, which uses passive

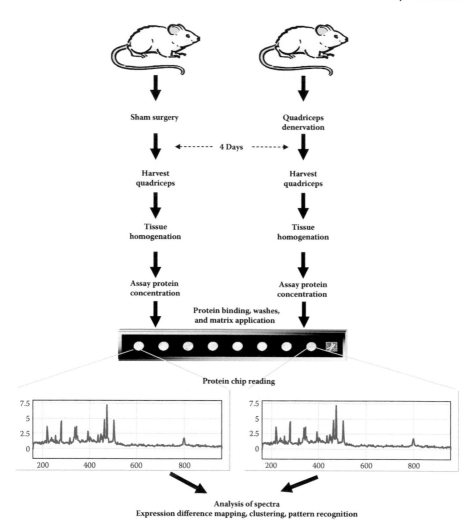

FIGURE 15.5 Experimental flow chart for the pilot protein expression data generated by the SELDI instrument.

surfaces. Like MALDI, SELDI also utilizes an energy-absorbing molecule (matrix) that absorbs laser energy, leading to the desorption and ionization of proteins on spot surfaces. Ciphergen provides chips with surfaces that are cationic, anionic, hydrophobic, normal phase, and capable of binding to metals, antibodies, or bait molecules.

The ProteinChip reader contains the SELDI mass spectrometer with a protein chip loader controlled by an external computer. Under computer control, a UV laser is fired at a limited area of each spot. The laser ionizes bound substances and causes them to fly from the surface. These ionized proteins are accelerated through an electrical field in a vacuum, and then are allowed to drift for a distance prior to encountering a multiple-stage electron multiplier ion detector (for more on ion detectors refer to Chapter 5).

The time of flight (TOF) of a substance depends on its charge and mass. Each laser shot generates a "spectrum" of data points that relate TOF to detector signal intensity. Spectra consist of intensity measures for each TOF value. For ease of interpretation, Ciphergen software converts TOF to the ratio mass/charge (m/z). Spectra can be pre-processed using ProteinReader software to adjust for baseline drift and to normalize across samples within a specified m/z range.

Operational parameters modifiable via SELDI ProteinChip Reader software include laser intensity, detector sensitivity, detector voltage, time-lag focusing, spot position, and number of laser shots per position. These are critical parameters that have a major impact on the quality and reproducibility of results. Our experiments include replicates of homogenized tissue samples for multiple animals within each experimental group. In pilot experiments, for example, 60 protein chip spots were required (10 animals × 2 experimental groups run in triplicate). There are 20 laser positions on each spot. Each of these positions is used to generate a spectrum that may be an average of as many as 99 laser shots. Thus, this experiment could generate 1200 spectra containing information from almost 120 laser shots. Obviously this is too much information to be processed manually, and automated analytic techniques were developed and utilized.

We adjusted for systematic variations across samples, spots, and spot positions with respect to TOF and intensity through the use of known protein standards and pooled experimental samples. Protein standard solutions are available from Ciphergen and are supported by software capabilities that adjust m/z estimates determined from TOF by applying correction factors based on proteins of known molecular weight. Experimental samples are pooled and applied to protein chips for analysis under a variety of possible combinations of laser intensity, detector sensitivities, and detector voltages.

We developed a SAS (SAS, Inc., Cary, NC) macro to generate spot and chip protocols that randomize each combination of laser intensity, detector voltage, and detector sensitivity to multiple laser positions, spots, and chips to identify optimum values of these parameters for pooled samples (i.e., all experimental samples combined). The protocols were generated in the XML format used by the ProteinChip reader. These protocols generate four spectra for each of 12 possible combinations of parameters. Data from the resulting 48 spectra are analyzed using SAS software to identify the parameter values that yield the best signal intensity and the least noise. These parameter values are then used in spot protocols to read protein chips containing individual (i.e., non-pooled) experimental samples. This approach substantially increases reproducibility among samples.

Data from the pre-processed spectra (i.e., baseline corrected and normalized) are analyzed by Ciphergen software to identify spectral peaks. Peaks that are common to a pre-specified proportion of spectra within a narrow molecular weight range are grouped by Ciphergen software into clusters (Figure 15.6). Clusters can be thought of as sets of measures of protein expression (i.e., intensity) across spectra for a substance of a given molecular weight.

The objective of cluster analysis is to identify significant differences in protein expression between experimental groups. In this manner, we have identified several potential biomarkers distinguishing denervated muscle from normal muscle, and these proteins would be candidates for purification using Ciphergen columns,

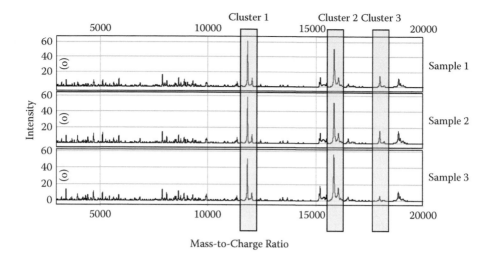

FIGURE 15.6 Example of reproducibility of SELDI spectra from triplicate spots on protein chip CM-10. Ciphergen software identifies common peaks across samples and groups them into "clusters." These clusters can then be analyzed across experimental samples.

combined with on-chip protein digestion to verify protein identity via peptide fingerprinting, as described in Chapter 5.

15.3.4 LESSONS LEARNED FROM SELDI PILOT STUDIES

We were early adopters of SELDI technology. There were few established protocols useful for our specific experimental conditions and tissues at the time of our initial experiments. As we learned more about the technology we found that results were highly sensitive to experimental conditions and variations in tissue sample characteristics and processing. We have found it essential to prepare tissue samples consistently with well-defined and uniform protein concentrations across samples. It is also important to ensure an appropriate experimental design with internal controls allowing the calibration of the instrument. We found substantial variability among individuals within experimental groups and among tissue samples within individuals, suggesting the need to increase experimental group size or refine experimental techniques to reduce that variability.

The resource commitments to refine protocols in order to achieve reproducible results were substantial and difficult to meet in a laboratory not specializing in this technique. We had available within our institution an established proteomics laboratory that was able to produce 2D-DIGE results at lower costs and higher quality than we could achieve with SELDI in our own laboratory. For these reasons we chose to complete this line of research using 2D-DIGE-based proteomics as an optimal choice. Nevertheless, our work with SELDI was informative and provided a foundation for our current work.

15.4 DISCOVERY OF TROPHOMORPHIC MEDIATORS WITH 2D-DIGE

15.4.1 OVERVIEW

Our pilot biomarker discovery investigations using SELDI suggested that there are protein expression differences comparing normal to denervated quadriceps muscle. This is consistent with experimental evidence such as our laboratory surgical findings showing that muscle tissue exerts strong trophomorphic effects on regenerating motor axons. We hypothesize the presence of soluble protein trophic factors that are released by denervated muscle tissue at the neuromuscular junction and diffuse in a retrograde fashion through the nerve tract to the site of peripheral nerve injury (Figure 15.7B). These factors then stimulate and direct the regeneration of motor axons to their appropriate target organ.

To identify candidate muscle proteins involved in axonal regeneration and PMR, we are implementing a series of biomarker discovery and protein identification experiments using two-dimensional difference gel electrophoresis (2D-DIGE). We have completed and describe here proof-of-concept pilot studies that demonstrate findings consistent with our expectations. Currently we are developing a series of comprehensive studies to identify candidate trophic factors using a more rigorous study design.

15.4.2 TWO-DIMENSIONAL DIFFERENCE GEL ELECTROPHORESIS

A major limitation of the 2-DE technique is gel-to-gel variation leading to poor reproducibility among gels. Two-dimensional difference gel electrophoresis (2D-DIGE) was developed to overcome this limitation (76). Rather than comparing samples run on separate gels as in 2-DE, in 2D-DIGE samples are run on the same gel and are distinguished using different fluorescent dyes. The different dyes are virtually identical with respect to characteristics that influence migration in the gel. Therefore, they do not differentially influence migration of their bound proteins. Identical proteins in one sample superimpose on their differentially labeled counterparts for the other sample in the gel (see Chapter 4).

Our biomarker discovery research to identify trophomorphic mediators in nerve regeneration uses 2D-DIGE–based proteomics. This is a powerful approach to evaluate differential protein expression in comparison tissue samples. Protein samples are covalently labeled with one of three fluorescent dyes. All three labeled samples are mixed and then run on a two-dimensional gel. An internal control is used to normalize the concentrations of each sample across gels and biological replicates. This control is usually a pooled mixture of equal aliquots of all samples (77). The relative abundance of each protein from samples is evaluated based on spot intensity for each CyDye.

Proteins with similar expression levels in the comparison samples will have the same relative concentrations and will thus create spots that appear white on the gel. Spots containing proteins that are expressed differently in one sample relative to the others will tend to take on the color of the dye associated with the highest expressed

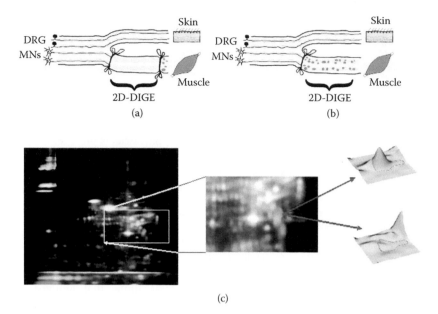

(c)

FIGURE 15.7 (See color insert following page 172.) 2D-DIGE results of individual terminal nerve branches of the rat femoral nerve. (a) and (b) Diagram of the two terminal nerve branches of the rat femoral nerve, showing the sensory cutaneous branch (continuing as the saphenous nerve and providing local cutaneous innervation) and the muscle branch to the quadriceps muscle. Some animals received two ligations: one just distal to the parent femoral nerve bifurcation and one just proximal to the quadriceps muscle (as shown in a). Other animals received a single ligation just distal to the parent femoral nerve bifurcation (as shown in b). Individual muscle branch samples were harvested 7 days post-ligation. Muscle-derived proteins (indicated by red dots in b) include trophomorphic factors. The translucent tube-like structures within the nerve represent Schwann cell tubes that remain following nerve degeneration. DRG = dorsal root ganglion, MNs = motor neurons. (c) Nerve samples were labeled with CyDyes, combined, and run on the same 2D-DIGE gel. The combined image of the gel is shown (all three CyDyes), with an enlarged view to the right. Red spots are proteins expressed more highly in single-ligated nerve samples. Spots that appear white are expressed equally in all three samples. Graphical images of spot intensities from the single-ligated nerve sample (*upper image*) and double-ligated nerve sample (*lower image*) illustrate an example of expressions differences for a candidate trophomorphic mediator. Spots showing large expression differences were subsequently selected for protein identification.

sample. Optical reading of the gel and evaluation with analytic software identify proteins and statistically evaluate expression differences.

15.4.3 PILOT EXPERIMENTS WITH 2D-DIGE

Our proof-of-concept pilot experiment with 2D-DIGE first required the development of protein isolation methods that would work with limited samples of peripheral nerve (e.g., single rat terminal muscle branches). Initial attempts at protein isolation from such small samples of nerve (<5 mg wet weight) proved to be problematic.

Peripheral nerve is notoriously difficult to homogenize due to the tough myelin and connective tissues, and the smaller the sample the more difficult it is to get consistent results. Quite a bit of pilot work was carried out with current popular protocols, and these were all disappointing for various reasons. An extensive protein extraction procedure was finally tailored after that of the 1979 procedure of Groswald and Luttges (78) and proved to be quite reliable. This procedure has probably fallen out of fashion because it is a multi-day, multi-step protocol; however, it proved to be quite robust and consistent.

A typical experiment involved rats in three experimental groups. Some animals received two nerve ligations, one just distal to the parent femoral nerve bifurcation and one just proximal to the quadriceps muscle (Figure 15.7A). Other animals received a single nerve ligation just distal to the parent femoral nerve bifurcation (Figure 15.7B). We also included a control group of naïve animals receiving no surgical procedures.

The rationale for this experimental design was that both sets of samples receiving ligation would have proteins from degenerating axons due to the proximal ligation, but only the samples from the group with a single ligation would contain muscle-derived proteins. Our hypothesis is that muscle-derived proteins are critical for determining the accuracy of regenerating motor neurons. These would migrate in a retrograde fashion from the denervated muscle via the neuromuscular junction to the site of the first ligation by a combination of diffusion and retrograde transport persisting in axons for a short time after injury.

Individual muscle branch samples (<5 mg wet weight) were harvested 7 days after the ligations were placed on the nerve. Nerve samples were harvested from naïve animals at that time as well. Nerve samples were placed in liquid N_2, and stored at $-80°C$ until they were transported to the proteomics laboratory where they were immediately processed. Total time from sacrifice to dissection is less than 10 minutes per rat.

For proteomics analysis nerves were rinsed in PBS, centrifuged, the supernatant discarded, and lysis buffer was added. The tissue was crushed and homogenized on ice with a sonicator, vortexed, centrifuged, and the protein supernatant was saved. The protein concentration was measured with a 2D-Quant kit (GE Healthcare, Piscataway, NJ, USA). We obtained a yield on average of 100 µg of protein per nerve, an amount sufficient for 2D-DIGE analyses. We labeled 40 µg each of the three samples (i.e., normal, single-ligation, double-ligation) with one of the CyDyes (Cy2, Cy3, Cy5, respectively). Two independent experiments were carried out with different nerve samples from different animals. Labeled samples from single animals in each of the three experimental groups were combined and run on the same gel.

The first dimension of the 2D-DIGE was run on an IPG strip 13 cm, pH 3–10 (GE Healthcare). We used a separation pH range from 3 to 10, and the conditions were as follows: active rehydration at 30 V, step voltage up to 500 V for 1 hour, step voltage up to 1000 V for 1 hour, finally a step voltage up to 8000 V until it reaches a total of 26,000 KVh. The second dimension (molecular weight) used a standard 12% polyacrylamide gel with sodium dodecyl sulfate (SDS-PAGE) stacking gel.

The fluorescently labeled spots (proteins) generated from the 2D-DIGE resolved gels were imaged with a Typhoon 9410 (GE Healthcare) Variable Mode Imager

(Figure 15.7). Because the excitation and emission energy for each dye is specific, the imager can digitally separate them, visualized as three colors (blue Cy2, green Cy3, and red Cy5). With these images, we performed automated analysis of individual spots as peaks characterized by their intensity and shape with the DeCyder (GE Healthcare) Differential Analysis software for proteomics (see Chapter 4). A ratio threshold is set by determining variation of comparable spots among the samples. We used a two-fold increase or decrease in spot intensity as a cutoff for pair-wise comparisons. Selected spots were cut from the gels with an Ettan spot picker, and then transferred into 96-well plates for further digestion and spotting for protein identification using liquid chromatography tandem mass spectrometry (LC-MS/MS) to identify protein fingerprints and peptide sequences (as described in Chapter 5).

The combined image of a representative gel from one experiment is shown in Figure 15.7C. One can quickly appreciate protein spots that are more highly expressed in either normal nerve (appears as blue), double ligated nerve (green), or single ligated nerve (red), as well as subsets of proteins from two samples in relation to the third, e.g., yellow equals more highly expressed in single- and double-ligated nerves compared to normal. Spots that appear white are equally expressed in all three samples.

Since our hypothesis centers on proteins originating from muscle, we focused our attention on spots that are more highly expressed in the single-ligated nerve with an intact connection to muscle (i.e., red spots in this gel). DeCyder 6.5 Differential Analysis software was used for automated analysis of individual relative spot intensities and shape. This system characterizes spots based on parameters such as area, maximum peak, maximum volume, and relative abundance. It produces quantitative (intensities) and qualitative (graphical) representations for protein expression comparisons.

Two independent sets of experimental groups in two 2D-DIGE gels were run and analyzed as described. About 2700 individual protein spots were resolved on each gel. Relative abundance measures based on spot volume ratios were used to identify proteins of interest. Expression was judged to be similar where the ratio of the highest spot volume to the lowest spot volume was 2.5 or less. More than 80% of the resolved spots were similar based on this criterion. Expression was elevated in double-ligated samples compared to single-ligated samples in about 6% of resolved spots, and the maximum spot volume ratio was less than seven for these. In about 10% of the spots, expression was higher in the single-ligated sample compared to the double-ligated sample. The magnitudes of the spot volume ratios in this case were much higher (maximum = 238). The spot volume ratio exceeded 100 for more than 10 of these single-ligated high expression spots.

The relatively high abundance of multiple proteins in single-ligated versus double-ligated nerve samples is consistent with our hypothesis that denervated muscle produces substances that migrate in a retrograde fashion from the muscle to the site of nerve injury. Any of these proteins can be considered a candidate mediator of axonal regeneration and PMR. The determination of which, if any, actually are mediators of trophomorphism is an active area of investigation in our laboratory.

A graphical example of a marked expression difference favoring single-ligated nerves is shown in two images of Figure 15.7C on the far right. A second independent gel with nerve samples from different animals verified the increased expression of this same spot in the single-ligated nerves. This spot was subsequently picked by a robot from both gels and subjected to protein sequencing with LC-MS/MS and identified as vimentin both times, a marker of reactive Schwann cells, which are known to have a large influence on nerve regeneration.

15.5 SUMMARY AND FUTURE DIRECTIONS

Our proof-of-concept pilot experiments using two prominent techniques in proteomics, SELDI and 2D-DIGE, have demonstrated that patterns of protein expression in skeletal muscle are influenced by denervation and that protein products from denervated muscle are conveyed in a retrograde fashion from muscle to the site of nerve injury. These findings are consistent with prior research in our laboratory and from other groups that muscles produce soluble protein mediators that control and direct axonal regeneration following nerve injury.

Ongoing and future research will utilize these and other proteomic techniques to identify and quantify all candidate biomarkers potentially related to axonal regeneration and PMR. Once candidates are identified, their individual effects will be tested in vitro and in experimental animals to evaluate the therapeutic potential of these factors in nerve repair. One can envision that this work might identify substances that could be introduced in muscle distal to recently injured nerve. These substances would enhance and guide axonal regeneration to appropriate end organs, thus reducing the substantial deficits experienced by persons with peripheral nerve injury. We are likely far from the realization of that dream, but neuroproteomics offers a promising pathway toward that goal.

ACKNOWLEDGMENTS

Supported by the NIH (to RDM, NS061106) and the Office of Research and Development, Biological Laboratory Research and Development (BLRD) Service, Department of Veterans Affairs (to RDM). RDM is a research career scientist for the BLRD service.

REFERENCES

1. Campbell, W. W. (2008) Evaluation and management of peripheral nerve injury, *Clin Neurophysiol* 119:1951–65.
2. Robinson, L. R. (2000) Traumatic injury to peripheral nerves, *Muscle Nerve* 23:863–73.
3. Noble, J., Munro, C. A., Prasad, V. S. and Midha, R. (1998) Analysis of upper and lower extremity peripheral nerve injuries in a population of patients with multiple injuries, *J Trauma* 45:116–22.
4. Sunderland, S. (1990) The anatomy and physiology of nerve injury, *Muscle Nerve* 13:771–84.
5. Ross, M. H., Reith, E. J. and Romrell, L. J. (1989) Nervous tissue. In *Histology*, 241–82 (Baltimore: Williams and Wilkins).
6. Farel, P. B. and Meeker, M. L. (1993) Developmental regulation of regenerative specificity in the bullfrog, *Brain Res Bull* 30:483–90.

7. Holmes, W. and Young, J. Z. (1942) Nerve regeneration after immediate and delayed suture, *J Anat* 77:63–96, 10.
8. Ide, C., Tohyama, K., Yokota, R., Nitatori, T. and Onodera, S. (1983) Schwann cell basal lamina and nerve regeneration, *Brain Res* 288:61–75.
9. Lee, M. T. and Farel, P. B. (1988) Guidance of regenerating motor axons in larval and juvenile bullfrogs, *J Neurosci* 8:2430–7.
10. Scherer, S. S. and Easter, S. S., Jr. (1984) Degenerative and regenerative changes in the trochlear nerve of goldfish, *J Neurocytol* 13:519–65.
11. Brown, M. C. and Hardman, V. J. (1987) A reassessment of the accuracy of reinnervation by motoneurons following crushing or freezing of the sciatic or lumbar spinal nerves of rats, *Brain* 110(Pt 3):695–705.
12. Brown, M. C. and Hopkins, W. G. (1981) Role of degenerating axon pathways in regeneration of mouse soleus motor axons, *J Physiol* 318:365–73.
13. Nguyen, Q. T., Sanes, J. R. and Lichtman, J. W. (2002) Pre-existing pathways promote precise projection patterns, *Nat Neurosci* 5:861–7.
14. Witzel, C., Rohde, C. and Brushart, T. M. (2005) Pathway sampling by regenerating peripheral axons, *J Comp Neurol* 485:183–90.
15. Grinnell, A. D. (1995) Dynamics of nerve-muscle interaction in developing and mature neuromuscular junctions, *Physiol Rev* 75:789–834.
16. Fox, M. A., Sanes, J. R., Borza, D. B. et al. (2007) Distinct target-derived signals organize formation, maturation, and maintenance of motor nerve terminals, *Cell* 129:179–93.
17. Chadaram, S. R., Laskowski, M. B. and Madison, R. D. (2007) Topographic specificity within membranes of a single muscle detected in vitro, *J Neurosci* 27:13938–48.
18. McCann, C. M., Nguyen, Q. T., Santo Neto, H. and Lichtman, J. W. (2007) Rapid synapse elimination after postsynaptic protein synthesis inhibition in vivo, *J Neurosci* 27:6064–7.
19. Seddon, H. (1943) Three types of nerve injury, *Brain* 66:237–88.
20. Sunderland, S. (1951) A classification of peripheral nerve injuries producing loss of function, *Brain* 74:491–516.
21. Brushart, T. M. (1998) Nerve repair and grafting. In *Green's operative hand surgery*, ed. Green, D., 1381–1403 (New York: Churchill Livingston).
22. Madison, R. D., Archibald, S. J. and Karup, C. (1992) Peripheral nerve injury. In *Wound healing: Biochemical and clinical aspects*, eds. Cohen, I. K., Diegelman, F. and Lindblad, W. J., 450–80 (Philadelphia: W. B. Saunders Co.).
23. Sunderland, S. (1991) *Nerve injuries and their repair: A critical appraisal* (New York: Churchill Livingstone.).
24. Weiss, P. and Edds, M. V. (1945) Sensory-motor nerve crosses in the rat, *J Neurophysiology* 8:173–93.
25. Brushart, T. M. (1988) Preferential reinnervation of motor nerves by regenerating motor axons, *J Neurosci* 8:1026–31.
26. Brushart, T. M., Gerber, J., Kessens, P., Chen, Y. G. and Royall, R. M. (1998) Contributions of pathway and neuron to preferential motor reinnervation, *J Neurosci* 18:8674–81.
27. Brushart, T. M. (1990) Preferential motor reinnervation: A sequential double-labeling study, *Restor Neurol Neurosci* 1:281–87.
28. Brushart, T. M. (1993) Motor axons preferentially reinnervate motor pathways, *J Neurosci* 13:2730–8.
29. Franz, C. K., Rutishauser, U. and Rafuse, V. F. (2005) Polysialylated neural cell adhesion molecule is necessary for selective targeting of regenerating motor neurons, *J Neurosci* 25:2081–91.
30. Madison, R. D., Archibald, S. J. and Brushart, T. M. (1996) Reinnervation accuracy of the rat femoral nerve by motor and sensory neurons, *J Neurosci* 16:5698–5703.

31. Mears, S., Schachner, M. and Brushart, T. M. (2003) Antibodies to myelin-associated glycoprotein accelerate preferential motor reinnervation, *J Peripher Nerv Syst* 8:91–9.

32. Robinson, G. A. and Madison, R. D. (2003) Preferential motor reinnervation in the mouse: Comparison of femoral nerve repair using a fibrin sealant or suture, *Muscle Nerve* 28:227–31.

33. Robinson, G. A. and Madison, R. D. (2005) Manipulations of the mouse femoral nerve influence the accuracy of pathway reinnervation by motor neurons, *Exp Neurol* 192:39–45.

34. Robinson, G. A. and Madison, R. D. (2004) Motor neurons can preferentially reinnervate cutaneous pathways, *Exp Neurol* 190:407–13.

35. Abercrombie, M. (1946) Estimation of nuclear population from microtome sections, *Anat Rec* 4:239–46.

36. Uschold, T., Robinson, G. A. and Madison, R. D. (2007) Motor neuron regeneration accuracy: balancing trophic influences between pathways and end-organs, *Exp Neurol* 205:250–6.

37. Hassig, R., Tavitian, B., Pappalardo, F. and Di Giamberardino, L. (1991) Axonal transport reversal of acetylcholinesterase molecular forms in transected nerve, *J Neurochem* 57:1913–20.

38. Miledi, R. and Slater, C. R. (1968) Electrophysiology and electron-microscopy of rat neuromuscular junctions after nerve degeneration, *Proc R Soc Lond B Biol Sci* 169:289–306.

39. Miledi, R. and Slater, C. R. (1970) On the degeneration of rat neuromuscular junctions after nerve section, *J Physiol* 207:507–28.

40. Smith, R. S. and Bisby, M. A. (1993) Persistence of axonal transport in isolated axons of the mouse, *Eur J Neurosci* 5:1127–35.

41. Stanley, E. F. and Drachman, D. B. (1980) Denervation and the time course of resting membrane potential changes in skeletal muscle in vivo, *Exp Neurol* 69:253–9.

42. Watson, D. F., Glass, J. D. and Griffin, J. W. (1993) Redistribution of cytoskeletal proteins in mammalian axons disconnected from their cell bodies, *J Neurosci* 13:4354–60.

43. Curtis, R., Scherer, S. S., Somogyi, R. et al. (1994) Retrograde axonal transport of LIF is increased by peripheral nerve injury: Correlation with increased LIF expression in distal nerve, *Neuron* 12:191–204.

44. McGraw, T. S., Mickle, J. P., Shaw, G. and Streit, W. J. (2002) Axonally transported peripheral signals regulate alpha-internexin expression in regenerating motoneurons, *J Neurosci* 22:4955–63.

45. McPhail, L. T., Oschipok, L. W., Liu, J. and Tetzlaff, W. (2005) Both positive and negative factors regulate gene expression following chronic facial nerve resection, *Exp Neurol* 195:199–207.

46. Murphy, P. G., Borthwick, L. S., Johnston, R. S., Kuchel, G. and Richardson, P. M. (1999) Nature of the retrograde signal from injured nerves that induces interleukin-6 mRNA in neurons, *J Neurosci* 19:3791–3800.

47. Richardson, P. M. and Verge, V. M. (1986) The induction of a regenerative propensity in sensory neurons following peripheral axonal injury, *J Neurocytol* 15:585–94.

48. Easter, S. S., Jr., Purves, D., Rakic, P. and Spitzer, N. C. (1985) The changing view of neural specificity, *Science* 230:507–11.

49. Levi-Montalcini, R. (1987) The nerve growth factor 35 years later, *Science* 237:1154–62.

50. Purves, D. (1988) *Body and brain* (Cambridge, MA: Harvard University Press).

51. Crutcher, K. A. and Saffran, B. N. (1990) Developmental remodeling of neuronal projections: Evidence for trophomorphism? *Comments Developmental Neurobiology* 1:119–41.

52. Saffran, B. N. and Crutcher, K. A. (1990) NGF-induced remodeling of mature uninjured axon collaterals, *Brain Res* 525:11–20.

53. Campenot, R. B. (1982) Development of sympathetic neurons in compartmentalized cultures. II. Local control of neurite survival by nerve growth factor, *Dev Biol* 93:13–21.

54. Campenot, R. B. (1994) NGF and the local control of nerve terminal growth, *J Neurobiol* 25:599–611.

55. Takahashi, Y., Maki, Y., Yoshizu, T. and Tajima, T. (1999) Both stump area and volume of distal sensory nerve segments influence the regeneration of sensory axons in rats, *Scand J Plast Reconstr Surg Hand Surg* 33:177–80.

56. Libelius, R., Lundquist, I., Templeton, W. and Thesleff, S. (1978) Intracellular uptake and degradation of extracellular tracers in mouse skeletal muscle in vitro: The effect of denervation, *Neuroscience* 3:641–7.

57. Vult von Steyern, F., Kanje, M. and Tagerud, S. (1993) Protein secretion from mouse skeletal muscle: Coupling of increased exocytotic and endocytotic activities in denervated muscle, *Cell Tissue Res* 274:49–56.

58. deLapeyriere, O. and Henderson, C. E. (1997) Motoneuron differentiation, survival and synaptogenesis, *Curr Opin Genet Dev* 7:642–50.

59. Henderson, C. E., Bloch-Gallego, E., Camu, W. et al. (1993) Motoneuron survival factors: Biological roles and therapeutic potential, *Neuromuscul Disord* 3:455–8.

60. Libelius, R. (1989) Lysosomes in skeletal muscle. In *Neuromuscular junction*, ed. Libelius, R., 481–85 (Elsevier).

61. Magnusson, C., Svensson, A., Christerson, U. and Tagerud, S. (2005) Denervation-induced alterations in gene expression in mouse skeletal muscle, *Eur J Neurosci* 21:577–80.

62. Thesleff, S. and Libelius, R. (1989) Some aspects of long term regulations of nerve-muscle relations. In *Neuromuscular stimulation: Basic concepts and clinical applications* (New York: Demos Publications).

63. Jennische, E., Ekberg, S. and Matejka, G. L. (1993) Expression of hepatocyte growth factor in growing and regenerating rat skeletal muscle, *Am J Physiol* 265:C122–8.

64. Lie, D. C. and Weis, J. (1998) GDNF expression is increased in denervated human skeletal muscle, *Neurosci Lett* 250:87–90.

65. Tonra, J. R., Curtis, R., Wong, V. et al. (1998) Axotomy upregulates the anterograde transport and expression of brain-derived neurotrophic factor by sensory neurons, *J Neurosci* 18:4374–83.

66. Wallenius, V., Hisaoka, M., Helou, K. et al. (2000) Overexpression of the hepatocyte growth factor (HGF) receptor (Met) and presence of a truncated and activated intracellular HGF receptor fragment in locally aggressive/malignant human musculoskeletal tumors, *Am J Pathol* 156:821–9.

67. Wehrwein, E. A., Roskelley, E. M. and Spitsbergen, J. M. (2002) GDNF is regulated in an activity-dependent manner in rat skeletal muscle, *Muscle Nerve* 26:206–11.

68. Yamaguchi, A., Ishii, H., Morita, I., Oota, I. and Takeda, H. (2004) mRNA expression of fibroblast growth factors and hepatocyte growth factor in rat plantaris muscle following denervation and compensatory overload, *Pflugers Arch* 448:539–46.

69. Zhao, C., Veltri, K., Li, S., Bain, J. R. and Fahnestock, M. (2004) NGF, BDNF, NT-3, and GDNF mRNA expression in rat skeletal muscle following denervation and sensory protection, *J Neurotrauma* 21:1468–78.

70. Lawoko, G. and Tagerud, S. (1995) High endocytotic activity occurs periodically in the endplate region of denervated mouse striated muscle fibers, *Exp Cell Res* 219:598–603.

71. Lefkovits, I., Kettman, J. R. and Frey, J. R. (2000) Global analysis of gene expression in cells of the immune system I. Analytical limitations in obtaining sequence information on polypeptides in two-dimensional gel spots, *Electrophoresis* 21:2688–93.

72. Harrison, P. M., Kumar, A., Lang, N., Snyder, M. and Gerstein, M. (2002) A question of size: The eukaryotic proteome and the problems in defining it, *Nucleic Acids Res* 30:1083–90.
73. Anderson, N. G., Matheson, A. and Anderson, N. L. (2001) Back to the future: The human protein index (HPI) and the agenda for post-proteomic biology, *Proteomics* 1:3–12.
74. Weinberg, A. M. (1999) The birth of Big Biology, *Nature* 401:738.
75. O'Farrell, P. H. (1975) High resolution two-dimensional electrophoresis of proteins, *J Biol Chem* 250:4007–21.
76. Viswanathan, S., Unlu, M. and Minden, J. S. (2006) Two-dimensional difference gel electrophoresis, *Nat Protoc* 1:1351–8.
77. Friedman, D. B. and Lilley, K. S. (2008) Optimizing the difference gel electrophoresis (DIGE) technology, *Methods Mol Biol* 428:93–124.
78. Groswald, D. E. (1979) Changes in sciatic nerve protein composition during postnatal development of mice, *Dev Neurosci* 2:51–64.

Index

A

Absolute quantitative analysis (AQUA), 37, 40–41
Acetone, 58
Acetylation, 95–96, *97*
Acrolein, 166
Actin, 182–183
Adenosine tri-phosphate (ATP), 275
Affinity-based separations, 31–35
Affinity chromatography, 26, 39–40
Aldolase, 178
Alpha-enolase, 176
Alpha-tocopherol, 167
Alzheimer's disease
    amyloid precursor protein (APP) in, 61
    domain-based similarity to other
        neurodegenerative diseases, 160–161
    KEGG Pathway for, 148, 152, 154
    MALDI profiling of, 122–125
    protein nitration in, 170
    signal transduction in, 183
    structural dysfunction in, 182–183
    synaptic dysfunction in, 60
Alzheimer's Disease Centers (ADCs), 17, 18
Amygdala, 21
Amyloid precursor protein (APP), 61
Amyotrophic lateral sclerosis (ALS), 148, 152, 155
    domain-based similarity to other
        neurodegenerative diseases, 160–161
Analytical ultracentrifugation (AUC), 6
Anatomical Therapeutic Chemical (ATC)
    classification, 152
Antibodies, targeted, 6
Antioxidants, 166–167
    defense, 180–182
Apoptosis, 156–157, 158, 161
Arrays, protein, 5
Ascorbic acid, 167
Assays, enzyme, 176
ATP molecules, 176–179
Auditory and vocal pathways, 240–241
    brain substrates for auditory processing, 266–267
    brain substrates of song learning and
        production, 264–266
    differences between, 252–254
    proposed hypothesis of singing-regulated
        genes and proteins, 254–256

RNA and protein concordance in, 251–252
singing-expression genes in, 242–251
Autopsy, 19–22
Avian species
    auditory and vocal pathways, 240–241
    brain substrates of song learning and
        production in, 264–266
    categories of singing-regulated genes in, 246–249
    differences between vocal and auditory
        pathways in, 252–254
    expression of singing-regulated genes and
        their proteins in, 242–251
    proposed hypothesis of singing-regulated
        genes and proteins, 254–256
    RNA and protein concordance in, 241–252
    social context regulation of gene expression
        in, 242–246
Avidin, 103
Axonal regeneration, 299–304

B

Banked tissue
    preservation methods and, 22–23
    procurement and storage, 18–22
    research importance of, 17–18
Basal ganglia, 21, 161
Benzyldimethyl-n-hexadecylammonium
    chloride-(16 BAC)/SDS-PAGE gel, 35
Biomarkers, 8, 41–42, 58, 305–309
Biotin, 103
Biotinylation, 103
Brain-derived neurotrophic factor (BDNF), 298
Brain regions
    development and, 61–62
    matrix-assisted laser desorption/ionization
        (MALDI) mass spectrometry of, 120–121
    specific protein markers, 61
    three-dimensional imaging mass
        spectrometry, 125–126
Brain stem, 20, 22
BRITE, KEGG, 149, 150–151

C

Calcium-driven plasticity, 279–282
Capillary electrophoresis (CE), 35–36, 39
Capillary zone electrophoresis (CZE), 39

315

T - #0159 - 171019 - C12 - 234/156/16 - PB - 9780367385019